S0-DUU-685

CHEMOSYSTEMATICS: PRINCIPLES AND PRACTICE

Proceedings of an International Symposium
held at the University of Southampton

THE SYSTEMATICS ASSOCIATION
SPECIAL VOLUME NO. 16

CHEMOSYSTEMATICS: PRINCIPLES AND PRACTICE

Edited by

F. A. BISBY

Department of Biology, University of Southampton

J. G. VAUGHAN

Department of Biology, Queen Elizabeth College, London

C. A. WRIGHT

British Museum (Natural History), London

1980

Published for the

SYSTEMATICS ASSOCIATION

by

ACADEMIC PRESS

LONDON NEW YORK TORONTO SYDNEY SAN FRANCISCO

ACADEMIC PRESS INC. (LONDON) LTD.
24–28 Oval Road
London NW1 7DX

U.S. Edition published by
ACADEMIC PRESS INC.
111 Fifth Avenue
New York, New York 10003

British Library Cataloguing in Publication Data

Chemosystematics. – (Systematics Association. Special volumes; no. 16 ISSN 0309–2593).
1. Biology – Classification
I. Bisby, F A II. Vaughan, J G
III. Wright, Christopher Amyas IV. Series
571′.01′2 QH83 80–41428

ISBN 0–12–101550–5

Printed in Great Britain at the Alden Press
Oxford London and Northampton

Contributors

Bell, Professor E. A., *Department of Plant Sciences, Kings College London, 68 Half Moon Lane, London, SE24 9JF, England*

Berry, Professor R. J., *Department of Genetics, Royal Free Hospital School of Medicine, University of London, 8 Hunter St, London, WC1N 1BP, England*

Bisby, Dr F. A., *Department of Biology, Building 44, University of Southampton, Southampton, SO9 5NH, England*

Boulter, Professor D., *Durham University, South Road, Durham, DH1 3LE, England*

Bradshaw, Dr J. W. S., *Chemical Entomology Unit, Departments of Biology and Chemistry, University of Southampton, Southampton, SO9 5NH, England*

Charlwood, Dr B. V., *Department of Plant Sciences, Kings College London, 68 Half Moon Lane, London, SE24 9JF, England*

Cristofolini, Dr G., *Istituto Botanica, Università di Trieste, Via A. Valerio 30, 34100 Trieste, Italy*

Cronquist, Dr A., *The New York Botanical Garden, Bronx, New York 10458, U.S.A.*

Dover, Dr G. A., *Department of Genetics, Cambridge University, Cambridge, CB2 3EH, England*

El-Tawil, Dr M. Y., *Marine Research Centre, Bab El-Bahr, P.O. Box 315, Tripoli, Libya.*

Estabrook, Professor G. F., *Department of Botany and University Herbarium, Michigan University, Ann Arbor, MI 48109, U.S.A.*

Estabrook, Dr G. F., *Department of Botany and University Herbarium, Michigan University, Ann Arbor, MI 48109, U.S.A.*

Friday, Dr A. E., *University Museum of Zoology, Downing Street, Cambridge, CB2 3EJ, England*

Gottlieb, Professor O. R., *Instituto de Quĩmica, Universidade de São Paulo, C. P. 2078, 05508 São Paulo, S. P., Brazil*

Gower, Dr J. C., *Rothamsted Experimental Station, Harpenden, Herts, AL5 2JQ, England*

Gray, Dr J. C., *Department of Botany, Cambridge University, Cambridge, CB2 3EH, England*

Harborne, Professor J. B., *Phytochemical Unit, Plant Science Laboratories, Reading University, Reading, RG6 2AS, England*

Harris, Dr J. A., *Bedford College, Regent's Park, London, NW1 4NS, England*

Howse, Dr P. E., *Chemical Entomology Unit, Departments of Biology and Chemistry, University of Southampton, Southampton, SO9 5NH, England*

Humphries, Dr C. J., *Department of Botany, British Museum (Natural History), Cromwell Road, London, SW7 5BD, England*

Hurka, Professor H., *Botanisches Institut und Botanischer Garten der Universität, Schlossgarten 3, D44 Münster, West Germany*

Joysey, Dr K. A., *University Museum of Zoology, Downing Street, Cambridge University, Cambridge, CB2 3EJ, England*

Miller, Dr P. J., *Department of Zoology, Bristol University, Bristol, BS8 1UG, England*

Richardson, Dr P. M., *Department of Plant Sciences, Reading University, Whiteknights, Reading, RG6 2AS, England*

Rollinson, Dr D., *Department of Zoology, British Museum (Natural History), London, SW7 5BD, England*

Sneath, Professor P. H. A., *Department of Microbiology, Leicester University, Leicester, LE1 7RH, England*

Thorpe, Dr R. S., *Department of Zoology, The University, Aberdeen, AB9 2TN, Scotland*

Webb, Dr C. J., *Department of Zoology, Bristol University, Bristol, BS8 1UG, England*

Wright, Dr C. A., *British Museum (Natural History), Cromwell Road, London, SW7 5BD, England*

Preface

This book contains all but two of the papers given at the Systematics Association's International Symposium "Chemosystematics: Principles and Practice" held at Southampton University from July 10th to July 12th, 1979. The Systematics Association had played a pioneer role in organizing a symposium on chemotaxonomy in 1967 and now wished to review the coming of age of a discipline that was then only just emerging. True to the Association's reputation for encouraging the study of the whole range of biological systematics as a broad cohesive discipline, the organizing committee wished to draw out some general principles and to bring into closer communication three remarkably separate subdisciplines: the working taxonomists, the chemosystematists and the numerical taxonomists. The former proved a somewhat elusive goal: the latter an exciting, even combustible, mixture!

To a large extent the more reasonable claims of the 1967 symposium are now accepted: a range of techniques is capable of producing chemically reliable comparative data on micromolecules and macromolecules for many groups of organisms, and these are of great systematic interest. The issues that came to the fore at the present meeting (excepting some important technical advances) relate to how systematists can use these data. We now appreciate that comparative chemical data have some of the same complexities as morphological data. We must check them not just for chemical consistency but also for normal taxonomic properties: authenticity of materials, variability within taxa, the possibilities of convergence and of mistaken homology. Secondly, we find that the fundamental debate current in taxonomy on the purpose and form of classifications naturally extends into chemotaxonomy. Do some aspects of chemical data give them special qualities whether in phenetic classification or in tracing phylogenies? Similarly the contribution that explicit numerical analyses can make extends to chemical as well as systematic data. Lastly, there are fundamental problems being caused by attempts to integrate chemical and morphological classifications. How common is incongruence? Does it bring into question the stability or naturalness of our morphological systems? How can we construct classifications or information systems that will give generalizations and retrieval to disparate data? By many accounts it was an interesting meeting and certainly it has been fun collecting the papers together for this volume.

I am very much indebted to my colleagues on the organizing committee, Dr J. G. Vaughan and Dr C. A. Wright for their help and advice in planning and executing the symposium and in editing this volume. I am most grateful to Mrs Yvonne Neville for the work she did as symposium secretary and to Mrs Barbara Otto, Mr R. Mothershaw and Mr F. A. Barrett for helping behind the scenes to run the symposium on the day. The sessions themselves all ran smoothly thanks to the chairmen – Professor M. R. House (President), Professor T. Swain, Dr A. Ferguson, Professor D. E. Fairbrothers, Professor P. H. A. Sneath and Dr J. G. Vaughan to whom I am most grateful. Several had travelled great distances to be with us.

I acknowledge with thanks the contributions made by the I.U.B.S., the British Council and the Systematics Association towards the travel and expenses of speakers coming from overseas.

September 1980 F.A.B.
Southampton

Contents

Systematics Association Publications

1. BIBLIOGRAPHY OF KEY WORKS FOR THE IDENTIFICATION OF THE BRITISH FAUNA AND FLORA
   3rd edition (1967)
   Edited by G. J. Kerrich, R. D. Meikle and N. Tebble
2. FUNCTION AND TAXONOMIC IMPORTANCE (1959)
   Edited by A. J. Cain
3. THE SPECIES CONCEPT IN PALAEONTOLOGY (1956)
   Edited by P. S. Sylvester-Bradley
4. TAXONOMY AND GEOGRAPHY (1962)
   Edited by D. Nichols
5. SPECIATION IN THE SEA (1963)
   Edited by J. P. Harding and N. Tebble
6. PHENETIC AND PHYLOGENETIC CLASSIFICATION (1964)
   Edited by V. H. Heywood and J. McNeill
7. ASPECTS OF TETHYAN BIOGEOGRAPHY (1967)
   Edited by C. G. Adams and D. V. Ager
8. THE SOIL ECOSYSTEM (1969)
   Edited by J. Sheals
9. ORGANISMS AND CONTINENTS THROUGH TIME (1973)†
   Edited by N. F. Hughes

LONDON. Published by the Association

Systematics Association Special Volumes

1. THE NEW SYSTEMATICS (1940)
   Edited by JULIAN HUXLEY (Reprinted 1971)
2. CHEMOTAXONOMY AND SEROTAXONOMY (1968)*
   Edited by J. G. Hawkes
3. DATA PROCESSING IN BIOLOGY AND GEOLOGY (1971)*
   Edited by J. L. Cutbill
4. SCANNING ELECTRON MICROSCOPY (1971)*
   Edited by V. H. Heywood
5. TAXONOMY AND ECOLOGY (1973)*
   Edited by V. H. Heywood
6. THE CHANGING FLORA AND FAUNA OF BRITAIN (1974)*
   Edited by D. L. Hawksworth
7. BIOLOGICAL INDENTIFICATION WITH COMPUTERS (1975)*
   Edited by R. J. Pankhurst
8. LICHENOLOGY: PROGRESS AND PROBLEMS (1977)*
   Edited by D. H. Brown, D. L. Hawksworth and R. H. Bailey

---

*Published by Academic Press for the Systematics Association

†Published by the Palaeontological Association in conjunction with the Systematics Association

*Published by Academic Press for the Systematics Association

1 | Chemistry in Plant Taxonomy: an Assessment of Where We Stand

A. CRONQUIST

The New York Botanical Garden, Bronx, New York 10458, USA

Abstract: Chemical characters are like other characters: they work when they work, and they don't work when they don't work. Like all taxonomic characters, they attain their value through correlation with other characters, and perfect correlations are the exception rather than the rule. Of all the sorts of chemical data, the structure of vital proteins may hold the most promise for helping to establish relationships among major groups of angiosperms. We do not know when or whether this promise will be fulfilled. The chemical compounds that have been taxonomically the most useful up to the present time are the secondary metabolites, perhaps simply because there are so many of them. The discovery and exploitation of major groups of new repellents may have played a decisive role in the origin and diversification of major groups of angiosperms. Serology has helped to resolve some major taxonomic problems in the past, and it may be on the threshold of a great leap forward, through the separation and individual study of the antigenic proteins. Accumulation of various minerals has a proper role to play in taxonomy, but I question the proposal that accumulation of aluminum is a fundamental primitive character in angiosperms. The nature of food storage products shows strong taxonomic correlations, sometimes at and above the level of families.

INTRODUCTION

At this stage in the development of taxonomy, it should not be necessary to expound the theoretical importance of chemical characters. We will all agree, I trust, that chemical data are potentially as important to the taxonomist as any other sort of data, and that they have a proper role to play in the development of taxonomic and

Systematics Association Special Volume No. 16, "Chemosystematics: Principles and Practice", edited by F. A. Bisby, J. G. Vaughan and C. A. Wright, 1980, pp. 1–27, Academic Press, London and New York phylogenetic schemes. My own position on this is expounded at some length in my paper published in Nobel Symposium 25 (Cronquist, 1973). Some further exposition of taxonomic theory may still be useful at this time.

THE VALUE OF CHARACTERS

One of the most important taxonomic principles is that characters attain their value through correlation with other characters, and that *a priori* assignment of value generally leads to artificial rather than natural classifications. Linnaeus was well aware of this principle, as shown by his aphorism in the Philosophia Botanica, "Scias characterem non constituere genus, sed genus characterem". Yet it is a lesson that comes hard, and it must be learned anew by each generation of taxonomists. After enough individual examples have been studied, one can make an educated guess as to the probable importance of a particular character is cases as yet unstudied, but it is still only an educated guess. In the next case the character may turn out to be either more important or less important that its average in previous cases.

Two hundred years ago, when the recognition of natural groups of plants above the rank of genus was still in a nascent state, there may have been more justification than now for efforts to set up systems on a few arbitrarily selected characters. For all his wise words about the theory, Linnaeus found it necessary to propose an admittedly artificial system for the placement of genera. Although the Adansonian method of assigning equal initial weight to all characters is conceptually sound, it is difficult or impossible in practice when one is beginning to construct a classification of a large group.

Things have changed since the time of Linnaeus. I trust that most of us will agree that the bulk of the families of angiosperms now recognized are truly natural groups. Their limits are subject to tinkering mainly by lumping or splitting, or by the transfer of aberrant genera. Although the orders are more debatable, considerable progress has been made on these also. The core groups of many of them, and the precise limits of some, now seem to be well established. I hope that the vigorous argument of the past decade about the limits of the Caryophyllales is now sufficiently well settled so that we can proceed to other matters.

Certainly there is still a lot of room for improvement in the general system of classification of flowering plants, but I think we can now properly be very cautious about using any one new source of data to make radical changes. Any new scheme should show more overall correlations and fewer anomalies than the scheme it proposes to replace. I have no patience with proposals to assemble all the iridoid-producing taxa into one group, in spite of their vast morphological diversity, or to associate the Asteraceae with the Apiaceae because of a degree of similarity in their secondary metabolites, when so many of their other features are very different.

At intervals some new morphological, anatomical, chemical or behavioral character, or set of characters, is touted as having the final solution to our taxonomic problems. Controversy ensues, with proponents and opponents giving ground reluctantly, until some sort of consensus is achieved on some intermediate level of value or reliability. Proponents of the new characters or approaches tend to feel that the orthodox or classical taxonomists are being mulishly obstinate in refusing to see the importance of the new things, whereas the classical taxonomists tend to feel that fools are rushing in where angels fear to tread. Enthusiastic proponents of the new beat the drums for their ideas, filling the air with what the more traditional taxonomists hear as cacophonous, infuriating noise that distracts them from contemplating the great symphony of data. But drums and even cannons have a proper place in some symphonies.

As a working taxonomist, a generalist not wedded to any particular approach or set of data, I often find it necessary to express some caution about the great importance of any particular sort of data or method of assessing the data. Thus it may seem to some that I and like-minded colleagues are kicking and screaming as we are being dragged into the present, whereas I feel that we are more nearly in the position of the girl who would like to be kissed but does not want to be raped.

Conceptually, there are two contrasting ways in which a given character or kind of character can be evaluated. One can assume nothing, and gradually build up a body of examples in which the character demonstrates its usefulness through correlation with other characters, and its limitations through the lack of such correlation. The limits of usefulness of the character are established from the centre outwards, and at any stage in the process the character is undervalued rather than overvalued. The opposite approach is to assume that new data of a particular sort have an overwhelming importance, and then chop away at the periphery by assembling examples in which the character goes contrary to the weight of the evidence. In this procedure, the character is at any stage of the process overvalued rather than undervalued.

My own philosophical and psychological bent is toward the first of these approaches rather than the second. My feeling (gut reaction, we would say in vernacular American) about this is strongly conditioned by the history of taxonomy, as recounted above. Faulty as it is, our general scheme is now good enough to warrant the greatest of caution in accepting radical changes.

My own appreciation of chemical characters has been expanded considerably by my efforts to improve the general system of classification of flowering plants. In arriving at conclusions I have tried to include chemical as well as other evidence, and in writing descriptions I have routinely tried to include information about secondary metabolites. In so doing, I have come to realize that there are many strong correlations between chemical and classical morphological characters, far too many to reflect mere coincidence. In my efforts to make sense out of the data, I have come to accept Fraenkel's view (1959) that a large proportion of secondary metabolites are basically defensive weapons, and the view of Ehrlich and Raven (1965) that the evolutionary origin and diversification of major groups may result from the adoption of new weapons. My paper of 1977 on this subject was an effort to apply the Ehrlich and Raven concept to the evolution and classification of dicotyledons.

At the same time I found many notable non-correlations, and I have tried to avoid basing conclusions on the chemistry alone, especially when it is not harmonious with the ensemble of morphological features. The most consistent single character marking the order Capparales is the uniform presence of mustard oils, an otherwise rare type of compound. On the other hand, it would make no sense to extract *Drypetes* from the Euphorbiaceae and put it into the Capparales merely because it produces mustard oils.

Evolutionary parallelism is so rampant in the angiosperms that taxonomists have learned to be suspicious of individual similarities or even syndromes of similarities as inevitable proof of relationships. No matter how we arrange things, there are generally some pieces of evidence that do not fit. We constantly face choices in which if the similarities between A and B are accepted as indicating a relationship, then other similarities between B and C must be dismissed as reflecting parallelism or convergence. An exposition of how to go about making such choices is beyond the scope of this paper. My point is that a taxonomist's dismissal of chemical similarities between two groups need not indicate an antichemical bias. Morphological similarities also must often be dismissed. One can still get a lively argument as to whether the pitchers of *Nepenthes* have anything to do with those of *Sarracenia.* Even if they do, probably neither has anything to do with the pitchers of *Cephalotus.*

It is perfectly clear, to me at least, that the pervasive parallelism which confounds our efforts to produce a natural system of classification of angiosperms extends to chemical as well as morphological characters. This does not mean that the chemical characters should be tossed aside, but only that they should be regarded with the same suspicion as morphological characters. This necessary caution has two important aspects. One is that we must resist the temptation to extrapolate broadly from studies of a limited number of species. The other is the principle, already noted, that characters attain their taxonomic value through correlation with other characters. I may be merely restating the obvious, but my observation is that these obvious problems are often ignored.

INTERPRETATION OF THE DATA

What I would like to do now is to evaluate the present state of the taxonomic use of chemical data, and consider what we might reasonably look forward to in the future. I speak as a classical taxonomist who is a consumer rather than the producer of chemical data. Although the photosynthetic pigments, food reserves and even the biosynthetic pathway of lysine have had a major impact on taxonomy and evolutionary thinking about the algae, I will here pass the algae by. In order to hold this commentary within reasonable bounds, and in order to stay closer to my home territory, I will limit my consideration to the angiosperms.

It may be useful to consider the chemical data under 5 headings: (1) the structure of vital proteins, (2) secondary metabolites, (3) serological reactions, (4) accumulation of minerals and (5) food storage products. I am well aware that there are other possibilities (such as the means of uptake of CO_2 and the form in which nutrients are transported), but I believe that these are the ones that have attracted the most taxonomic attention and have had the most impact to date. Obviously I cannot here review all the significant papers; I choose some that hold the most interest for me.

1. The Structure of Vital Proteins

Of all the sorts of chemical data, the structure of vital proteins may prove to be the most useful in the long run in helping to establish relationships among the major groups of angiosperms. However, we are still only on the threshold of making use of this set of features. Nothing that is more than minimally helpful has come out of it so far.

Several years ago I explored at some length the potential usefulness and limitations of amino acid sequences as a botanical taxonomic tool (Cronquist, 1976). At that time virtually the only data that merited critical attention pertained to cytochrome *c*. Since then a number of sequences for plastocyanin have also been presented (Boulter, 1979; Boulter *et al.*, 1979). All, or nearly all, of the data and interpretations concerning plants continue to come from Dr Donald Boulter and his associates. Thus it is difficult to avoid a degree of *ad hominem* colouration in my comments; all I can do is assure you that I am concerned with the results, and that I do not feel any personal bias.

As I pointed out in 1976, the phylogenetic tree presented by Boulter in 1973, based on the cytochrome *c* of 25 species in 22 families of seed plants, is hopelessly flawed. Since nothing more has been presented, no further analysis is now warranted.

More recently, Boulter *et al.* (1979) have presented a scheme based on partial amino acid sequences of plastocyanin in 40 members of 10 families of flowering plants. Since only three of the species are the same as those used in the tree based on cytochrome *c*, no comparison of the two sets of results is possible. Based on its degree of correspondence with existing concepts, the plastocyanin tree is a good deal better than the cytochrome *c* tree, but still not good enough to be taxonomically helpful.

Among the good features of Boulter's plastocyanin tree, from my standpoint, are the following: all seven tested members of the Asteraceae are associated, and fairly well removed from anything else; all of the other tested members of the Asteridae come out on one major branch of the tree, and the three tested members of the Caprifoliaceae come out on the same side branch; the four tested members of the Apiaceae all come out on the same branch; the three tested members (two species) of Brassicaceae all come out on the same branch; the two tested members of the Magnoliaceae (the only tested members of the Magnoliidae) both come out on the same branch, fairly well removed from anthing else.

Among the bad features of the same tree are the following: the Asteraceae are about as far separated as can be from other tested Asteridae; two genera (*Phaseolus* and *Vigna*) of the Fabaceae are well removed from the seven other tested genera of the family, with *Capsella, Brassica* and *Crataegus* coming between them; the two tested genera of Rosaceae, *Prunus* and *Crataegus,* are well separated, with most of the Fabaceae coming between them; the five tested genera of Solanaceae come out on three separate branches of the main branch on which they are more loosely associated, and the branch for the Caprifoliaceae separates one of these solanaceous branches from the other two. These are serious flaws indeed. It is scarcely to be believed that *Phaseolus* and *Vigna* are more closely related to the Brassicaceae than to other genera of the Fabaceae, or that *Capsicum* is more closely related to the Caprifoliaceae than to *Lycopersicon* and *Solanum.*

Boulter (1979) has also presented an affinity tree of plastocyanin for 21 members of the Asteraceae. One might have hoped that such a tree for a single large family would have fewer anomalies than one for more distantly related groups. Unfortunately, this is not the case. The five tested members of the Lactuceae (surely a natural group) are scattered from one end of the tree nearly to the other, and the nearest neighbour of *Lactuca* in the scheme is *Senecio.* One climbs such a tree at mortal risk.

I hear from Dr Uwe Jensen that some sites on the plastocyanin molecule are very labile (in an evolutionary sense), varying within a genus or even within a species, whereas others are more stable, varying only between genera or families (so far as present evidence shows). Thus it might be possible and indeed necessary to ignore the more labile sites in the construction of affinity diagrams, and to rely on the more stable ones. This sort of weighting, necessary though it may be, has its own problems.

At this point we must directly consider some of the problems in assessing the data on amino acid sequences. A more extended discussion is presented in my 1976 paper, but some recapitulation is necessary here.

The construction of phylogenetic trees or affinity diagrams from the amino acid sequence data requires complex calculations, based on mathematical programs whose validity for the purpose is not immediately obvious and can be tested only by the results. Simple one to one comparisons, with a count of the number of differences, were early given up as leading to palpably untenable conclusions. (As Dr Uwe Jensen pointed out in discussion at Nobel Symposium 25 in Stockholm in 1973, there are more differences in cytochrome *c* between maize and wheat than between maize and some dicotyledons.) More complicated, and correspondingly more arcane, methods were devised by people considering cytochrome *c* in various groups of animals. The different methods have been judged, in effect, by their success in producing plausible phylogenetic schemes. One of these methods, hopefully called the ancestral sequence method, gives mostly tolerable results when applied to groups of animals. It has been less successful, to say the least, when applied to plants. Furthermore, when any large number of groups is considered, the method requires excessive calculation time even for a computer. Thus, in 1979, Boulter *et al.* had to say, "The ancestral sequence method was applied only to a limited set of plastocyanin data owing to the excessive computing time which would have otherwise been required; even so, many different equally parsimonious affinity trees were found. However, a method based on the identification of the 'compatible' residues reduced the ambiguity in the results and gave the relationships summarized in figure 1".

Thus in an effort to handle the sequence data we are forced to use increasingly complicated methods, of uncertain validity, in order to get results we can live with. The conclusions, such as they are, are progressively further removed from the data, allowing progressively more room for the operation of Murphy's Law (Anything that can go wrong will go wrong).

Even if (I hesitate to say when) we eventually find a generally suitable method to handle the data, we will unfortunately have lost one of the things that initially made the amino acid sequence approach attractive: its independence from the traditional sorts of data. Different ways of handling the data are judged by how well they correlate with what else we know or believe. This does not make the sequence data inherently any worse than other kinds of data, but neither does it make them any better. No radical restructuring of the system can be based on the sequence data, but when we find a way to get results that are concordant with the more firmly established parts of the present system, the remaining discordances can become suggestions for possible change.

In concluding this section, I should point out that in their most recent paper Boulter *et al.* (1979) are appropriately cautious about the present significance of the sequence data. I can do no better than to quote them: "...at this stage in the investigation it is not possible to say to what extent the amino acid sequence data give new insights into the phylogeny of Flowering Plants since 'random' evolutionary amino acid substitutions lead to distortions when 'trees' are constructed from amino acid sequence data using present data handling methods. The distinction between useful new information and the distortions will only become fully apparent when more protein data sets are assembled".

With all its problems, I have great hopes for the amino acid sequence approach in the future. Flying machines were strictly visionary in 1900.

2. *Secondary Metabolites*

The chemical compounds that have been taxonomically the most useful up to the present time are the secondary metabolites, perhaps simply because there are so many of them. They form a great horde of chemically unrelated substances that do not appear to be necessary for ordinary metabolic functions. They include such diverse things as alkaloids, anthraquinones, betalains, cyanogens, many (not all) flavonoids, iridoid compounds, mustard oils, polyacetylenes, saponins, sesquiterpene lactones, tannins, triterpenoids – the list seems endless.

Each kind of secondary metabolite, and each set of kinds, occurs in some quantity in a limited array of plant taxa, and is wanting (or nearly so) from others. The distribution is not at all random, but shows varying degrees of correlation with groups recognized on other bases. I do not know of any perfect correlations of secondary metabolites with taxonomic groups larger than a good-sized genus, but within the angiosperms imperfect correlations occur at all levels up to and including the two classes (monocots and dicots).

Like other chemical features, secondary metabolites have had their greatest acceptance as taxonomic characters when they have been used in conjunction with other characters as part of a comprehensive re-evaluation. Taxonomy depends on multiple correlations, and we want to make sure that any new scheme shows a better set of correlations and fewer anomalies than the one it replaces.

Two of the groups of secondary metabolites that show the best correlation with major taxa are the betalains and the mustard oils. Within the angiosperms, betalains are restricted to the Caryophyllales, where they are widespread, but missing from two families, the Caryophyllaceae and Molluginaceae (Mabry, 1977 and many other papers). These latter two families produce anthocyanins, which are missing from all the betalain families. This degree of correlation requires that *Gisekia,* which is in some respects transitional between the Phytolaccaceae and Molluginaceae, be assigned to the Phytolaccaceae. The Molluginaceae have been only very sparsely sampled for pigments, and I am not convinced that we will not eventually have to recognize both betalain and anthocyanin genera in the family.

Betalains have played an important role in the now generally accepted assignment of the Cactaceae and Didieriaceae to the Caryophyllales. Prior to study of the pigments, the position of these two families had been debated, and they were not widely considered to belong to the Caryophyllales. Their production of betalains made a position in the Caryophyllales seem plausible, and subsequent studies of sieve tube plastids (Behnke, 1976) and pollen (Straka, 1965; Nowicke, 1975; Skvarla and Nowicke, 1976) provide strong support for this assignment.

Here we have one of the best examples of the successful use of chemical data to help solve a major taxonomic problem in flowering plants. It is worth noting that it took a decade of controversy for botanists to come to a general agreement on how best to adjust the system to provide for the data, and that the eventual accord was based on multiple correlations rather than on an arbitrary assignment of a particular weight to a particular chemical or other character. These events hold a lesson, if we will heed it.

The taxonomic use of mustard oils presents a problem opposite to that of betalains. Betalains are restricted (within the angiosperms) to the Caryophyllales, but not everything in the order has them. Mustard oils occur throughout the Capparales, but also in a sprinkling of other groups. In addition to the 4000 or so species of Capparales, mustard oils occur in a total of a little more than 300 species that are variously distributed in different schemes of classification. No one that I know of puts all species with mustard oil into the Capparales, although Dahlgren (1977) comes close, taking in everything but *Drypetes* and the Caricaceace. In my own scheme, the non-capparalean producers of mustard oil are distributed in nine families, belonging to six orders: Bataceae and Gyrostemonaceae, collectively making up the order Batales; Limnanthaceae and Tropaeolaceae, both in the Geraniales; Bretschneideraceae (consisting of a single species) and some Zygophyllaceae, in the Sapindales; Caricaceae, in the Violales; some Salvadoraceae, in the Celastrales; *Drypetes*, in the Euphorbiaceae, Euphorbiales. These groups are also widely scattered in the systems of Takhtajan (1966) and Thorne (1976).

It now seems to be generally agreed that at least the following families belong to the order Capparales: Capparaceae, Brassicaceae, Resedaceae, Moringaceae and Tovariaceae (if distinguished from the Capparaceae). Although there are other reasons for associating these families, their consistent production of mustard oil is in my opinion the best single unifying feature. I do not myself find it useful to include any other families in the order. I take some comfort from the fact that one of the other mustard oil families (Caricaceae) is in the closely related order Violales, and the presence of mustard oils is one of the reasons why I assign the Batales to a position next to the Capparales, but I simply have to accept the fact that the other mustard oil families are scattered amongst several orders of Rosidae. In summary, it now seems to be generally agreed that the vast majority of species with mustard oils being together in one order, but that a small proportion of mustard oil species must be excluded from the order. There is still active disagreement on how many such species should be excluded, and how they should be distributed in the system.

Chemical characters played a large role in the now generally accepted dissolution of the traditional order Rhoeadales into the orders Capparales and Papaverales. The present agreement arises from a general reconsideration of both morphological and chemical features, but we are here concerned only with the chemical ones. Serological studies (Frohne, 1962; Jensen, 1967), as well as the secondary metabolites, support the present view. The alkaloids of the Papaverales suggest that the order belongs in the Magnoliidae *sens. lat.* (under whatever name), close to the Ranunculaceae and Berberidaceae. Here Dahlgren, Taktajan, Thorne and I are all in basic accord.

The widespread (though far from universal) occurrence of the otherwise rare benzyl-isoquinoline and the related aporphine alkaloids in the Magnoliidae is one of the generally recognized features that helps to mark the group. Except for the addition of the Papaverales, however, the alkaloids have had little influence on efforts to define the group. For the record, I should add that Takhtajan (e.g. 1973) now segregates the Ranunculidae as a subclass distinct from the Magnoliidae, but his concepts of the evolutionary relationships remain unchanged and similar to my own and those of Thorne and (to a perhaps lesser extent) Dahlgren.

Some of the Rutaceae also contain benzyl-isoquinoline alkaloids, and some authors (e.g. Kubitzki, 1969) have seized upon this to suggest that the Rutaceae are related to the Magnoliidae (under whatever name). The Rutaceae have so many different kinds of alkaloids, however, that it scarcely seems fair to choose only one of these, at the expense of the others, as a guide to affinity. In some other chemical features, such as the abundant triterpenoid bitter substances, the Rutaceae are perfectly at home in the Sapindales as I define that order. It is also significant, though not conclusive, that a serological study by Jensen (1974) revealed no affinity between *Phellodendron* (Rutaceae) and *Berberis*.

From the standpoint of a generalist like myself, one of the most useful papers suggesting taxonomic reconsiderations to account for chemical evidence is that of Hegnauer (1969), who has been extraordinarily perspicacious in such matters – which is not to say that I always agree with him. His comments on the Papaverales and Caryophyllales (Centrospermae) in the cited paper are in accord with the views presented here, and they are indeed now widely accepted. His views on the Cornales and Apiales, noted in the following paragraph, are more controversial.

There is a long-standing difference of opinion as to whether the Apiales (Araliales, Umbellales) and Cornales make up one order or two. In my opinion they form two separate orders, which relate to each other only through the ancestral Rosales. Hegnauer strongly supports the separation of the two orders on the basis of their secondary metabolites. For instance, the Cornales frequently accumulate iridiod compounds and ellagic acid, which the Apiales do not, and the Apiales characteristically accumulate polyacetylenes, which the Cornales do not. Heganuer would even put the eventual connection between the Cornales and Apiales one step further back than I, in the "Ranalian stock" (= Magnoliidae). I welcome Hegnauer's support, but I cannot say that many people on the other side have yet been convinced. There is in fact even some chemical evidence that points the other way. *Garrya* and *Aucuba,* in the Cornales, store petroselenic acid in seeds. Petroselenic acid is a characteristic seed oil of the Apiales and is also found in *Picrasma,* of the Simaroubaceae. No matter what sort of phylogenetic scheme one proposes for whatever group, there are generally some resemblances or shared characters that can only be explained by postulating parallelism or convergence. Evolutionary parallelism is a way of life for the angiosperms.

Chemistry has played a role in the now widely accepted removal of the Callitrichaceae from the vicinity of the Haloragaceae, the still controversial removal of the Hippuridaceae from the same area and the assignment of both to a position as florally reduced aquatic derivates of the higher sympetalous families (Hegnauer, 1969). I find it useful to recognize an order Callitrichales in the Asteridae, to include the Callitrichaceae, Hippuridaceae and the chemically uninvestigated Hydrostachyaceae. Here again the taxonomic changes were not based solely on the chemistry, but on a reconsideration of all the evidence. The unitegmic, tenuinucellar ovules in these families are also important in this reassessment.

I must disagree completely, however, with Hegnauer's suggestion (1969 and elsewhere) that the Asterales originated from the Apiales. I would not address myself to the matter in a formal, public way, except for the fact that some people seem to be taking the suggestion seriously.

There are indeed some chemical similarities (as well as some differences) between the two groups. Both accumulate polyacetylenes, which are not otherwise very common, but those of the Asterales characteristically have cyclic end-groups, in contrast to the mainly aliphatic polyacetylenes of the Apiales (Sørensen, 1977). Sesquiterpene lactones, widespread in the Asterales, have now been found in a few Apiaceae, but I believe not as yet in the Araliaceae; certainly they are not typical for the order. The pyrrolizidine alkaloids of the Senecioneae (and some Eupatorieae) and the latex of the Lactuceae have no parallel in the Apiales. The two orders also differ in their storage products. The Asterales characteristically store carbohydrate as inulin, which the Apiales do not. Seeds of the Apiales characteristically contain large amount of petroselenic acid, which is wanting from the Asterales. The chemical similarities between the two orders are strong enough to warrant a further comparison, but they are certainly not overwhelming.

Comparison of morphological features of the two orders discloses many important differences. The Asterales are morphologically at home in the subclass to which they give their name, even though it is difficult to decide which order is most closely related. Furthermore, they are advanced within the group; their immediate ancestors must also have been Asteridae. If one seeks an eventual Rosidan ancestor for the Asterales, one must seek something that could be ancestral to the whole subclass, not just to one highly advanced order. Surely the ancestors of the subclass Asteridae had a superior ovary, which would remove them at once from the Apiales. And of course the chemical similarity, such as it is, between the Asterales and Apiales is dissipated when one must take in the whole subclass Asteridae.

Even if one assumes that the Asterales can somehow be dissociated from their subclass, formidable problems remain. The pattern of relationships within the Asterales clearly indicates (to me, at least) that their ancestors must have been woody plants with simple, opposite leaves, a cymose inflorescence, a sympetalous corolla with epipetalous stamens and two median carpels united to form a compound, unilocular ovary with a single bifid style. The Apiales simply do not qualify. All other features aside, the specialized gynoecium of the Apiaceae and even the Araliaceae is totally unsuited to evolve into that of the Asterales.

Without an adequate fossil record one cannot be mathematically certain that any particular group of angiosperms could not possibly be ancestral to another, but surely the Apiales must rank very far down on the list of logically possible ancestors for the Asterales.

Iridoid compounds have received justified attention in recent years as possible taxonomic markers. Jensen *et al.* (1975) have presented an admirable review of the taxonomic distribution and possible chemical evolution of iridoids. Dahlgren (1975 *et seq.*) goes further and presents a taxonomic scheme in which all plants with iridoids are considered to belong to a single evolutionary line. He admits, however, that not every member of an iridoid group produces these compounds; presence of iridoids is a ticket of admission, but absence is no barrier if the candidate is otherwise associated with a ticket-holder.

I believe that Dahlgren's assignment of *a priori* value to iridoids is destructive rather than helpful. I see no reason why iridoids or any other group of chemicals should escape the pervasive parallelism among the angiosperms and arise only once. I see no value in trying to associate such dissimilar things as *Liquidambar,* Cornaceae, Ericaceae and Scrophulariaceae merely because they have iridoid compounds. In my view, the early Asteridae probably inherited their compounds from ancestral Rosidae, but *Liquidambar,* the Ericaceae and some other groups acquired them independently.

Giannasi (1978a) and Crawford (1978) have separately reviewed the status and future prospects of the use of flavonoids in reaching taxonomic conclusions. They agree that up until now such taxonomic attention as these compounds have received has been mainly at the level of species within a genus. The possible future use of flavonoids at higher taxonomic levels awaits the accumulation of enough data so that correlations with other characters can be observed and verified.

The Ulmaceae provide one of the few examples in which flavonoids have been used to help clarify relationships among genera. The family has traditionally been broadly defined to include a subfamily Celtidoideae as well as a subfamily Ulmoideae. (The status of *Barbeya* is not considered here.) More recently, Grudzinskaja (1965) has redefined the two groups and restored the Celtidoideae to the status of a distinct family. Giannasi (1978b) finds that the flavonoid chemistry corresponds well with Grudzinskaja's definition, although some discrepancies remain. The taxonomic rank at which the two groups should be received is still a matter of opinion.

Flavonoids have also contributed to taxonomic understanding of the position of the Julianiaceae. Although this small group has traditionally been associated with the Juglandaceae in the "Amentiferae", it has in recent years come to be appended to the Anacardiaceae as a florally reduced derivative. Takhtajan (1966) considers that the Julianiaceae find their closest relatives in the tribe Rhoideae of the Anacardiaceae. More recently, Young (1976) has found that the flavonoid constituents are most similar to those of the Rhoideae. Young even proposes to reduce the Julianiaceae to subtribal status within the Rhoideae, which is a more subjective and debatable matter.

The Simaroubaceae are a loosely knit family of uncertain limits. I currently recognize about 25 genera, after excluding *Suriana, Cadellia* and *Guilfoylia,* which together with *Stylobasium* form a small family Surianaceae. Some authors would also exclude several other genera, notably *Kirkia* and *Irvingia.* A characteristic group of triterpenoid lactones called simaroubalides has been found in at least 15 genera of Simaroubaceae (Polonsky, 1973, and subsequent authors) and is not known outside the family. Thus these compounds may eventually aid the definition of the group. *Suriana* does not have them, but so far as I know they have not been sought in *Cadellia, Guilfoylia, Stylobasium, Kirkia,* or *Irvingia.*

Bate-Smith has noted in a series of papers (e.g. 1973) that ellagic acid is found mainly in a limited set of families and orders, to the exclusion of others. In my scheme nearly all of the species with ellagic acid go into three subclasses, the Hamamelidae, Rosidae and Dilleniidae. Even with the latter two subclasses, ellagic acid is widespread in some orders and wanting from other. In spite of these correlations, the distribution of ellagic acid has not had much influence on taxonomic schemes. It is considered along with other characters, but the detectable effect has been minimal. No one has tried to use it to define a major taxonomic group, although I suggested (1977) that it probably originated at about the same time in the Hamamelidae, Rosidae and Dilleniidae when these groups were proliferating during the Upper Cretaceous.

In an effort to provide a rationale for intergrating the data from secondary metabolites into the general system of classification, I recently (1977) presented a paper in which major groups of angiosperms were visualized as originating through the discovery and exploitation of new groups of repellents. The same discovery might be made by two or more quite different groups at about the same time, but once the discovery had been effectively exploited, its potential value for subsequent exploitation by other groups would be curtailed. The theoretical basis for my approach had already been established by others, notably Ehrlich and Raven (1965) and Levin (1976). I merely tried to put the ideas into practice for the dicotyledons. My success, or lack of it, in this effort must be assessed by others. People are notoriously not the best judges of their own work.

A very similar interpretation for a more limited group of dicotyledons was independently advanced at about the same time by Gardner (1977). Gardner visualizes a "gradual replacement of a defence based on tannin and crystals (primitive Rosidae) by defences based on a variety of toxic and repellent substances (advanced Asteridae)".

One could go on with a long list of secondary metabolites whose distribution is correlated to some degree with taxonomic groups recognized on other bases, but I believe the ones we have discussed are the ones that have had the most influence on major taxonomic schemes. Cyanogens, once thought to be too irregularly distributed to be useful, may soon achieve taxonomic significance. There is real merit in Hegnauer's suggestion (1977) that the correlations are much better when differing biogenetic pathways are taken into account. Until now, however, the taxonomic possibilities have not been exploited.

3. Serological Reactions

Like so many other things, serology works when it works, and doesn't work when it doesn't. Serological reactions correlate with other features often enough so that they must be taken seriously by taxonomists, but they so often fail to correlate that they can never be taken as definitive. They provide a partially independent check on taxonomic systems that have been developed on morphological bases. If the serological data fit, fine; if not, one should take notice and see if there is cause to reconsider. Do I begin to sound like a broken record? I must accept the hazard in order to make my point.

As I have previously (1973) noted, serological methods have the special problem that the data exist only as one to one comparisons, gathered with much time and effort. Even after you have checked *A* against *B*, *C* and *D*, you do not know how *B* will react with *C* or *D*, etc. The problem seems small when we are studying a small number of species, but it becomes progressively more limiting as more species are added to the sample. If we consider that there are about 375 families of flowering plants, then it would take about 70 000 tests to compare just one sample of one species in each family with one sample of one species in each of all the other families. When one thinks of comparing all the species of angiosperms, the figures become astronomical. Thus the sample must always be relatively small, and its selection is inevitably biased by previous taxonomic concepts. In spite of these and other problems, serology can make a useful contribution to taxonomy.

Probably the most significant examples in which serology has helped to resolve taxonomic problems are those of the Caryophyllales, Rhoeadales (= Papaverales and Capparales), and the genus *Paeonia*. We have already commented on the first two groups, noting that serology placed a significant but not dominant role in the taxonomic reconsideration.

Hammond's studies (1955) of *Paeonia* and the Ranunculaceae merit special comment. *Paeonia* has traditionally been included in the Ranunculaceae, but it is discordant there in many features. Difference after difference was pointed out by a series of botanists, but the genus was still routinely included in the Ranunculaceae. After Hammond showed that *Paeonia* is also serologically isolated from the Ranunculaceae, the evidence could no longer be blithely ignored. Virtually everyone now treats *Paeonia* in its own family, allied to the Dilleniaceae and well removed from the Ranunculaceae. So far as I know, no one has yet checked the serological reaction of *Paeonia* with anything in the Dilleniaceae.

Serology is now being brought to bear on the long-disputed position of the Juglandaceae. Preliminary results (Peterson and Fairbrothers, 1978) support the placement of the Juglandaceae in the subclass Hamamelidae, and their wide separation from the Anacardiaceae. The study is still proceeding, and we should be cautious not to anticipate its conclusions. I can say that I find the results to date comforting, but their ultimate impact on the taxonomic system cannot yet be predicted.

Chupov (1978) has given the serological method a new twist by using protein from pollen instead of seeds. He too sees the Juglandaceae as closely linked to members of the Hammamelidae (*sensu* Cronquist and *sensu* Takhatajan). The taxonomic significance of some of the weaker reactions that he reports is, to say the least, subject to challenge.

A recent serological study of the Geraniales *sens. lat.* by Rullmann (1978) has produced interesting but scarcely startling results. The close affinity of the Linaceae (including *Hugonia*) and Erythroxylaceae is strongly supported, as is their relative isolation from most other tested families. This should surprise no one. Some of us maintain the Linales as a separate order. The linkage of three of the twelve tested genera of Euphorbiaceae with the Linaceae and Erythroxylaceae is a bit unexpected, but not difficult for those such as myself who have the Euphorbiales in the subclass Rosidae along with the Linales. Those who put the Euphorbiales in the Dilleniidae near the Malvales will have to dismiss this linkage as just another of the not infrequent wrong connections suggested by serology. Rullmann found only weak links among the other tested families, although members of a given family generally reacted well together. Even the Geraniaceae (four genera tested) and Oxalidaceae (three genera tested) did not show strong cross-links. Although it does not resolve any previously difficult problems, Rullmann's study contributes appropriately to the data base for the taxonomic organization of the Rosidae.

A study by Lee and Fairbrothers (1972) directed at elucidating the relationships of the taxonomically isolated order Typhales unfortunately disclosed nothing. *Typha* and *Sparganium* react well with each other, and neither reacts well with any of a wide range of other monocotyledons, including members of the Agavaceae, Araceae (including *Acorus*), Arecaceae, Commelinaceae, Cyperaceae, Juncaceae, Liliaceae, Pandanaceae and Poaceae.

Extensive serological studies of the Cornales and Caprifoliaceae by Fairbrothers and his students (Hillebrand and Fairbrothers, 1970a, b; Brunner and Fairbrothers, 1978) have produced interesting results, but these have had no major impact on the taxonomic system at least as yet.

One of the problems of the serological method has been that we do not really know what we are measuring. In a study centred on the Ranunculales, Dr Uwe Jensen is now refining the method so as to reduce the uncertainty. Instead of using a mixture of all extracted proteins, he is separating and isolating them. He tells me that roughly half of the material extracted from the seed will be one protein, about a quarter may be another and several others make up the remainder. Each protein has several (generally seven or more) active sites. Testing the individual proteins separately, it is possible to ignore the results from those with a wide pattern of reactivity, and concentrate on those that are more selective. From my viewpoint, this is all to the good. We shall, of course, have to await a formal presentation of the data.

4. Accumulation of Minerals

Several different mineral elements are accumulated by some kinds of plants in quantities far greater than is usually required for ordinary metabolism. Only two of these elements, aluminum and selenium, have attracted more than minimal taxonomic attention. I pass over the accumulation of silicon by many plants, because this has a direct morphological expression that has already been considered in classical taxonomy.

At least some members of nearly 40 families of dicotyledons are known to accumulate aluminum. These are distributed among more than 20 orders, including members of five of the six subclasses. Only the subclass Caryophyllidae has not been found to include aluminum accumulators.

In spite of this scattered taxonomic distribution, Chenery and Sporne (1976) maintain that accumulation of aluminum is a primitive feature in dicotyledons. Although they may possibly be right, I am not convinced by their complicated statistical argument. The same distribution could just as easily be accounted for in other ways.

I can see easily enough that some way to tolerate a superabundance of aluminum might have survival value for many plants especially in tropical soils, but evidently there is more than one way to achieve such a tolerance. Obviously there are many plants that tolerate aluminum without accumulating it. If aluminum accumulation is a primitive feature among angiosperms, then a great many species even on aluminum-rich soils have been able to dispense with this ancestral adaption.

It may be significant that extensive testing in the larger aluminum-accumulating families, such as the Euphorbiaceae, Melastomataceae, Proteaceae, Rubiaceae and Theaceae, turns up significant numbers of non-accumulators as well (Chenery, 1948). Given the pervasive parallelism among the angiosperms in other features, I see no *a priori* reason when the accumulation of aluminum might not have arisen several or many times. It should not be surprising if this method of meeting the aluminum problem were adopted more often by some groups than by others.

Fortunately, no one is proposing radical taxonomic changes on the basis of aluminum accumulation. The existence of accumulators and non-accumulators in the same family seems to be viewed with equanimity by all.

Selenium is accumulated by a number of species of plants in several families. Some of these require it for normal growth; others merely tolerate it. Most of the species known to require selenium belong to only a few genera: *Stanleya* (Brassicaceae), *Astragalus* (Fabaceae) and *Haplopappus* and *Machaerantherea* (Asteraceae). All tested species of *Stanleya* are seleniferous, but only certain species or sections of the other genera.

Selenium accumulation has originated at least three times in *Astragalus* (Barneby, 1964 and personal communication), but it is not randomly distributed. In North America it is restricted to and consistently present in six rather small sections. Five of these sections (Albuli, Bisulcati, Oceati, Oocalyces and Pectinati) belong to one group, whereas the sixth (Preussiani) stands apart from the others. Some of the Asiatic species of *Astragalus* also accumulate selenium. I do not know whether or not these form a coherent group, but in any case they do not belong to the North American groups of selenium accumulators.

Although the distribution of the seleniferous habit adds to the data base for the taxonomy of *Astragalus*, it has not been critical to the formulation of present taxonomic concepts.

5. Food Storage Products

Inulin may be the food storage product that has attracted the most taxonomic attention. It is the typical reserve carbohydrate of the Asterales, Campanulales, Calycerales, Boraginaceae and Polemoniaceae, among the Asteridae, and is found here and there in various groups of Rosidae (e.g. Crossosomataceae, Cornaceae, Greyiaceae, Hippocastanaceae, Malpighiaceae, Melianthaceae, Simaroubaceae), Dilleniidae (e.g. Marcgraviaceae, Pyrolaceae), Magnoliidae (e.g. Lauraceae) and a few liloiid monocots. It is generally accompanied by isokestose, a trisaccharide that may be an intermediate stage in the formation of inulin; or one member of a group may have inulin, and others may have isokestose without inulin.

The production of inulin is one of the features frequently cited as linking the Asterales with the Campanulales. It also provides evidence for keeping *Donatia* alongside the Stylidiaceae in the Campanulales, instead of transferring it to a position near the Saxifragaceae. Certainly the inulin favours these conclusions, but it can scarcely be decisive by itself. No one would propose to link the Asteraceae to the Lauraceae or Pyrolaceae by their common possession of inulin. Inulin takes its place along with other features, both chemical and morphological, that merit consideration in the formation of taxonomic schemes.

It is interesting that although the Polemoniaceae have inulin, their most immediate relatives (Hydrophyllaceae and Convolvulaceae) apparently do not. Some people (including myself) believe that the ancestry of the Campanulales is to be sought in or near the Solanales, perhaps not far from the Polemoniaceae. The presence of inulin in the Polemoniaceae would of course be highly compatible with such a relationship, but not of itself decisive.

Seeds of angiosperms most often have food reserves consisting largely of oil or fat, and commonly some protein as well. Differences in the kind of fatty acids making up the fats or oils frequently have a strong taxonomic correlation, as does the introduction of starch or hemicellulose as a major storage product.

The seed fats of the four core families of the Malvales (Tiliaceae, Sterculiaceae, Bombacaceae, and Malvaceae) characteristically contain some fatty acids of an unusual sort, with a three membered (thus cyclopropene) ring. These cyclopropenoid fatty acids have not been found, so far as I know, outside the Malvales, nor yet in the basal family of the order, Elaeocarpaceae, which connects the

Malvales to the Flacourtiaceae in the order Violales. Thus it would appear that the cyclopropenoid fatty acids evolved fairly early in the history of the Malvales and have been retained throughout their subsequent evolution.

The absence of cyclopropenoid fatty acids is one of several reasons why I do not accept Baas' proposal (1972) to include the distinctive small family Huaceae in the Malvales. Incidentally, the Huaceae have a strong garlicky odour that sets them off from all other dicotyledons. This helps to link the two genera (*Hua* and *Afrostyrax*), but it provides no guide to the relationships of the family.

Erucic acid is a characteristic major component of the seed fats of the Brassicaceae, and appears to be of very limited occurrence outside the Capparales. Is it significant that the seeds of Tropaeolaceae and Limnanthaceae also have erucic acid? These two small families also produce mustard oil, but otherwise they have little in common with the Capparales. It would be easy to dismiss one or the other chemical similarity here as a mere coincidence, but the two together are a bit worrisome, or should be, to those of us who do not associate these families with the Capparales. Still, their transfer to the Capparales would created more anomalies than it removes. Even the chemistry alone does not unequivocally favour such an association, since the cytochrome *c* of *Tropaeolum* is very different from that of *Brassica* (Boulter, 1973).

Petroselenic acid has already been mentioned as a major component of the seed fats of the Apiales and just three genera (as far as is known) in other groups of Rosidae.

Studies of many features during the past two decades have contributed to a sharp definition of the Caryophyllales, and emphasized their isolation from other orders. Questions now arise as to whether the Polygonales and Plumbaginales, associated with the Caryophyllales in some systems, might better be removed altogether from the vicinity. There is at least one chemical feature that supports the continued association of the three orders. All have starch as a major or principal component of the food reserves. But so do the Tamaricaceae and Frankeniaceae, which no one wants to associate with the Caryophyllales.

Two major groups of monocotyledons, the Commelinidae and Zingiberidae, have substituted starch for oil as the principal food reserve of the seed. Often the starch grains are clustered or compound, and the endosperm has a mealy texture. The taxonomic correlation is, expectably, not perfect. We should all be aware of the amount of oil that can be extracted from seeds of maize, a member of the Commelinidae; and four small families of the Liliales (Philydraceae, Pontederiaceae, Haemodoraceae and Cyanastraceae) have starch as the principal food reserve in the seed.

Most of the remaining families of Liliales have hemicellulose as a very important food reserve, along with oil and some protein. The hemicellulose is stored in thickened, often pitted, cell walls and the endosperm is very hard. This type of endosperm serves as a unifying feature for the bulk of the Liliales, but it is also characteristic of the palms. The Cyclanthaceae, the only near relatives of the palms, have a very ordinary, mostly rather soft, endosperm, with reserves mainly of oil and protein.

EPILOGUE

In closing this review and prospect, I must acknowledge the debt we all owe to two great gatherers, compilers and interpreters of chemical data, R. Darnley Gibbs and Robert Hegnauer. Their direct contributions to the data are enormous, and their interpretive compendia (Gibbs, 1974; Hegnauer, 1962–1973) are indispensible. I, at least, would be lost without them.

REFERENCES

Baas, P. (1972). Anatomical contributions to plant taxonomy. II. The affinities of *Hua* Pierre and *Afrostyrax* Perkins et Gilg. *Blumea* **20**, 161–192.

Barneby, R. C. (1964). Atlas of North American *Astragalus*. *Mem. N.Y. bot. Gdn.* **13**, 1–1188.

Bate-Smith, E. C. (1973). Systematic distribution of ellagitannis in relation to phylogeny and classification of the angiosperms. *In* "Chemistry in Botanical Classification. Nobel Symposium 25" (G. Bendz and J. Santesson, eds), pp. 93–102. Academic Press, New York and London.

Behnke, H.–D. (1976). Ultrastructure of sieve-element plastids in Caryophyllales (Centrospermae), evidence for delimination and classification of the order. *Pl. Syst. Evol.* **126**, 31–54.

Boulter, D. (1973). The use of amino acid sequence data in the classification of higher plants. *In* "Chemistry in Botanical Classification. Nobel Symposium

25" (G. Bendz and J. Santesson, eds), pp. 211–216. Academic Press, New York and London.

Boulter, D. (1979). Present status of the use of amino acid sequence data in plant phylogenetic studies. *In* "Evolution of protein molecules" (H. Matsubara and T. Yamanaka, eds) pp. 243–250. Japan Scientific Societies Press. Tokyo.

Boulter, D. *et al.* (1979). Relationships between the partial amino acid sequences of plastocyanin from members of ten families of flowering plants. *Phytochemistry* **18**, 603–608.

Brunner, F. and Fairbrothers, D. E. (1978). A comparative serological investigation within the Cornales. *Bull. serol. Mus., New Brunsw.* **53**, 2–5.

Chenery, E. M. (1948). Aluminum in the plant world. Part 1. General Survey in dicotyledons. *Kew Bull.* **1948**, 173–183.

Chenery, E. M. and Sporne, K. R. (1976). A note on the evolutionary status of aluminum-accumulators among dicotyledons. *New Phytol.* **76**, 551–554.

Chupov, V. S. (1978). Comparative immuno-electrophoretic studies of pollen proteins of some amentiferous taxa. *Bot. Zh.SSSR* **63**, 1579–1585 (in Russian).

Crawford, D. J. (1978). Flavonoid chemistry and angiosperm evolution. *Bot. Rev.* **44**, 431–456.

Cronquist, A. (1973). Chemical plant taxonomy: A generalist's view of a promising specialty. *In* "Chemistry in Botanical Classification. Nobel Symposium 25" (G. Bendz and J. Santeson, eds), pp. 29–39. Academic Press, New York and London.

Cronquist, A. (1976). The taxonomic significance of the structure of plant proteins: a classical taxonomist's view. *Brittonia* **28**, 1–27.

Cronquist, A. (1977). On the taxonomic significance of secondary metabolites in angiosperms. *Pl. Syst. Evol.*, Suppl. 1, 179–189.

Dahlgren, R. (1975). A system of classification of the angiosperms to be used to demonstrate the distribution of characters. *Bot. Notiser* **128**, 119–147.

Dahlgren, R. (1977). A commentary on a diagrammatic presentation of the angiosperms in relation to the distribution of character states. *Pl. Syst. Evol.*, Suppl. 1, 253–283.

Ehrlich, P. R. and Raven, P. H. (1965). Butterflies and plants: a study in co-evolution. *Evolution, Lancaster,* Pa. **18**, 586–608.

Fraenkel, G. S. (1959). The *raison d'être* of secondary plant substances. *Science, N.Y.* **129**, 1466–1470.

Frohne, D. (1962). Das Verhältnis von vergleichender Serobotanik zu vergleichender Phytochemie, dargestellt an serologischen Untersuchungen im Bereich der "Rhoeadales". *Planta med.* **10**, 283–297.

Gardner, R. O. (1977). Systematic distribution and ecological function of the secondary metabolites of the Rosidae – Asteridae. *Biochem. Syst. Ecol.* **5**, 29–35.

Giannasi, D. E. (1978a). Systematic aspects of flavonoid biosynthesis and evolution. *Bot. Rev.* **44**, 399–429.

Giannasi, D. E. (1978b). Generic relationships in the Ulmaceae based on flavonoid chemistry. *Taxon* **27**, 331–344.

Gibbs, R. D. (1974). *In* "Chemistry of flowering plants", Vol. 4 McGill-Queen's Universities Press, Montreal and London.

Grudzinskaja, I. A. (1965). The Ulmaceae and reasons for distinguishing the Celtidoideae as a separate family Celtidaceae Link. *Bot. Zh.SSSR.* **52**, 1723–1749 (in Russian).

Hammond, H. D. (1955). Systematic serological studies in Ranunculaceae. *Bull. serol. Mus., New Brunsw.* **14**, 1–3.

Hegnauer, R. (1962–73). Chemotaxonomie der Pflanzen, 6 Vols. Birkhäuser Verlag, Basel and Stuttgart.

Hegnauer, R. (1969). Chemical evidence for the classification of some plant taxa. *In* "Perspectives in Phytochemistry" (J. B. Harborne and T. Swain, eds), pp. 121–138 Academic Press, London and New York.

Hegnauer, R. (1977). Cyanogenic compounds as systematic markers in Tracheophyta. *Pl. Syst. Evol.*, Suppl. 1, 191–209.

Hillebrand, G. R. and Fairbrothers, D. E. (1970a). Serological investigation of the systematic position of the Caprifoliaceae. I. Correspondence with selected Rubiaceae and Cornaceae. *Am. J. Bot.* **57**, 810–815.

Hillebrand, G. R. and Fairbrothers, D. E. (1970b). Phytoserological systematic survey of the Caprifoliaceae. *Brittonia* **22**, 125–133.

Jensen, S. R., Nielsen, B. J. and Dahlgren, R. (1975). Iridoid compounds, their occurrence and systematic importance in the angiosperms. *Bot. Notiser* **128**, 148–180.

Jensen, U. (1967). Serologische Beiträge zur Frage der Verwandtschaft zwischen Ranunculaceen und Papaveracee. *Ber. dt. bot. Ges.* **80**, 621–624.

Jensen, U. (1974). Close relationships between Ranunculales and Rutales? Systematic considerations in the light of new results of comparative serological research. *Bull. serol. Mus., New Brunsw.* **50**, 4–7.

Kubitzki, K. (1969). Chemosystematische Betrachtungen zur Grossgliederung der Dicotylen. *Taxon* **18**, 360–368.

Lee, D. W. and Fairbrothers, D. E. (1972). Taxonomic placement of the Typhales within the monocotyledons: preliminary serological investigation. *Taxon* **21**, 39–44.

Levin, D. A. (1976). The chemical defenses of plants to pathogens and herbivores. *Ann. Rev. Ecol. Syst.* **7**, 121–159.

Mabry, T. J. (1977). The order Centrospermae. *Ann. Mo. bot. Gdn.* **64**, 210–220.

Nowicke, J. (1975). Pollen morphology in the order Centrospermae. *Grana* **15**, 51–77.

Peterson, F. and Fairbrothers, D. E. (1978). A serological investigation of selected amentiferous taxa. *Bull. serol. Mus., New Brunsw.* **53**, 10.

Polonsky, J. (1973). Chemistry and biogenesis of the quassinoids (simaroubolides). *Recent Adv. Phytochem.* **6**, 31–64.

Rullmann, H.–D. (1978). Phytoserologische Untersuchungen zur Gliederung der Geraniales. Dissert. Doktorgrades Christ-Albrecht-Universität, Kiel.

Skvarla, J. J. and Nowicke, J. W. (1976). Ultrastructure of pollen exine in the centrospermous families. *Pl. Syst. Evol.* **126**, 55–78.

Sørensen, N. A. (1977). Polyacetylenes and conservatism of chemical characters in the Compositae. *In* "The Biology and Chemistry of the Compositae" (V. H. Heywood *et al.*, eds), 2 Vols, pp. 385–409. Academic Press, London and New York.

Straka, H. (1965). Die Pollenmorphologie der Didiereaceen. *Sber. heidelb. Akad. Wiss., Math. Naturwiss. Kl.* **11**, 435–443.

Takhtajan, A. (1966). *In* "Systema et Phylogenia Magnoliophytorum" (in Russian). Soviet Sciences Press, Moscow and Leningrad.

Takhtajan, A. (1973). Evolution und Ausbreitung der Blütenpflanzen. VEB Gustav Fischer Verlag, Jena.

Thorne, R. F. (1976). A phylogenetic classification of the Angiospermae. *Evol. Biol.* **9**, 35–106.

Young, D. A. (1976). Flavonoid chemistry and the phylogenetic relationships of the Julianiaceae. *Syst. Bot.* **1**, 149–162.

2 Chemosystematics: Perspective, Problems and Prospects for the Zoologist

C. A. WRIGHT

British Museum (Natural History), London SW7 5BD, England

> Taxonomy is at the same time the most elementary and most inclusive part of zoology, most elementary because animals cannot be discussed or treated in a scientific way until some taxonomy has been achieved, and most inclusive because taxonomy in its various guises and branches gathers together, utilizes, summarizes and implements everything that is known about animals, whether morphological, physiological, psychological or ecological.

This succint summary of vitally important principles was published by G. G. Simpson in 1945 but did he, at that time, realize just what he was asking of those of us who work in the field of taxonomy?

PERSPECTIVE

From many points of view 1945 was the beginning of a new era. It certainly marked the start of an explosive phase in technological development, and scientific progress has always been dependent upon technical innovation. While the principles upon which many of our present techniques in biological research are founded had been known for some time, it was the development of easily applied technology which made them widely available. Since the work of Tswett in 1903 chromatography had been little more than an

Systematics Association Special Volume No. 16, "Chemosystematics: Principles and Practice", edited by F. A. Bisby, J. G. Vaughan and C. A. Wright, 1980, pp. 29–38, Academic Press, London and New York.

interesting toy until Consden, Gordon and Martin introduced the use of filter paper as a medium in 1944. Tiselius and his co-workers in the 1930s had shown that proteins can be separated in an electrical field and König's (1937) adaptation of the method to filter paper as a supporting medium was largely overlooked until, again, Consden and his colleagues publicized the technique in 1946. Nuttall at the turn of the century had developed a relatively crude ring-test method for determining immunological relationships between animals. Boyden and his co-workers in the 1930s improved upon the quantitative analysis of antigen-antibody precipitin reactions and started a revival of interest in this field, but it was the development of gel-diffusion techniques by Oudin (1946) and Ouchterlony (1949) which provided the kind of simplified technology needed to make immunological methods accessible to large numbers of investigators.

General application of these techniques to a broad spectrum of topics led to the development of improvements in resolving power and adaptations allowing the use of smaller quantities of material for analysis. A range of new chromatographic media appeared, enabling very precise purification of biological substances. In the field of electrophoresis, starch gel was introduced by Smithies in 1955, cellulose acetate by Kohn in 1957, polyacrylamide by Ornstein and Davis in 1964 and now the general availability of ampholines has given us isoelectric focusing. Grabar and Williams (1953) combined the advantages of antigen separation by gel electrophoresis with immunodiffusion to provide more critical analysis of protein relationships. These techniques and their derivatives suddenly made available to taxonomists a formidable array of potential tools for adding to our knowledge of animals. However, the possibilities presented have not been without their problems, many of which are still with us.

Animal taxonomists have tended to be traditionalists, basing their studies on series of preserved specimens curated in museum collections. The impact of the new methods predictably had a mixed reception, but a few workers, frustrated in the resolution of particularly intractable problems within their respective groups, welcomed the possibility of alternative approaches. At first, the need for fresh or living specimens for most of the techniques proved to be a considerable discouragement. Consequently, before launching into the complications of obtaining and maintaining appropriate material, some of those interested enough to pursue the matter carried out a few pilot experiments on locally available species. Unfortunately, many of these attempts were based upon inadequate technical knowledge (techniques were often selected more or less at random, on the basis of the most recently publicized method for which claims were being made) and the results proved to be disappointing. There is no doubt that some of the disappointment stemmed from the realization that biochemical analyses were not necessarily a panacea for resolution of their problems and that the philosophers' stone of taxonomy remained to be discovered.

In the face of difficulties such as these, the "experimental" taxonomist needed to be strongly motivated to pursue his objectives. Such motivation usually had two sources. Either it came from the frustrated taxonomists attempting to supplement equivocal morphological data, or it came from biochemists wanting to use their techniques to explore biological diversity. The first group rarely had much experience in biochemistry and few of the biochemists were conversant with the methods and traditions of classical taxonomy. Inevitably, the early, tentative contacts between two long-separated biological disciplines led to some misunderstandings. There was a certain amount of "new" taxonomy supported by indifferent biochemical data and rather more bad taxonomy based upon elegant technical procedures. Both still exist but, on the whole, the union is proving fruitful and showing signs of heterosis. Indeed, in the mid 1960s, when I reviewed the development of experimental techniques in animal taxonomy, it was possible to give reasonable coverage to the subject in a single chapter (Wright, 1966). By the early 1970s it was necessary to call upon a number of experts to review progress and the resultant volume, although treating the major groups in some detail, still omitted important contributions in many of the invertebrate phyla (Wright, 1974). If nothing else these facts refute the uncharitable remark about taxonomists by J. B. S. Haldane that "Species-making is a nice job for biologists who have an eye for small differences but lack that peculiar quality which makes the good experimenter".

PROBLEMS

1. Communication

While vigorous growth in science is highly desirable, there is often an unfortunate by-product in the form of jargon. This phenomenon is even more pronounced in activities which involve the convergence of previously divergent disciplines. Attempts to communicate between the practitioners of the different arts can lead to the development of hybrid terminology and this can be exacerbated by various factors. Within every close-knit research group a certain amount of "domestic" phraseology evolves to deal with day-to-day situations. In long-term programmes this phraseology becomes so familiar to the members of the group that it eventually finds its way into the published results. Sometimes this can help to enrich the language of science but all too often it only adds unnecessary new terms to the literature. In the analysis of protein sequence data in systematic studies one of the fashionable principles often evoked is that of "maximum parsimony". (Why not "greatest economy"?) It is a matter of regret that similar economical principles are not always applied in the written presentation of the results. If biological research is to progress and provide information of value to the world at large then it is essential that intelligible lines of communication between the constituent sub-disciplines be maintained. In taxonomy the need for clarity of communication is even greater. If the results of taxonomic research are to serve their integrative function and be properly utilized they must be readily accessible to a wide audience, not all of whom will be conversant with the finer points of the subject.

2. Quality of Data

A significant part of this book is being devoted to modern methods of analysing chemical data in systematic work. This is a most important activity in view of the increasing complexity and volume of information which is becoming available. However, we must be sure that the results which are generated are of a quality that justifies the treatment accorded to them. There is danger that the aura of exactness which mathematical analysis confers upon biological data may dignify the conclusions which are reached with a status which they scarcely deserve. To the inexperienced in the early stages of this kind of biochemical investigation the techniques themselves suggest that the results from standardized apparatus and procedures are likely to be of a more objective nature than those derived from visual comparisons. This is, of course, generally true but there is still a subjective element in the choice of the technique to be used, the material to be analysed and in the interpretation of the results. Immunological and micro-complement fixation studies have done much to dispel naive assumptions that similarity of electrophoretic mobility is necessarily indicative of the identity of proteins. The refinements of enzyme electrophoresis provide information of much greater accuracy than the earlier general protein methods but we are only just beginning to appreciate the possible effects of environmental influences, disease and ontogenetic development on enzyme systems. Attempts to quantify immunodiffusion data may be fraught with difficulties because of the ways in which antigen concentrations can alter the number and patterns of precipitin lines. Ideally the protein content of test antigens should be determined accurately in order to avoid this kind of problem but this is not always possible in practice and the results must therefore be treated with appropriate caution. The high resolution methods such as protein sequencing lend themselves best to detailed systematic analyses, but these are time consuming techniques which are scarcely practical on a large scale, and therein lies one of their possible disadvantages in that the numbers of individuals which can be examined may be limited.

3. Utilization of Data

While it is relatively easy to obtain biochemical data of some kind, it is not always so simple to decide upon the proper application of these data in a taxonomic context. Where the original objective of the investigation was to determine the possible identity or otherwise of a few previously defined taxa, the problem is not too great, provided that a sufficient number of characteristics have been examined in adequate samples of the animals concerned. This is, after all, no more than normal taxonomic practice. However, the answer begs the question by using the adjectives sufficient and adequate, neither of which can be readily or generally quantified. What is absolutely clear is that examination of a single character in one individual of each taxon is not enough. It should not be necessary to make such an obvious statement but the regrettable truth is that some experimental studies have come close to this level of inadequacy. The numerical deficiencies may be masked in various groups. It is possible to examine a large number of progeny derived from a single, self-fertilizing hermaphrodite parent or, in groups where a phase of asexual multiplication occurs in the life-cycle, it may be that the individuals examined all originate from one zygote. In either case, however large the number of individuals subjected to analysis, the sample can scarcely be said to be adequately representative. Definitions of what constitute adequate sample size must be contingent upon the size of the population being sampled and upon the reproductive characteristics of the animals concerned. No absolute criterion is possible.

Where the purpose of the investigation is more extensive and involves large complexes of ill-defined forms, lacking in distinctive morphological characters, the problem of integrating biochemical data becomes more difficult. The poor structure of the basic taxonomic framework in such complexes makes decisions concerning the level at which a biochemical trait has significance far from easy. Despite the existence of a number of guide lines established from studies on some insects and a few vertebrates, the worker in other invertebrate groups will rarely find these of much help and he must try to establish his own relevant principles. This is best achieved by first applying some technique of relatively low resolving power, possibly a general protein electrophoresis, in order to find diagnostic break-points in what at first appears to be a continuum of morphological characters. When some general groupings are established it is advisable to reassess the morphological background because it often transpires that once discontinuities have been shown to exist they can be identified morphologically. At this stage it is helpful if the nominal taxa can be grouped within the preliminary framework before proceeding to the use of methods offering higher resolution. The most usual technique of choice here is enzyme electrophoresis but the potential scope is vast and precipitate decisions on the systems to be employed can lead to enhancing the confusion rather than helping to resolve it. Experience shows that within any particular complex of species some systems will exhibit very little variation while others will prove to be extremely polymorphic. In most cases there is no way of predicting the behaviour of a particular system in an animal group and extensive preliminary trials will be needed to establish a reasonable limited basis for comparative work. If the comparative study is to contribute substantially to strengthening the framework, an effort should be made first to analyse samples from a wide geographical range and subsequently to concentrate upon intensive sampling in more restricted areas. This process of attrition should reduce the overall problem to manageable proportions and reveal those aspects in need of particular attention. To take short cuts and apply high resolution methods without the preliminary stages is similar in effect to using a scanning electron microscope for detailed structural examination before making a general appraisal of the whole organism under a dissecting microscope. Such action only leads to the development of a separate biochemical classification independent of the established system.

PROSPECTS

The technological advances which made possible our new approaches to taxonomy had an even greater impact on other fields of biology. As a result taxonomy passed through a period of recession during which it became unfashionable and lost many potential recruits to other, apparently more exciting, activities. Inevitably the proportion of resources allocated for taxonomic research declined so that even people with a primary interest in identification and classification of animals had difficulty in obtaining support. Equally inevitably, the fundamental importance of taxonomy once again asserted itself, particularly as a result of intensive environmental and biological resource investigations. These activities are now placing an enormous work load on the depleted numbers of taxonomic research workers and, as a result, there is increasing emphasis on the need for improving taxonomic services.

Resolution 9 of the eighteenth General Assembly of the International Union of Biological Sciences (IUBS) in 1973 urged all governments "to strengthen taxonomy and especially the taxonomic infrastructure required for training taxonomists". Three years later the nineteenth General Assembly of the IUBS, meeting in Bangalore, adopted Resolution 4 which "resolved that IUBS will, as a matter of urgency, tell agencies responsible for the provision of technical assistance, particularly those of the United Nations Organisation, the International Development Banks and Governments and Foundations providing multilateral and bilateral aid, of the desperate need for resources to enable taxonomic knowledge and competence to be increased in the developing world". Since the publication of the first of these IUBS resolutions, several reports on the present status and needs of taxonomy have appeared. In 1975 the North West European Working Group on Cooperation in Zoo-taxonomy (NOS) published its report, followed in 1976 by that of the Natural Environment Research Council on "The Role of Taxonomy in Ecological Research". In 1977 the European Science Foundation produced Review No. 13 from the Committee of European Science Research Councils on "Taxonomy in Europe" and this year, at last, the review group on taxonomy set up by the Advisory Board for the Research Councils has reported under the title "Taxonomy in Britain". In all of these documents the need for increased support for taxonomic research is recognized. However, although the NERC report recommends that encouragement should be given to the use of mathematical, chemical and computer techniques in the taxonomy of ecologically important groups, the ESRC review does not even mention chemotaxonomy in its zoological section. The ABRC report, which is the most extensive of the four, does give greater coverage to chemotaxonomy, but its list of applications in the zoological field appear as a poor third to those in microbiology and botany. The report recognizes that biochemical and immunological techniques are in many cases essential tools for the taxonomist, that there is a need for improved equipment and instrumentation in taxonomic laboratories and that the cost of maintaining living cultures is high compared with curation of non-living collections.

To some extent the apparent low level of zoological activity in chemotaxonomy in Britain is because much of the work is concerned with applied problems in the medical and veterinary fields. The resultant papers are published in specialist journals, where the results will be seen by those concerned with the problems, and seldom in the more traditional outlets for taxonomic work. Despite these contributions the fact remains that the number of zoological studies in which new techniques have contributed to the solution of defined problems, of the kind which stimulated the recent resolutions and reports, are few. The old criticism that modern methods have done little more than confirm that which was already known still holds in some quarters.

Although the climate of scientific opinion appears to be favourable for increased activity in chemotaxonomy, the economic forecasts are less cheerful. Many of the methods employed are expensive when compared with traditional comparative morphology but they are nevertheless cheap by comparison with the costs of "big science". Furthermore, in major programmes of environmental research, resource development and pest and disease control, where sound taxonomy is vitally important, the proportion of overall project costs, which needs to be devoted to providing good taxonomic baselines, is trivial. Not all of the problems encountered, particularly by ecologists, will merit the immediate application of chemotaxonomy until the more intractable aspects have been identified by morphological study. At that stage, when the precise problems have been defined, we must be prepared to apply appropriate techniques in order to provide the answers which are needed. In the face of continuing financial constraints we cannot expect unlimited support for esoteric studies of our own choice, but opportunities do exist for carrying out useful and exciting work. We should not ignore Humboldt's advice that the man of science must also be a man of the world!

The argument that solutions to applied problems often have their origins in academic studies has some validity. However, there is an equally valid point that many academic studies could be enhanced by an infusion of new information from a wider data base. The potential feedback of sound data from a more diverse spectrum of the animal kingdom can do nothing but good by providing broader perspectives within which theories of speciation and evolution can be synthesized.

Whatever the eventual purpose of our studies, chemotaxonomy is essentially concerned with gathering additional data, of adding to the sum "of everything that is known about animals". The methods of observation which we use may be less direct than those of our predecessors but our objectives are similar. It is particularly appropriate that this meeting is being held in Southampton, within a few miles of the home of Gilbert White of Selborne, one of the greatest biological observers that ever lived. Over two hundred years ago, without the aid of binoculars, he discovered that swifts mate on the wing and, without a tape-recorder or methods of sound spectrograph analysis, he was able to show, by differences in their songs, that the bird then known as the willow wren is in fact as assemblage of three species, the chiffchaff, the willow warbler and the woodwarbler. If the observations which we make and the conclusions which we reach stand the test of time as well as those of Gilbert White, then we shall indeed have made some useful contributions to animal taxonomy.

REFERENCES

Consden, R., Gordon, A. A. and Martin, A. J. P. (1944). Qualitative analysis of proteins: a partition chromatographic method using paper *Biochem. J.* **38**, 224–232.

Consden, R., Gordon, A. A. and Martin, A. J. P. (1946). Ionophoresis in silica jelly. *Biochem. J.* **40**, 33–41.

Davis, B. J. (1964). Disc electrophoresis. Methods applicable to human proteins. *Ann. N.Y. Acad. Sci.* **121**, 404–427.

Grabar, P. and Williams, C. A. (1953). Méthode permettant l'étude conjuguée des propriétés électrophorétiques et immunochimiques de melange de proteins. Applications au sérum sanguin. *Biochim. biophys. Acta* **10**, 193–194.

Kohn, J. (1957). A new supporting medium for zone electrophoresis. *Biochem. J.* **65**, 9P

König, P. (1937). *Actas e trabalhos do terceiro Congresso Sud-Americano de Chimica* 2, 334.

Ornstein, L. (1964). Disc electrophoresis. Background and theory. *Ann. N.Y. Acad. Sci.* **121**, 321–349.

Ouchterlony, O. (1949). Antigen-antibody reactions in gels. *Acta Path. Microbiol. Scand.* **26**, 507–509.

Oudin, J. (1946). Méthode d'analyse immunochimique par precipitation spécifique en milieu gélifié. *C.r.hebd. Seanc. Acad. Sci., Paris* **222**, 115–116.

Simpson, G. G. (1945). The principles of classification and a classification of mammals. *Bull. Am. Mus. nat. Hist.* **85**, 1–350.

Smithies, O. (1955). Zone electrophoresis of serum in starch gels. *Biochem. J.* **61**, 629–641.

Wright, C. A. (1966). Experimental Taxonomy: a review of some techniques and their applications. *Int. Rev. gen. exp. Zool.* **2**, 1–42.

Wright, C. A. (1974) (ed.). *"Biochemical and Immunological Taxonomy of Animals"* Academic Press, London and New York.

3 | New Experimental Approaches to Plant Chemosystematics

J. B. HARBORNE

Phytochemical Unit, Plant Science Laboratories, Reading University Reading RG6 2AS, England

INTRODUCTION

Since the Systematics Association's previous symposium on chemotaxonomy and serotaxonomy (Hawkes, 1968), very considerable progress has been made with the study of secondary compounds as taxonomic markers in plants. Indeed, the enthusiasm shown by plant taxonomists for utilizing chemical characters derived from plant flavonoids, alkaloids and terpenoids has become almost an embarrassment to phytochemists, since it is rarely possible to predict with certainty which particular set of chemical characters is likely to yield new information for systematic evaluation in a given plant group. Several reviews considering the impact of secondary chemistry on plant taxonomy have appeared recently, covering much of the progress in the field since 1967 (see for example Cronquist, 1977; Harborne, 1975; Swain, 1979). In the present review, attention will be focused more on some of the newer approaches to chemosystematics, involving low molecular weight substances, which are open to the plant systematist.

Originally, the secondary characters used as taxonomic markers in plants were essentially the natural products of the organic chemist (Swain, 1963). They were largely thought of as being waste products

Systematics Association Special Volume No. 16, "Chemosystematics: Principles and Practice", edited by F. A. Bisby, J. G. Vaughan and C. A. Wright, 1980, pp. 39–70, Academic Press, London and New York.

of metabolism, which accumulated in plant tissues because there was no efficient method by which the plant could excrete them. These organic molecules fell into three main groups, according to their biogenetic origin: non-protein amino acids, cyanogens and the amino acid based alkaloids; the terpenoids, including essential oils, resin acids, saponins, phytosterols and carotenoids; the phenolics, especially the flavonoids. Other substances more directly related to primary metabolism were also examined at this time, including oligosaccharides, sugar alcohols, unusual fatty acids and the fatty acid derived polyacetylenes (see Swain, 1963).

Since 1967, two main changes have taken place. First, further chemical characters have emerged which are either primary metabolites or which are so closely related to primary metabolites that they cannot be separated from primary metabolism. Secondly, more has become known about the ecological role of secondary compounds in plants, particularly in relationship to insect feeding. An important pioneering review which stimulated much research in this area was that of Ehrlich and Raven (1965) on plant–butterfly co-evolution. While much of the evidence regarding the function of alkaloids, terpenoids and flavonoids is still very circumstantial, it is nevertheless now impossible to regard them as "neutral" substances of no possible value in the plant which synthesizes them. Evidence for their undoubted survival value to the plant in terms of both plant–plant and plant–animal interactions has recently been reviewed (see for example Swain, 1977).

In the present chapter, therefore, some of these newer ideas on the relationship between taxonomy and ecology will be presented. Some of the new types of low molecular weight substance available in plants for taxonomic consideration will be mentioned, including both physiological and pathological agents. The importance of variations in metabolism and biosynthesis of secondary constituents will be discussed. Finally, the interaction between taxonomy and ecology will be evaluated in the case of both primary and secondary constituents present in the floral tissues of plants.

PHYSIOLOGICAL FACTORS

1. Photosynthetic Variation

The idea that characters derived from such a basic universal process as photosynthesis should be of assistance in systematics is, on the face of it, a very unexpected one. And yet, recent studies of plants with C_4 photosynthesis (as distinct from the more common C_3 plants) indicate that such plants are taxonomically related to each other. As is now well known, plants with C_4 photosynthesis have a distinctively different carbon pathway, because of the presence of an additional biochemical cycle, the so-called Hatch-Slack Pathway. In such plants, CO_2 from the atmosphere is first incorporated into the synthesis of two organic acids, oxalacetate and malate, before being "passed on" into sugar and starch, via the Calvin (C_3) cycle. C_4 photosynthesis represents an adaptation to tropical climates and is restricted to a relatively small number of angiosperm species.

C_4 plants are also anatomically distinct from C_3 plants, the various anatomical features distinguishing such plants being referred to as the Kranz syndrome. Depending on one's viewpoint, the two classes of plant are termed C_4 or Kranz and C_3 or non-Kranz. Indeed, both biochemical measurements and microscopic techniques can be used to determine the condition. Furthermore, both methods can be applied successfully to small fragments of herbarium tissue, so that wide surveys can be readily achieved.

As a result of such surveys, it is now known that C_4 plants are found, and often consistently, in about 11 angiosperm families (Table I). No less than six of the 11 families belong to the same natural order, the Centrospermae, a group of plants already distinguished by their unusual pigmentation based on betalains and their distinctive sieve-tube plastids. Furthermore, two of the remaining families – the Gramineae and the Cyperaceae – are recognized as being closely related in many other features. All families with C_4 plants are relatively advanced and herbaceous, so that the characteristic is apparently a specialized feature within the angiosperms.

While the overall distribution of C_4 photosynthesis in plants, as presently understood, is clearly not of major taxonomic significance, it is apparent that at the narrower confines of subfamily, tribe and

Table I. The Major Plant Families with Members which Possess the C_4 Photosynthetic Cycle[a]

Aizoaceae	Gramineae
Amaranthaceae	Cyperaceae
Chenopodiaceae	–
Molluginaceae	Euphorbiaceae
Nyctaginaceae	Zygophyllaceae
Portulaccaceae	–
	Compositae

[a] There are also isolated records of C_4 plants in seven other families: Acanthaceae, Asclepiadaceae, Boraginaceae, Capparaceae, Polygalaceae, Scrophulariaceae and Liliaceae. In all, at least 943 spp. are known to be C_4. Eighteen genera have both C_3 and C_4 members and a few (e.g. *Panicum*) have species which are intermediate in character.

genus the character becomes of interest. Thus, in the family Gramineae, it occurs in the whole of the subfamily Eragrostoideae without exception (over 60 taxa surveyed). It also occurs regularly in the subfamily Panicoideae, being represented in all tribes except for the small group Isachneae and some genera of the Paniceae. Within Paniceae, the genera considered to be the most primitive on other grounds are nearly all non-Kranz. At the generic level in *Panicum,* the character is of considerable interest in relation to recent taxonomic revision. Because of heterogeneity within this genus, as defined by Linneaus, many taxa have been separated and described under new generic names, such as *Dichanthelium* and *Hymenache.* The correctness of these decisions is nicely reflected in the fact that all recently removed species are non-Kranz, while all remaining true *Panicums* are Kranz (Brown and Smith, 1975).

Surveys for the C_4 condition in the Cyperaceae (Raynal, 1973), Compositae (Smith and Turner, 1975) and the Euphorbiaceae (Webster *et al.*, 1975) have all yielded results of taxonomic value. This character, then, which is basically an adaptive feature to tropical and subtropical environments, is nevertheless a useful taxonomic marker in plants. Further surveys are clearly called for, because of the systematic potential.

Another photosynthetic adaptation in xeric plants, which is associated with the need to conserve moisture in desert habitats, is the so-called Crassulacean Acid Metabolism (abbreviated to CAM). Such CAM plants accumulate organic acids, especially malic acid, during the night and then convert them to sugars via the Calvin cycle during the daytime. As a result, there is a large diurnal variation in organic acid in the plant leaves and the condition can be recognized by recording an extremely acid pH in the cell sap at dawn (Kluge and Ting, 1978).

First described in *Bryophyllum* and *Kalanchoe* (Crassulaceae), this photosynthetic modification is widespread, but not always present, in succulent plants. According to a recent survey (Szarek and Ting, 1977), CAM has been reported in 18 families of angiosperms (109 genera). It is also present in one gymnosperm, *Welwitschia mirabilis,* and in two epiphytic Filicophyta. Not surprisingly in view of its close biochemical similarity to C_4 photosynthesis, the presence of Crassulacean Acid Metabolism is correlated to some extent with the presence of C_4 photosynthesis at the ordinal level (Table II).

Table II. Taxonomic Distribution of CAM Plants

Succulents	
Caryophyllales•	Saxifragales
Euphorbiales•	Celastrales
Asterales•	Liliales
Rhamnales	Orchidales
Gentianales	
Non-succulents	
Thymelaeales	Bromeliales

• = C_4 plants

The fact that CAM has been found in several non-succulents is perhaps a most significant recent finding for plant systematists. It suggests that the character may turn out to be of chemotaxonomic value. No deliberate surveys, outside succulent groups, have yet been attempted; these new findings hint that such surveys might well yield data of systematic interest.

2. *Adaptation to Salinity*

Biochemical aspects of the processes by which plants adapt to the stress conditions of growing in salt marshes and near the sea have recently been the subject of intensive experimental investigations and the results again indicate a significant chemotaxonomic element in the way plants respond to NaCl stress. Plants in fact adapt in various ways to the presence of high concentrations of NaCl in the soil: they may accumulate salt in the vacuole, resist its entry into the plant or dilute it after entry. Whatever the process adopted, the presence of NaCl within the plant has the possibility of disturbing the osmotic balance within the cell. A feature of biochemical adaptation is the accumulation within the cytoplasm of low molecular weight constituents, which act as osmotica to balance any harmful influx of sodium cations.

Such substances which accumulate are usually zwitterions but a variety of structures may achieve the same result. The first osmotic agent identified in this way was proline, which accumulates to the extent of 10-20% dry weight in shoots of the halophyte, *Triglochin maritima.* More recently, a range of quaternary ammonium salts have been recognized in many halophytes, particularly the substance glycine betaine. It is interesting that glycine betaine, which is biosynthetically related to the lipid base choline, was regarded as a typical "waste product of metabolism" until this function was discovered for it. Some indication of the variety of organic molecules which have been implicated as cytoplasmic osmotica in halophytes to the present time are shown in Table III. The subject has recently been reviewed by Stewart *et al.* (1979).

Although taxonomic aspects have not yet been fully explored, it is apparent already that there is a taxonomic element (Table III) in the ability of higher plants to tolerate salt stress. Thus, all Chenopodiaceae studied appear to respond by the synthesis of glycine betaine. By contrast, grasses may produce a sulphonium salt such as S-dimethylsulphonium propanoic acid. The most recent finding in the Plantaginaceae is of the simple sugar alcohol, sorbitol, as the osmoticum of *Plantago maritima* (Ahmad *et al.,* 1979). The possibility of using these osmotica for taxonomic purposes is apparent from work on the Plumbaginaceae, where two different responses have already been recorded in this family. Thus, proline accumulates in *Armeria,* whereas a glycine betaine analogue is found in *Limonium* (Table III). Since relatively few halophytes have been screened so far for their response to NaCl entry, much further work is needed before a representative picture is available. Nevertheless, there is the promise here of a new chemotaxonomic approach to plant systematics.

Table III. Low Molecular Weight Substances which Accumulate in Salt–tolerant Higher Plants

Cytoplasmic osmotica	*Formulae*	*Halophytes known to accumulate it*[a]
Proline[b]	$\overset{+}{N}H_2-(CH_2)_3-CHCO_2^-$ (ring closed between N and CH)	*Aster tripolium* (Compositae) *Armeria maritima* (Plumbaginaceae) *Juncus roemerianus* (Juncaceae) *Triglochin maritima* (Juncaginaceae)
Glycine betaine	$Me_3\overset{+}{N}-CH_2-CO_2^-$	*Atriplex spongiosa* *Beta vulgaris* ssp.*maritima* *Salicornia* spp. (all Chenopodiaceae)
β–Trimethylaminopropanoic acid	$Me_3\overset{+}{N}-CH_2-CH_2-CO_2^-$	*Limonium vulgare*, (Plumbaginaceae)
S–Dimethylsulphonium propanoic acid	$Me_2\overset{+}{S}-CH_2-CH_2-CO_2^-$	*Spartina anglica* and other *Spartina* spp. (Gramineae)
S–Dimethylsulphonium pentanoic acid	$Me_2\overset{+}{S}-(CH_2)_4-CO_2^-$	*Diplotaxis tenuifolia* (Cruciferae)
Sorbitol	$CH_2OH-(CHOH)_4-CH_2OH$	*Plantago maritima* (Plantaginaceae)

[a] For references, see Stewart *et al.*, 1979, Ahmad *et al.*, 1979 and Cavalieri and Huang, 1979.

[b] As a protein amino acid, proline is present in low amounts in all plants; it is present at much higher levels in these halophytes.

3. Hormonal Variation

The possibility that certain classes of growth hormones in plants might be useful in systematic investigations was pointed out earlier (Harborne, 1971b), but at that time too little was known about the distribution of different hormone types to be able to substantiate the claim. The main problem in screening tissues for hormones is their extremely low levels in plants, so that highly sophisticated techniques are needed for their separation and subsequent identification (Hillman, 1978). The fact that one class of hormone, the gibberellins, accumulate in seeds has meant that such tissues have been more widely surveyed than other plant parts. Evidence that gibberellin patterns in seeds are correlated with taxonomy at family and tribal level is just beginning to emerge (Sponsel *et al.*, 1979).

Thus, data obtained on seed gibberellins indicate that fairly complex mixtures may occur in any one species. However, patterns found in representatives of the Leguminosae and the Cucurbitaceae indicate distinct differences in structural type. Thus, 13–hydroxy gibberellins predominate in the Leguminosae but are not found at all in the Cucurbitaceae. Again, within the Leguminosae, significant differences have been found at the tribal level. The gibberellins present in *Vicia* and *Pisum* (both Vicieae) lack hydroxylation at C–3 and the compounds do not normally occur in conjugated form with sugar attachments. By contrast, in *Phaseolus* and *Vigna* (Phaseoleae), the gibberellins are commonly hydroxylated at C–3; furthermore, they are often present as glucosides or as glucose esters. Typically members of Vicieae contain a hormone such as GA_{20}, whereas members of Phaseoleae might have GA_8 2–glucoside (see Fig. 1).

Another situation where hormones may accumulate in greater amounts than usual is in stressed plants. This is so, for example, for the growth hormone abscisic acid, which is present at 40 times the normal level in wheat leaves subjected to drought conditions (Milborrow and Noddle, 1970). The enhanced synthesis of abscisic acid in the wilted leaves is related to the role of this hormone in bringing about closure of the stomata, thus reducing transpiration. Again, structural variation is possible, without affecting the hormonal function. Thus, Fenton *et al.* (1977) have found that in *Sorghum* leaves *trans*-farnesol takes over the role of abscisic acid in drought stress, while Loveys and Kriedemann (1974) have reported

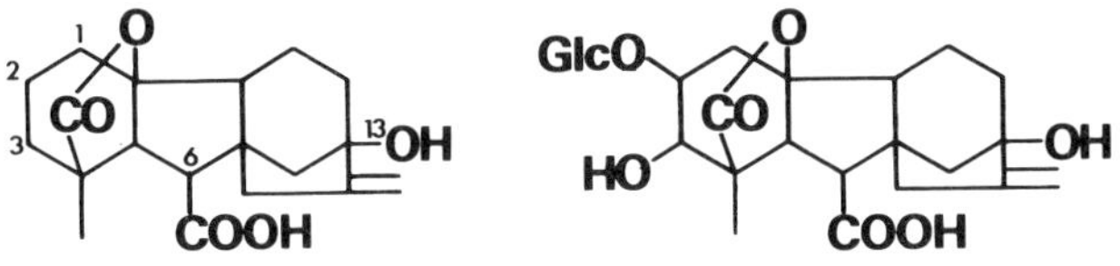

GA_{20}
(*Pisum, Vicia*)

GA_8 2–glucoside
(*Phaseolus*)

Fig. 1. Different gibberellins found in members of the Vicieae and Phaseoleae. (For simplicity, stereochemistry is omitted.)

that phaseic acid controls stomatal closure in *Vitis vinifera.* Taxonomically, the fact that *Triticum* and *Sorghum*, two members of the same family (the Gramineae), have different hormone regulators in the leaves (Fig. 2) is clearly of significance and this suggests that wider studies of this phenomenon might be rewarding from a chemotaxonomic view point.

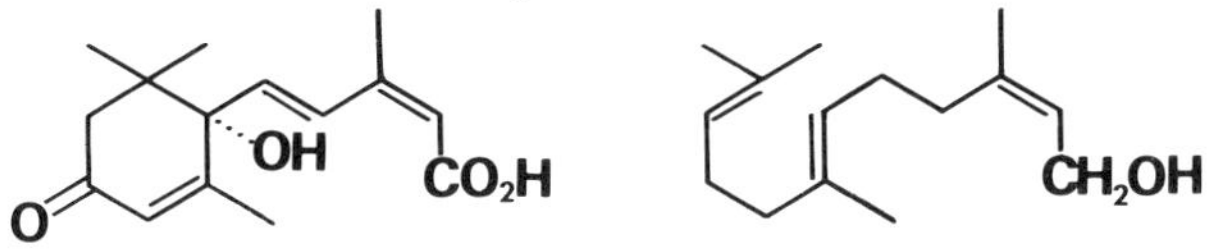

Abscisic Acid:
wilting hormone
of *Triticum*

t–Farnesol:
wilting hormone
of *Sorghum*

Fig. 2. Variation in wilting hormones in the Gramineae.

DISEASE RESISTANCE FACTORS IN PLANTS

1. *Phytoalexin Induction*

It is now well established that many higher plants respond to microbial invasion by the *de novo* production, around the site of infection, of organic compounds called phytoalexins. These "warding off" substances, although typically secondary in terms of biosynthesis, are absent from healthy plants. Although occasionally formed in plants by stress situations of a non-microbial nature, they are only produced consistently and in high concentration in plants in response to fungal invasion. While there is good evidence that phytoalexins are of importance in the protection of higher plants from fungal colonization, their production does not necessarily limit the invasion of every pathogenic organism. Indeed, phytoalexin synthesis is only one of a number of barriers to microbial attack that are present in higher plant tissues, so that their importance in disease control varies from species to species.

Although most early studies of phytoalexins concentrated on crop plants (e.g. beans, peas, potatoes), it soon became evident that a taxonomic element was present in the type of phytoalexin produced by a plant. Thus different families tended to accumulate their own distinctive classes of phytoalexin molecule: the Leguminosae in general produced isoflavonoids, the Solanaceae diterpenoids, the Compositae polyacetylenes, the Orchidaceae dihydrophenanthrenes and so on (Ingham, 1972). Anomalies were rare, for example the furanoacetylene wyerone is produced atypically in *Vicia faba,* a member of a family normally synthesizing isoflavonoid phytoalexins.

The possibility of actually turning disease resistance investigations to taxonomic advantage was first considered at Reading and successful results were achieved at the generic level in the legume genus *Trigonella* (Ingham and Harborne, 1967). The results obtained since then encourage us to suggest that this represents a valid new approach to plant relationships. The method specifically uncovers for comparative purposes the synthesis of organic molecules, which are not normally produced as part of the secondary metabolism of a given plant.

Experimentally, the procedures are very simple, but they do require living plants, normally as leaf tissue. Phytoalexins can be usually obtained by the drop diffusate technique, in which droplets containing a non-pathogenic fungus (such as *Helminthosporium carbonum*) are placed on the surface of the leaves, which are floated on water in a suitable container and left for 48 hours in diffuse light. After a few hours of incubation, the spores germinate and the fungus starts penetrating the leaf surface. This triggers phytoalexin synthesis in the tissues at the site of invasion. Quite massive amounts are produced and much is exuded from the leaf surface and is pushed into the droplet. After 48 hours, the phytoalexins, now present in the droplet in quantity, can be collected, isolated and identified by standard phytochemical techniques. Bioassays for antifungal activity are carried out at the same time, since phytoalexins by definition are significantly fungitoxic. It is also essential to run water controls in parallel and show that no phytoalexin is produced in the absence of the microbial trigger.

It is not possible here to give more than a couple of examples of the chemotaxonomic results achieved with this phytoalexin approach. Most of our effort has been devoted to the family Leguminosae, and these data have been summarized elsewhere (Harborne and Ingham, 1978; Ingham, 1980). The results of a study of phytoalexin production in the tribe Vicieae (Table IV) can be

Table IV. Phytoalexin Differences within the Tribe Vicieae

Genus	*Number of species studied*	*Phytoalexin class*	*Compounds identified*[a]
Vicia	27	furanoacetylene	Wyerone and wyerone epoxide widespread
Lens	2		
Pisum	4	6a–hydroxypterocarpan	pisatin in all *Pisum* and in 29 of 31 *Lathyrus* spp. representing 10 sections
Lathyrus	31		
Cicer	1	pterocarpan	medicarpin and maackiain

[a] Data mainly from Robeson and Harborne (1980); *Cicer,* see Ingham (1976). Other compounds besides those mentioned were found in individual species. Traces of medicarpin have been found in *Vicia faba*, but the major response is furanoacetylene production (Hargreaves *et al.*, 1976). The genus *Vicia* was screened mostly using cotyledon rather than leaf tissue.

taken as representative of the data that can be obtained by these procedures. Typical structures of phytoalexins formed in the Vicieae are illustrated in Fig. 3.

Here, the most striking result from phytoalexin comparisons is the dichotomy in the tribe, which cuts across previous views of the relationships between the four main genera. Thus *Vicia* and *Lens* are grouped together, since they both produce furanoacetylenes, a type not known anywhere else in the Leguminosae, in spite of a wide search (Ingham, 1980). By contrast, *Pisum* and *Lathyrus* are united in forming the usual legume-type phytoalexin, that is based on pterocarpan. Species of these latter genera regularly produce pisatin as well as a number of other related isoflavonoids. This chemical separation of *Vicia* and *Lathyrus* thus disclosed is unexpected, since these two genera are morphologically close; they are also similar in this ability to accumulate large amounts of non-protein amino acids in the seeds (Bell, 1966).

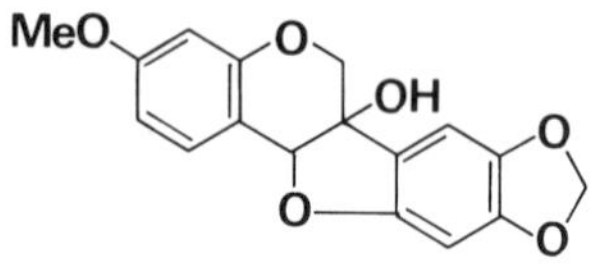

pisatin
(*Pisum, Lathyrus*)

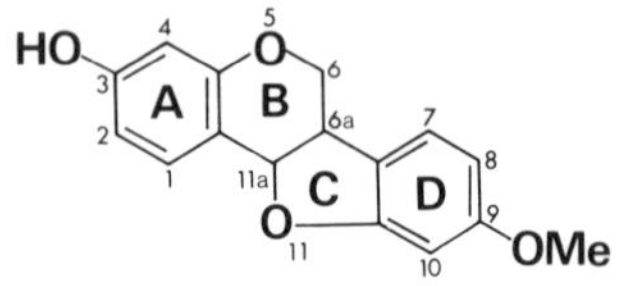

Medicarpin
(*Cicer*)

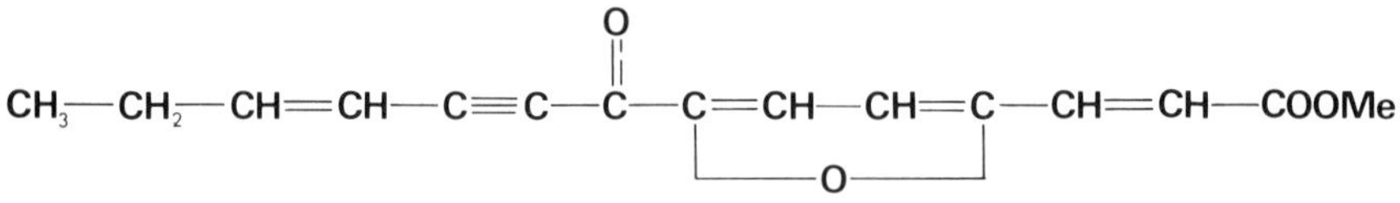

Wyerone (*Vicia, Lens*)

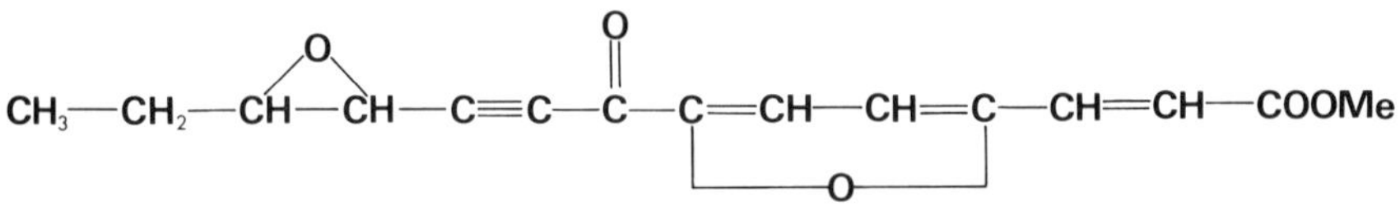

Wyerone epoxide (*Vicia, Lens*)

Fig. 3 Phytoalexin structural variation in the *Vicieae.*

As indicated in Table IV, *Cicer* is somewhat distinct from *Pisum* or *Lathyrus* in making a simple pterocarpan lacking the 6a–hydroxyl present in pisatin, namely medicarpin. This chemical difference supports the separation of *Cicer* into a separate tribe, the Cicereae, as recently proposed by Kupicha (1977) on the grounds of serology and pollen morphology.

It is interesting to compare these phytoalexin results with chemical data obtained from surveying the plants for conventional secondary products. Anthocyanin patterns have been examined in a representative number of these and related taxa (Table V). Here, it may be noted that the pattern throughout the Vicieae is distinct from that elsewhere in the Leguminosae, in the presence of anthocyanins in which the 3–substituted sugar is rhamnose rather than the much more common glucose. In fact anthocyanins are commonly present in the Vicieae as the 3-rhamnoside-5-glucosides,

Table V. Distribution of Anthocyanins in Different Legume Tribes

		Anthocyanin 3–O–sugar[a]	
Tribe	*Genus*	*Rhamnose*	*Glucose*
Vicieae	*Vicia*	+	—
	Lathyrus	+	—
	Pisum	+	—
Cicereae	*Cicer*	+	—
Trifolieae	*Parochetus*	+	—
	Trigonella	+	—
	Medicago	—	+
	Trifolium	—	+
	Ononis	—	+
Abreae	*Abrus*	—	+

[a] Data, from Harborne (1971b) and unpublished results, refer mainly to surveys of floral tissues. A rare glycosidic type, the 3–lathyroside (3–xylosylgalactoside) has been detected in flowers of *Lathyrus odoratus* and leaves of *Parochetus communis*.

although the simpler 3-rhamnoside type does also appear. This Vicieae anthocyanin type has recently been found (Harborne, unpublished results) in several Trifolieae, so that it provides a link between the two tribes. Furthermore, the distribution of 3–rhamnoside–5–glucosides fits in well with Kupicha's recent reclassification of the group, in which *Cicer* is regarded as a bridge taxon between the two tribes.

While phytoalexin induction as a technique has found its major application so far at the lower levels of plant classification, e.g. within the family Leguminosae, it is a tool that might also be useful for distinguishing different familes. Some preliminary experiments have been carried out at Reading (Richardson, unpublished results) indicating its potential with various families within the order Centrospermae. Thus an isoflavone betavulgarin has been confirmed as the typical phytoalexin response in *Beta* (Chenopodiaceae) and the same isoflavone has been provisionally detected in *Dianthus* (Caryophyllaceae). Again, unidentified orange coloured phytoalexins have been noted in representatives of two other families of the order, the Didieriaceae and the Portulacaceae. These results hint that a range of phytoalexin types might be expected within the order and that their different distributions might eventually be of significance in analysing family relationships within this natural group of angiosperms.

2. *Preformed Antimicrobial Compounds*

Other chemical barriers against the penetration of fungi into higher plant tissues may be present at the leaf surface in fungitoxic lipid soluble substances of the wax or cutin. Again, plants may contain within their cells tissue-bound toxins, which are hydrolysed during fungal attack by anzymes of the plant or of the attacking fungus to yield free toxins. Such substances are sometimes termed *post inhibitins.* Chemical variation has been noted in both of these classes of natural antifungal agents (Harborne and Ingham, 1978). Their production is, of course, part of normal secondary metabolism and they might be identified during routine investigations of the chemistry of a given plant. However, for comparative purposes, it could be useful to concentrate on these functional molecules and study their distribution in particular within a given plant group. This would appear to be as valid a chemotaxonomic approach as any other.

In the case of bound toxins, two related aliphatic compounds, tuliposides A and B, are associated in the tulip plant with the resistance of the bulbs to the wilting pathogen, *Fusarium oxysporum.* On enzyme release, the two tuliposides undergo hydrolysis, rearrangement and cyclization to yield tulipatins A and B, the true toxins (see Fig. 4). The tuliposides are not confined to cultivated tulips but occur in wild species and also in many related taxa, where they presumably have the same function. Indeed, a survey of 200 taxa in the Liliiflorae, to which *Tulipa* belongs, has been carried out (Slob *et al.,* 1975) with interesting taxonomic results. The tuliposides occur exclusively in all species surveyed of five genera: *Erythronium, Tulipa* and *Gagea* of the Lilioideae; *Bomarea* and *Alstroemeria* of the Alstroemeriaceae. These results indicate a close chemical relationship

Bound toxins
tuliposide A, R = H
tuliposide B, R = OH

Free toxins
tulipatin A, R = H
tulipatin B, R = OH

Fig. 4. Bound and free antifungal toxins of tulip bulb.

between Lilioideae (of the Liliaceae) and the Alstroemeriaceae, which is not in conflict with overall morphological resemblances between the two groups. Furthermore, they support phyletic arguments advanced on other grounds that the latter family evolved from within the Liliaceae.

There are also chemotaxonomic possibilities in surveying leaf waxes for fungitoxins, although little deliberate work has yet been done in this direction. In *Lupinus*, the isopentenyl isoflavone luteone has recently been detected in the leaf wax as a pre-infectional antifungal agent. In a study of some ten species, it was found to be accompanied by at least two related compounds, so that there was some variation between species (Harborne *et al.*, 1976). Since *Lupinus* is a taxonomically difficult genus with many species, a wider survey of surface antifungal agents would seem to be worthwhile. In other plant groups, structurally different wax constituents have been designated as antifungal agents (e.g. the diterpenes sclareol and episclareol in *Nicotiana*, see Bailey *et al.*, 1974), so that this approach is one of some general application among the angiosperms.

METABOLISM OF FOREIGN COMPOUNDS

It has been appreciated for some years (see for example Towers, 1964) that plants vary in the way they may detoxify or conjugate substances which are fed to them through petiole or stem or by spraying on to the leaf. The growth hormone, indole–3–acetic acid, may be conjugated either as the glucose ester or as the aspartic acid derivative. 6,7–Dihydroxycoumarin (aesculetin) is converted to the 6–glucoside, to a mixture of 6– and 7–glucoside, or to the 6–diglucoside. Herbicides are also known to be variously metabolized *in vivo* by plants (Naylor, 1976). Chemotaxonomic programmes in which plants are deliberately challenged with foreign compounds in order to obtain differential responses have, however, been rather few.

Until recently, the only taxonomic approach by feeding experiments appears to have been that with unnatural D–isomers of protein amino acids. For example, it has been found that D–methionine is acylated in vascular plants to the N–malonyl derivative, while in fungi and lichens it is converted to the

N–acetate. By contrast, in bacteria and algae, it is not conjugated at all but metabolized immediately by deamination. While all the above phyla are uniform in the way they dispose of D–methionine, the bryophytes are exceptional in having all three routes. Thus some bryophytes deaminate it, some conjugate it with acetic acid and others with malonic acid. The results of a feeding survey in the Bryophyta have yielded valuable data of chemotaxonomic significance, particularly for subclass classification (Pokorny, 1974).

Here, I wish to point in detail to one recent example, of possible chemotaxonomic importance in higher plants, which uses tissue culture feeding for the first time. Willeke *et al.* (1979) have demonstrated that when nicotinic acid is fed to suspension cultures, one of two alternative pathways come into operation. It is *either* converted by methylation to trigonelline *or* it is conjugated with sugar and bound as the N–arabinoside (Fig. 5). Only when conjugated as the arabinoside does the molecule then undergo further metabolism and degradation.

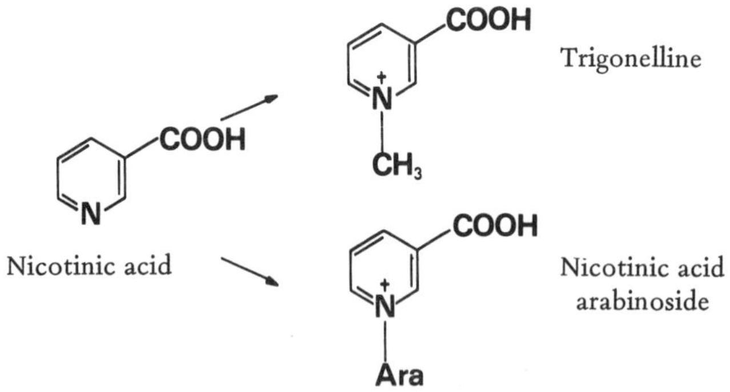

COOH
Trigonelline
N
CH_3
COOH
N
COOH
Nicotinic acid
N
Nicotinic acid
arabinoside
Ara

Fig. 5. Alternative pathways of nicotinic acid conjugation in higher plant suspension cultures.

Some fifty species of higher plant, that were available in cell culture, were surveyed by these workers for their ability to metabolize nicotinic acid and it was found that the arabinoside is, with few exceptions (Table VI), formed exclusively in members of the subclass Asterideae – all other plant groups generally follow the trigonelline route. This metabolic character thus appears to be a marker at the level of order or subclass. The production of the arabinoside in a few

Table VI. Distribution of Different Metabolic Pathways of Nicotinic Acid in the Dicotyledons

	Trigonelline production	*Arabinoside production*
Magnoliidae	*Papaver*	–
Hamamelididae	*Cannabis*	–
Rosidae	*Sedium, Rosa, Mucuna, Phaseolus, Cicer, Glycine, Arachis, Drosophyllum, Ruta, Aesculus, Parthenocissus, Euphorbia*	*Cornus, Daucus, Petroselinum*
Dilleniidae	*Sinapis, Cucumis, Bryonia*	*Anagallis*
Caryophyllidae	*Chenopodium*	–
Asteridae	–	*Galium, Catharanthus, Nicotiana, Duboisia, Digitalis, Tectona, Mentha, Ocimum, Tagetes, Haplopappus, Tanacetum*

Data modified from Willeke *et al.,* 1979. In most cases, only single species were studied in any given genera. The three gymnosperms and seven monocotyledons studied all gave trigonelline.

members of the Rosidae and Dilleniidae (see Table VI) supports the view of some taxonomists of the presence of a phylogenetic line linking these orders with the Asteridae. The fact that there are these exceptional taxa makes the character a more interesting one, since its distribution pattern questions the validity of the existing system of plant classification above the family level.

It is important to realize that this character is strictly a property of suspension cultures. While the arabinoside has not yet been demonstrated unequivocally as a product of intact plants, the methyl derivative trigonelline is a well known natural product. In fact, it occurs in whole plants of several members of the Asteridae, e.g. in *Coffea arabica* (Rubiaceae), and *Solanum tuberosum* (Solanaceae), the tissue cultures of which synthesize the arabinoside. The difference in metabolism is thus something that can be observed in cell culture. This limits its widespread use as a taxonomic tool, since there are still only relatively few plants which have been grown successfully in suspension culture. However, in the future, more and more species are likely to be cultivated in this way for other purposes. A more representative sample of angiosperm species should thus become available for exposure to foreign compounds in order to test their metabolic abilities and hence yield results of possible taxonomic value.

BIOSYNTHETIC PATHWAYS

1. Pathway Variations

It has been axiomatic in chemotaxonomic studies of secondary constituents when finding the same complex molecule, say an isoquinoline alkaloid, in two unrelated plant groups that the character can only really be considered homologous if the pathway of biosynthesis is shown to be the same in representative species from both groups. The importance of checking the biosynthetic origin of secondary compounds was established early during the development of modern chemical plant taxonomy (see Swain, 1963). The fact that one of the basic protein amino acids, namely lysine, can be formed by two alternative pathways, via α-aminoadipic acid or diaminopimelic acid, was known even before this (see Vogel, 1959).

Subsequent work on plant alkaloids has shown that in general the same pathway is followed to a particular structure, whatever the plant being studied. However, differences in the pathway do exceptionally appear. One of the best examples of this is the synthesis of simple piperidine bases. In *Conium* (Umbelliferae), such alkaloids are known to be of polyketide origin, derived from acetate units, while in *Punica* (Punicaceae), similar structures are derived from the amino acid lysine. Perhaps the class of secondary substance showing most biosynthetic variation are the naphthoquinone pigments, which may be formed, depending on the plant studied, by any one of four routes. A useful summary of biosynthetic pathways to all the major classes of secondary compound has been provided by Mann (1978).

In the case of higher plant flavonoids, radiocarbon feeding experiments have shown that these substances are always formed by the same route. Furthermore, some of the key enzymes along the pathway have also been characterized and found to have similar properties from several plant sources. Thus it can be assumed that the same flavonoid, wherever it may occur within the vascular plants, has the same taxonomic value for comparative purposes.

Although flavonoids are normally thought of as being exclusive to the vascular plant kingdom, there have been a number of reports of their occurrence in bacteria and fungi. While some of these reports have subsequently been questioned, one of them is undoubtedly correct. The presence of the flavonol chlorflavonin as an antibiotic produced by the fungus *Aspergillus candidus* is very well established. This chlorflavonin is very similar in structure to a higher plant flavonol, such as quercetin (see Fig. 6), the only real difference being the presence of a chlorine substituent. No chlorinated flavonoid has yet been reported in a vascular plant. In spite of this, the question has remained whether chlorflavonin production in *Aspergillus* is homologous or not with flavonol synthesis in higher plants.

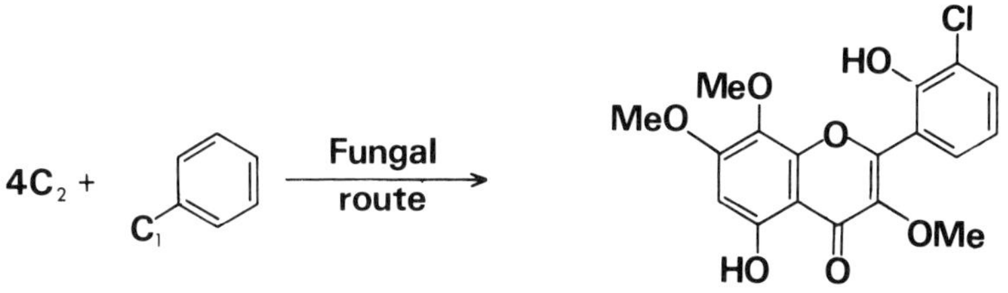

$4C_2$ + C_1 —Fungal route→

Chlorflavonin
(*Aspergillus candidus*)

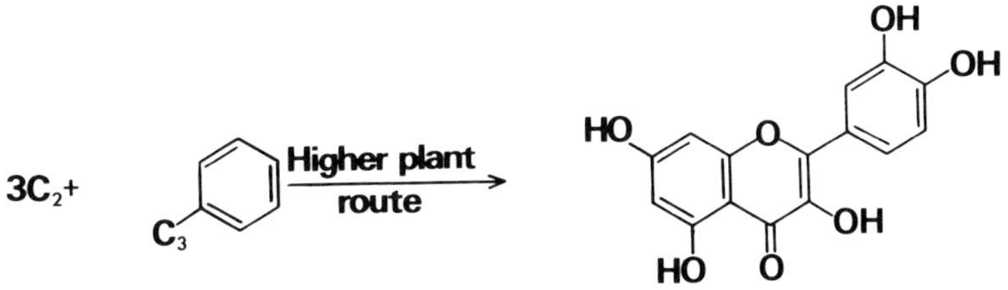

$3C_2$+ C_3 —Higher plant route→

quercetin
(widespread in higher plants)

Fig. 6. Variations in biosynthetic pathways to flavonoids in fungi and in higher plants.

An answer to this dilemma has fortunately just been provided by Burns *et al.* (1979), who have shown that chlorflavonin is actually formed by a special fungal route. Thus, the carbon skeleton of this pigment is built up from the condensation of a $C_6 - C_1$ (benzoic acid) unit with four acetate residues (as malonyl CoA). This contrasts with the higher plant route, in which quercetin is derived from the condensation of a $C_6 - C_3$ (cinnamate) unit with three acetates (see Fig. 5). Considering the wide separation between higher plants and fungi in morphology, it is very satisfying to find that there is a real diversity in the biosynthetic origin of flavonoids in the two phyla. This new discovery is of some general importance since it can be said to strengthen the validity of flavonoid comparisons within different higher plant groups.

2. *Isoenzyme Variation*

In general, the pathways of secondary metabolism within a given phylum are stable and unchanging. Differences in pathways of the type mentioned in the previous section with regard to piperidine alkaloids and naphthoquinones are relatively few and far between. A deliberate search of angiosperms for biosynthetic variants is unlikely to be rewarded with many new examples. However, even if the basic pathway is practically always the same, variation may exist at a different level: in the composition of the enzymes catalysing the various steps along the pathway. Indeed, such variation has been uncovered in the shikimate pathway, the universal route in micro-organisms and all plants for aromatic synthesis. It is also the route by which all phenolic compounds, including the flavonoids, are formed.

One type of variation is in the way the enzymes of this pathway are separate proteins or are linked together in the same protein as multi-enzyme complexes. This varies mainly according to whether the enzyme source is a bacterium, a fungus, an alga or a higher plant (see Harborne, 1980). Within higher plants, another type of variation is present in the number and type of isozymes. Thus any given enzyme of the pathway may, in some organisms, occur in more than one form. Such multiplicity in enzyme forms (or isozymes) may be needed for feedback control through product inhibition, one isozyme being specifically inhibited by one product, another isozyme be a second product and so on. Even if this does not happen, the presence of several forms of the same enzyme must surely be a safety factor and of some survival value to the organism.

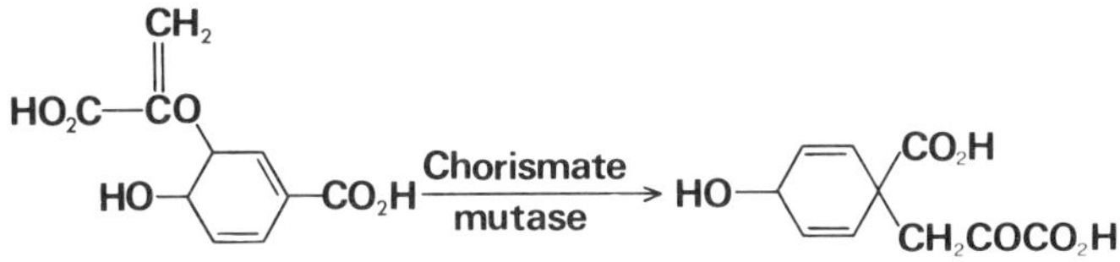

Chorismate $\xrightarrow{\text{Chorismate mutase}}$ Prephenate

2 isozymes: *Selaginella*, fern, pine, 3 legumes
3 isozymes: all other angiosperms

Fig. 7. Enzymic diversity in phenolic biosynthesis: chorismate mutase.

Two examples may be quoted from recent enzymic studies of the shikimate pathway, illustrating the taxonomic potential of isozyme variation. One is of the enzyme chorismate mutase (Fig. 7), of which three isozymes exist in vascular plants. Only two of these are formed in more primitive groups, e.g. in representative pteridophytes (*Selaginella,* a fern) and gymnosperms (*Pinus*). In contrast, most angiosperms studied have three isozymes. The only exception to this rule so far is the Leguminosae, where the primitive two isozyme situation holds (Woodin *et al.*, 1978). Further surveys are obviously of interest to see whether Leguminosae is the only "more primitve" family in the Angiospermae.

A second example of isozyme variation is in the enzyme, dehydroquinate hydrolase, which catalyses the production of dehydroshikimate from dehydroquinate (Fig. 8). Two forms have been described, the second differing from the first in that it is specifically activated by shikimic acid (Boudet *et al.*, 1977). This second enzyme is restricted in its occurrence to the monocotyledons and within the monocotyledons it occurs almost exclusively in Juncaceae, Gramineae and Cyperaceae. These are fairly closely linked families. There are also occasional occurrences in Liliaceae and Iridaceae. Clearly, this is a character that could be useful in determining the position within the monocotyledons of organisms of uncertain or disputed affinities.

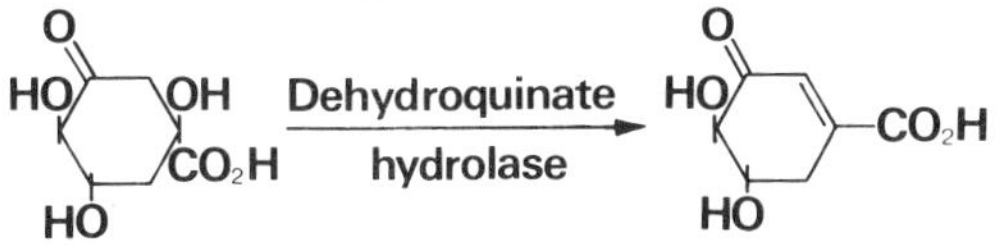

Dehydroquinate $\xrightarrow{\text{Dehydroquinate hydrolase}}$ Dehydroshikimate

Second isozyme restricted to:
Juncaceae – Gramineae – Cyperaceae

Fig. 8. Enzymic diversity in phenolic biosynthesis: dehydroquinate hydrolase.

CHEMOTAXONOMY AND POLLINATION ECOLOGY

1. *Anthocyanin Pigments*

Anthocyanins are widely distributed in the floral tissues of higher plants. Here they make an important contribution to flower colour and thus provide much of the visual attraction through which pollinating animals are drawn to plants. Since they vary structurally, particularly in the nature of the sugars attached to the anthocyanidin chromophore, these pigments are potentially useful as chemotaxonomic markers in the angiosperms. It was earlier pointed out (Harborne, 1963) that the glycosidic pattern is more useful than the anthocyanidin type, since it is very consistent within a given genus, tribe or even family. The anthocyanidin type has to be used with more caution, since it is very variable genetically and is highly selected for in relation to the differing colour preferences of pollinating animals.

Since 1963, anthocyanins have been utilized as chemotaxonomic markers in a range of different plant groups. The available data on anthocyanin distribution are summarized by Harborne (1967) and by Timberlake and Bridle (1975). In general, however, such studies have not been related directly to the ecological importance of anthocyanins as flower pigments. The opportunity was therefore taken recently to examine critically the taxonomic value of anthocyanin characters in relation to pollination mechanisms in the family Polemoniaceae. This is one of the very few angiosperm groups where detailed information is available on pollination vectors. Furthermore, many different insects and also birds are known to be pollinators within this family (Grant and Grant, 1965).

A representative sample of 34 species from 14 genera were surveyed for anthocyanin pigments, including plants pollinated by hummingbirds, bees, beeflies, flies, hawkmoths and various butterflies (Harborne and Smith, 1978a). Only three anthocyanidins were detected: the scarlet pelargonidin, the crimson cyanidin and the mauve delphinidin. As expected, the anthocyanidin or anthocyanidins present in a particular polemoniad was not related to its taxonomic position, but instead was directly correlated with the pollinator (Table VII): hummingbird-pollinated species examined contain pelargonidin or sometimes cyanidin, while bee- and beefly-pollinated

Table VII. Correlation between Anthocyanidin Type, Flower Colour and Pollinator in Polemoniaceae

Species	*Flower colour*	*Anthocyanidin*[a]
Hummingbird-pollinated species		
Cantua buxifolia	Scarlet	Cy
Loeselia mexicana	Orange red	Pg
Ipomopsis aggregata ssp. *aggregata*	Bright red	Pg
I. aggregata ssp. *bridgesii*	Red to magnenta	Pg/Cy
I. rubra	Scarlet	Pg
Collomia rawsoniana	Orange red	Cy
Bee-pollinated species		
Polemonium caeruleum	Blue	Dp
Gilia capitata	Blue–violet	Dp
G. latiflora	Violet	Dp/Cy
Eriastrum densifolium	Blue	Dp
Langloisia matthewsii	Pink	Dp/Cy
Linanthus liniflorus	Lilac	Dp/Cy
Lepidoptera-pollinated species		
Phlox diffusa	Pink to lilac	Dp/Cy
P. drummondii	Pink to violet	Dp/Cy (Pg)
Ipomopsis thurberi	Violet	Dp
Leptodactylon californicum	Bright rose	Dp/Cy
L.pungens	Pink to purple	Dp/Cy
Linanthus dichotomus	Reddish–brown	Cy

a Pg = pelargonidin; Cy = cyanidin; Dp = delphinidin. From Harborne and Smith (1978a).

species usually contain delphinidin. Lepidopteran species, on the other hand, have cyanidin or mixtures of cyanidin and delphinidin. In fact, the correlation between anthocyanidin type, flower colour and pollinator is remarkably regular (Table VII), particularly considering the fact that other biochemical factors (e.g. copigmentation with flavone) can have significant effects on the colour of anthocyanidin pigments *in vivo*.

In the Polemoniaceae, the anthocyanidins are present as the 3–glucoside or 3,5–diglucoside. In addition, the 3–sugar may be acylated with *p*–coumaric acid, so that two series of acylated derivative are found: the 3–(*p*–coumarylglucoside) and the 3–(*p*–coumarylglucoside)–5–glucoside. Acylation of the 3–sugar with *p*–coumaric acid and attachment of a glucose in the 5–position are two independent characters and they do not appear to have any direct relation to pollinating mechanisms. For example, hummingbird-pollinated species have both acylated and unacylated pigments. Furthermore, the presence/absence of acylation and 5–glucosylation is completely consistent at the species level within the various genera studied in the family.

Although these two characters vary at the generic and tribal level (Table VIII), they are of some taxonomic interest. For example, *Microsteris gracilis* has variously been considered either a *Phlox* or a

Table VIII. Distribution of Acylation and 5–Glucosylation in the Anthocyanins of the Polemoniaceae

Tribe	Genus (no. of species surveyed)[a]	Presence/Absence of	
		Acylation	5–Glucosylation
Cantueae	*Cantua* (1)	–	–
Bonplandieae	*Loeselia* (1)	–	–
Polemonieae	*Polemonium* (2)	+	+
	Allophyllum (1)	+	–
	Collomia (2)	+	+
	Phlox (5)	–	+
	Microsteris (1)	–	+
Gilieae	*Gilia* (4)	+	+
	Ipomopsis (4)	+	+
	Eriastrum (3)	+	–
	Langloisia (3)	+	+
	Navarretia (1)	+	+
	Leptodactylon (2)	–	+
	Linanthus (4)	–	+

[a] Flowers were obtained for pigment analysis from plants of the Californian flora; in some cases, collections were made in the same localities as studied by Grant and Grant (1965).

Gilia on morphological grounds. The absence of acylation in its anthocyanins would suggest it is, in fact, nearer to *Phlox* (acylation absent) than to *Gilia* (acylation present). Again, *Leptodactylon* and *Linanthus* are the only two genera in the tribe Gilieae lacking acylation. Should they, therefore, perhaps be placed nearer to the Polemonieae, where acylation is also absent in *Phlox* and *Microsteris*? In the case of *Leptodactylon,* other data from leaf flavonoids (Smith *et al.*, 1977) also indicate that it may be misplaced in the Gilieae and should be near *Phlox* (Polemonieae). These are all suggested rearrangements and clearly much more work is needed with these plants before such data can be incorporated into revised classification. This example from the Polemoniaceae, however, does reinforce again the importance of assessing the ecological significance of a particular set of chemical characters before utilizing them for chemotaxonomic purposes.

2. Ultraviolet Patterning in Flowers

The presence of ultraviolet (UV) patterning due to the differential distribution of pigment in the flowers was recognized by Daumer (1958) and Kugler (1963). Such patterns are invisible to the human eye but can be seen by bees, whose sight extends into the UV range. These authors showed that the patterning was of importance in attracting bees to flowers as pollinators. Chemical investigations of UV patterning were first stimulated by the discovery of Thompson *et al.* (1972) that the yellow flowers of *Rudbeckia hirta* (Compositae) contain two types of pigment, carotenoid in the chromoplast and yellow flavonol in the vacuole, with quite different functions. The carotenoid, uniformly present throughout the ray, is responsible for the UV reflection of the outer ray and provides visual colour to attract the bee from a distance. By contrast, the yellow flavonol is specifically located in the inner ray, where it is strongly UV absorbing, and attracts the bee to land near the centre of the inflorescence. Bees are thus able in such flowers to locate the nectar and pollen more effectively.

The role of yellow flavonoids in pollination ecology is not, however, restricted to *Rudbeckia.* Subsequent studies have shown that other yellow flowered species in the Compositae and in certain other families (e.g. Onagraceae) similarly contain yellow flavonoids which are specifically responsible for UV patterning in the flower. In fact, UV absorption may also be produced by colourless flavonols and such has been detected in *Helianthus annuus* (Harborne and Smith, 1978b). However, colourless flavonols are less advantageous, since they are not able to contribute to visual colour and thus reinforce the visual impact of the carotenoid yellow.

Although UV patterning has an important function in increasing the efficiency of pollination by some insect vectors, there is clearly much structural variation in the nature of the flavonoid responsible and hence UV patterning is chemotaxonomically of interest. A recent study of the flavonoids responsible for UV absorption within a single tribe of the Compositae has indicated the chemotaxonomic potential. No less than five different flavonoid systems were detected in these flowers: ordinary flavonol glycosides, 6–substituted flavonols, 8–substituted flavonols, chalcones and aurones (Harborne and Smith, 1978b). Some of these structures are illustrated in Fig. 9. In *Helianthus*, there was a clear taxonomic separation between the perennial and annual species in the agents responsible for UV absorption. In the perennial species, the yellow chalcone coreopsin and aurone sulphurein were widely detected. By

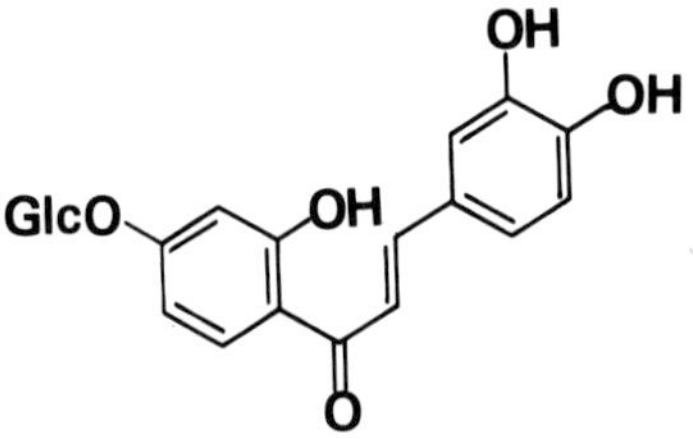

Coreopsin
perennial *Helianthus* spp.

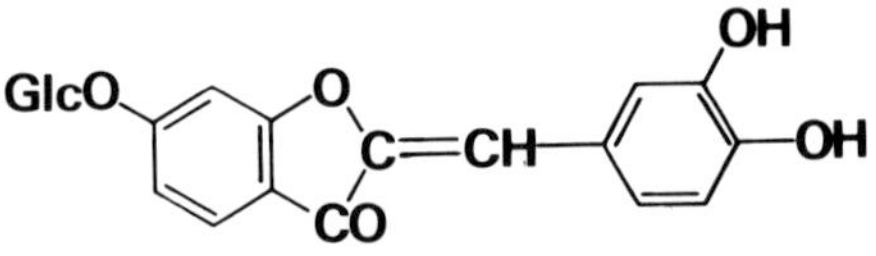

Sulphurein
perennial *Helianthus* spp.

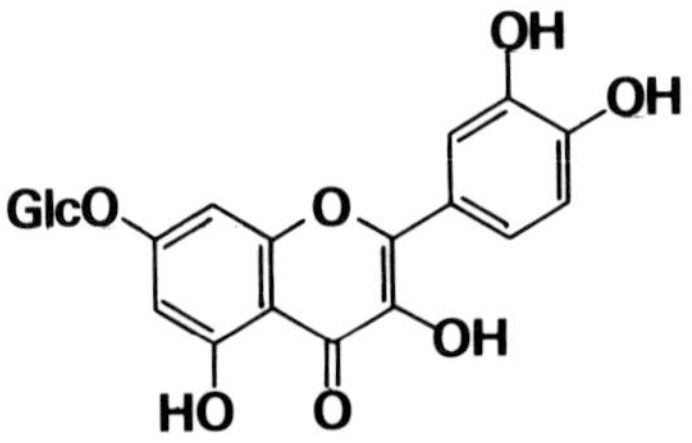

Quercetin 7–glucoside,
annual *Helianthus* sp.

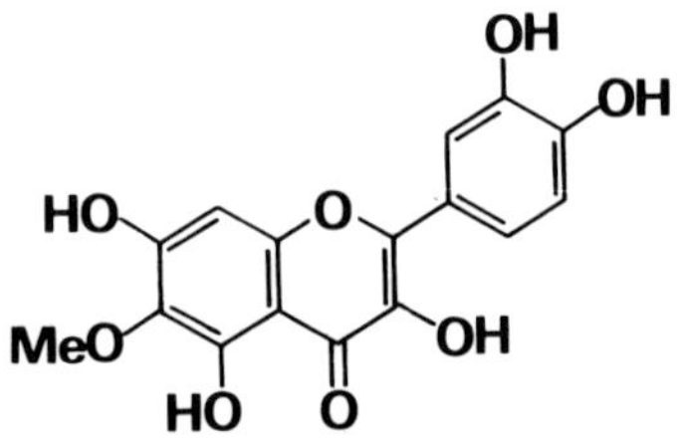

Patuletin,
Eriophyllum, Rudbeckia

Fig. 9. Some flavonoids responsible for UV patterning in composite flowers.

contrast, in the annual species *H. annuus*, the UV absorbing material of the inner ray was identified as a mixture of the 7– and 3–glucosides of quercetin. This difference may be very great to an insect pollinator. It could explain the very low incidence of inter-sectional hybridization compared to the contrastingly high incidences of hybridization among annual species or among perennial species.

Another possible taxonomic use of UV patterning is in *Eriophyllum*, a genus where the pigments responsible are the 7–glucosides of quercetagetin and patuletin (see Fig. 9). The taxonomic problem here is one of tribal affinities, the genus having originally been placed in the Helenieae and then moved more recently to the Senecioneae. The presence of patuletin and quercetagetin as the UV guides firmly links *Eriophyllum* with the Heliantheae, where these two pigments occur frequently (e.g. in *Rudbeckia*). In this case, the chemical evidence is strengthened by data from the ligule microcharacters, which also strongly favour its placement in the Heliantheae (Baagøe, 1977).

3. Nectar Constituents

Although the pattern of sugars in plant nectars is a simple one, based on glucose, fructose and sucrose, there are sufficient quantitative differences from species to species to warrant chemotaxonomic consideration. The differences in the proportion of the three main sugars are consistent and vary only slightly with physiological changes in the flower (Percival, 1961). Such differences in sugar contents have been applied to the taxonomy at the species level in several plant groups (for *Rhododendron* nectars, see Harborne 1977). Less well known is the fact that flower nectars contain trace amounts of several protein amino acids. These are of nutritional importance, especially to butterfly and bird nectar-gatherers. Again, there are significant variations in the pattern of amino acids between species. The phyletic value of analysing the amino acids in nectars has recently been demonstrated by Baker and Baker (1976). Two illustrations from their work indicate the chemotaxonomic potentialities of such an approach.

First, in a study of *Silene alba* and *S. dioica* (Caryophyllaceae), these authors showed that the nectar amino acid pattern is consistently different between the two species. Furthermore, hybrid plants can be detected (Table IX), because the amino acid patterns of the two parents are additive in the F_1 generation. Secondly, an analysis of amino acids in *Geranium* nectars was shown to confirm one of several possible origins for the octaploid *G. rubescens*. Here, the amino acids (Table X) fitted in well with the hypothesis that the octaploid was derived by autopolyploidy from *G. robertianum*, a tetraploid which itself was formed by alloploidy from the diploid *G. purpureum* and some other now extinct diploid species. The data argue against an alternative suggestion that allopolyploidy is involved

Table IX. Amino Acid Variation in Plant Nectars in Silene

Amino acid	Silene dioica	*Hybrid*	Silene alba
ASP	+	+	—
HIS	+	+	—
THR	—	++	++
VAL	+	+	—

ALA, ARG, GLU, GLY, ILE, LEU, LYS, PRO, SER common to all three plants.

Table X. Nectar Amino Acids: Confirmation of the Auto-Octaploid Origin of *Geranium rubescens*

Species Ploidy	*G. purpureum* Diploid	allo →	*G. robertianum* Tetraploid	auto →	*G. rubescens* Octaploid
Amino Acid Pattern	4 Amino Acids		identical:		14 Amino Acids[a]
	ARG		ALA	GLY	PRO
	ASP		ARG	HIS	SER
	SER		ASN	ILE	THR
	THR		ASP	LEU	VAL

[a] Also, γ—aminobutyric acid and an unknown.

at both stages of doubling the chromosomes to give *G. rubescens*. These two examples indicate that this new chemical approach is one of considerable promise at the lower levels of plant classification.

CONCLUSION

Experimental techniques have been described in this paper which have exposed new chemical characters as stable markers in plant systematics. Some of these characters are associated with the physiological adaptation of plants to hostile climatic or soil factors. Some depend on a dynamic analysis of the response of the plant to microbial invasion or to the presence of foreign compounds. Others are associated with floral ecology and the adaptation of plants to their animal pollinators. Most of these new techniques have yet to be fully developed and it is too early to assess their impact on conventional plant chemosystematics.

Undoubtedly, the main source of chemical characters for taxonomic comparison will continue in the future to be provided by the isolation and identification of the many and various secondary constituents, obtained by the fractionation of a whole plant extract. It is hoped, however, that some of the new techniques mentioned here will be used in conjunction with established procedures to yield a wider array of low molecular weight compounds than was previously available. The ideal, of course, is to employ as many kinds of chemical data as possible for the assessment of taxonomic relationships in a given plant group. Only by such means is chemistry likely to have a real impact on classical taxonomic practices.

REFERENCES

Ahmad, I., Larher, F. and Stewart, G. R. (1979). Sorbitol, a compatible osmotic solute in *Plantago maritima*. *New Phytol.* **82**, 671–678.

Baagøe, J. (1977). Microcharacters in the ligules of the Compositae. *In* "The Biology and Chemistry of the Compositae" (V. H. Heywood, J. B. Harborne and B. L. Turner, eds), pp. 119–140. Academic Press, London and New York.

Bailey, J. A., Vincent, G. G. and Burden, R. S. (1974). Diterpenes from *Nicotinia glutinosa* and their effect on fungal growth. *J. Gen. Microbiol.* **85**, 57–64.

Baker, I. and Baker, H. G. (1976). Analyses of amino acids in flower nectars of hybrids and their parents, with phylogenetic implications. *New Phytol.* **76**, 87–98.

Bell, E. A. (1966). Amino acids and related compounds. *In* "Comparative Phytochemistry" (T. Swain, ed.), pp. 195–210. Academic Press, London and New York.

Boudet, A. M., Boudet, A. and Bouysou, H. (1977). Taxonomic distribution of isoenzymes of dehydroquinate hydrolyase in the angiosperms. *Phytochemistry* **16**, 919–922.

Brown, W. V. and Smith, B. N. (1975). The genus *Dichanthelium* (Gramineae). *Bull Torrey bot. Club* **102**, 10–13.

Burns, M. K., Coffin, J. M., Kurobane, I. and Vining, L. C. (1979). Biosynthesis of chlorflavonin in *Aspergillus candidus:* a novel fungal route to flavonoids. *J. C. S. Chem. Comm.*, 426–427.

Cavalieri, A. J. and Huang, A. H. C. (1979). Evaluation of proline accumulation in the adaptation of diverse species of marsh halophytes to the saline environment. *Am. J. Bot.* **66**, 307–312.

Cronquist, A. (1977). On the taxonomic significance of secondary metabolites in angiosperms. *Pl. Syst. Evol.*, Suppl. 1, 179–189.

Daumer, K. (1958). Blumenfarben wie sie die Bienen Sehen. *Z. vergl. Physiol.* **41**, 49–110.

Ehrlich, P. R. and Raven, P. H. (1965). Butterflies and plants: a study in coevolution. *Evolution* **18**, 586–608.

Fenton, R., Davies, W. J. and Mansfield, T. A. (1977). Role of farnesol as a regulator of stomatal opening in *Sorghum*. *J. Exp. Bot.* **28**, 1043–1053.

Grant, V. and Grant, K. (1965). "Flower Pollination in the Phlox Family". Columbia University Press, New York.

Harborne, J. B. (1963). Distribution of anthocyanins in higher plants. *In* "Chemical Plant Taxonomy" (T. Swain, ed.), pp. 359–388. Academic Press, London and New York.

Harborne, J. B. (1967). "Comparative Biochemistry of the Flavonoids", 383 pp. Academic Press, London and New York.

Harborne, J. B. (1971a). Distribution of flavonoids in the Leguminosae. *In* "Chemotaxonomy of the Leguminosae" (J. B. Harborne, D. Boulter, and B. L. Turner, eds), pp. 31–72. Academic Press, London and New York.

Harborne, J. B. (1971b). Terpenoid and other low molecular weight substances of systematic interest in the Leguminosae. *In* "Chemotaxonomy of the Leguminosae" (J. B. Harborne, D. Boulter and B. L. Turner, eds), pp. 257–284. Academic Press, London and New York.

Harborne, J. B. (1975). Biochemical Systematics of flavonoids. *In* "The Flavonoids" (J. B. Harborne, T. J. Mabry and H. Mabry, eds), pp. 1056–1095. Chapman and Hall, London.

Harborne, J. B. (1977). "Introduction to Ecological Biochemistry", 243 pp. Academic Press, London and New York.

Harborne, J. B. (1980). Phenolic compounds derived from shikimate. *In* "Biosynthesis, volume 6" (J. Bu-Lock, ed.). The Chemical Society, London. In Press.

Harborne, J. B. and Ingham, J. L. (1978). Biochemical aspects of the coevolution of higher plants with their fungal parasites. *In* "Biochemical Aspects of Plant and Animal Coevolution" (J. B. Harborne, ed.), pp. 343–405. Academic Press, London and New York.

Harborne, J. B. and Smith, D. M. (1978a). Correlations between anthocyanin chemistry and pollination ecology in the Polemoniaceae. *Biochem. Syst. Ecol.* **6**, 127–130.

Harborne, J. B. and Smith, D. M. (1978b). Anthochlors and other flavonoids as honey guides in the Compositae. *Biochem. Syst. Ecol.* **6**, 287–291.

Harborne, J. B., Ingham, J. L., King, L. and Payne, M. (1976). The isopentenyl isoflavone luteone as a preinfectional agent in the genus *Lupinus*. *Phytochemistry* **15**, 1485–1487.

Hargreaves, J. A., Mansfield, J. W. and Coxon, D. T. (1976). Identification of medicarpin as a phytoalexin in the broad bean plant, *Vicia faba*. *Nature (Lond.)* **262**, 318–319.

Hawkes, J. G. (1968) (ed.). "Chemotaxonomy and Serotaxonomy". Systematics Assoc. spec. vol. 2, 299 pp. Academic Press, London and New York.

Hillman, J. R. (1978) (ed.). "Isolation of Plant Growth Substances", Soc. Exp. Biol. Seminar 4, 157 pp. Cambridge University Press.

Ingham, J. L. (1972). Phytoalexins and other natural products as factors in plant disease resistance. *Bot. Rev.* **38**, 343–424.

Ingham, J. L. (1976). Isoflavonoids from stems of *Cicer arietinum* inoculated with the spores of *Helminthosporium carbonum. Phytopath. Z.* **87**, 353–367.

Ingham, J. L. (1980). Chemosystematic survey of phytoalexin induction in the Leguminosae. *In* "Proceedings of the International Legume Conference, Kew, 1978". In press.

Ingham, J. L. and Harborne, J. B. (1976). Phytoalexin induction as a new dynamic approach to the study of systematic relationships among higher plants. *Nature (Lond.)* **260**, 241–243.

Kluge, M. and Ting, I. P. (1978). "Crassulacean Acid Metabolism", Ecological Studies 30, 209 pp. Springer-Verlag, Berlin.

Kugler, H. (1963). UV – Musterungen auf Bluten und Ihr Zustande Kommen. *Planta* **59**, 296–329.

Kupicha, F. K. (1977). The delimitation of the tribe Vicieae and the relationship of *Cicer. Bot. J. Linn. Soc.* **74**, 131–162.

Loveys, B. R. and Kriedemann, P. E. (1974). Stomatal regulation and associated changes in endogenous levels of abscisic and phaseic acids. *Aust. J. Pl. Physiol.* **1**, 407–415.

Mann, J. (1978). "Secondary Metabolism", 316 pp. Clarendon Press, Oxford.

Milborrow, B. V. and Noddle, R. C. (1970). Conversion of 5–(1,2–epoxy–2,6,6–trimethylcyclohexyl)–3–methylpenta–*cis*–2–*trans*–4–dienoic acid into abscisic acid in plants. *Biochem. J.* **119**, 727–734.

Naylor, A. W. (1976). Herbicide metabolism in plants. *In* "Herbicides" (L. J. Audus, ed.), Vol. 1, pp. 397–426. Academic Press. London and New York.

Percival, M. S. (1961). Types of nectar in angiosperms. *New Phytol.* **60**, 235–281.

Pokorny, M. (1974). D–Methionine metabolic pathways in Bryophyta: a chemotaxonomic evaluation. *Phytochemistry* **13**, 965–972.

Raynal, J. (1973). Notes Cyperologiques: contribution à la classification de la sous-famille des Cyperoideae. *Adansonia,* Ser. 2, **13**, 145–171.

Robeson, D. J. and Harborne, J. B. (1980). *Phytochemistry* **19**, 2359–2366

Slob, A., Jekel, B., Jong, B. de and Schlatmann, E. (1975). On the occurrence of tuliposides in the Liliiflorae. *Phytochemistry* **14**, 1997–2005.

Smith, B. N. and Turner, B. L. (1975). Distribution of the Kranz syndrome among Asteraceae. *Am. J. Bot.* **62**, 541–545.

Smith, D. M., Glennie, C. W., Harborne, J. B. and Williams, C. A. (1977). Flavonoid diversification in the Polemoniaceae. *Biochem. Syst. Ecol.* **5**, 107–115.

Sponsel, V. M., Gaskin, P. and MacMillan, J. (1979). Identification of gibberellins in immature seeds of *Vicia faba* and some chemotaxonomic considerations. *Planta* **146**, 101–106.

Stewart, G. R., Larher, F., Ahmad, I. and Lee, J. A. (1979). Nitrogen metabolism and salt-tolerance in higher plant halophytes. *In* "Ecological Processes in Coastal Environments" (R. L. Jefferies and A. J. Davy, eds), pp. 211–228. Blackwell, Oxford.

Swain T. (1963) (ed.) "Chemical Plant Taxonomy", 543 pp. Academic Press, London.

Swain, T. (1977). Secondary compounds as protective agents. *Ann. Rev. Pl. Physiol.* **28**, 479–501.

Swain, T. (1979). Flavonoids as chemotaxonomic markers in plants. In Press.

Szarek, S. R. and Ting, P. I. (1977). The occurrence of Crassulacean Acid Metabolism among plants. *Photosynthetica* **11**, 330–342.

Thompson, W. R., Meinwald, J., Aneshansley, D. and Eisner. T. (1972). Flavonols: pigments responsible for ultraviolet absorption in nectar guide of flowers. *Science* **177**, 528–530.

Timberlake, C. F. and Bridle, P. (1975) Anthocyanins. *In* "The Flavonoids" (J. B. Harborne, T. J. Mabry, and H. Mabry, eds), pp. 214–266. Chapman and Hall, London.

Towers, G. H. N. (1964). Metabolism of phenolics in higher plants and micro-organisms. *In* "Biochemistry of phenolic compounds" (J. B. Harborne, ed.), pp. 249–294. Academic Press, London and New York.

Vogel, H. J. (1959). On biochemical evolution: lysine formation in higher plants. *Proc. Nat. Acad. Sci. U.S.* **45**, 1717–1721.

Webster, G. L., Brown, W. V. and Smith, B. N. (1975). Systematics of photosynthetic carbon fixation pathways in *Euphorbia. Taxon* **24**, 27–33.

Willeke, U., Heeger, V., Meise, M., Neuhann, H., Schindelmeiser, I., Vordemfelde, K. and Barz, W. (1979). Mutually exclusive occurrence of trigonelline and nicotinic acid arabinoside in plant cell cultures. *Phytochemistry* **18**, 105–110.

Woodin, T. S., Nishioka, L. and Hsu, A. (1978). Comparison of chorismate mutase isozyme patterns in selected plants. *Pl. Physiol.* **61**, 949–952.

4 Chemical Systematics of Social Insects with particular reference to Ants and Termites

P. E. HOWSE

and

J. W. S. BRADSHAW

Chemical Entomology Unit, Departments of Biology and Chemistry, Southampton University, Southampton, England

Abstract: There have been detailed chemical studies on the exocrine gland secretions of ants and termites during the last few years, but in general the behavioural role of components of such secretions is poorly known. The chemical nature of defensive secretions used by termites may be related to the mode of application of the secretion, and the susceptibility of the principal predators. Many species use terpenes as repellents or toxins in which the mixture of diterpenes present appears to be species-specific. However, until the ecological significance and the mode of action of such secretions is better understood their importance in systematics can be only very subsidiary.

Ants produce a considerable variety of volatile compounds and attempts have already been made to use these for taxonomic purposes, and to reconstruct phylogenies. However, the relationships of the important primitive subfamily Ponerinae with other subfamilies are not illuminated by consideration of their exocrine chemistry. Examples from the tribe Attini and the genera *Oecophylla* and *Myrmicaria* are used to indicate the complexity of chemical characters, at genus and species level, and the variations that frequently occur within species, related to caste polymorphism, and intercolony differences.

Systematics Association Special Volume No. 16, "Chemosystematics: Principles and Practice", edited by F.A. Bisby, J.G. Vaughan and C.A. Wright, 1980, pp. 71–90, Academic Press, London and New York.

THE BASIS FOR CHEMOTAXONOMY IN SOCIAL INSECTS

The literature on insect pheromones has grown at a considerable pace during the last decade, and it is this that provides us with information of potential use as an aid to systematics. However, unlike their colleagues in the field of botany, entomologists have been motivated from the beginning by a desire to determine the functions of chemical secretions, so that chemical identification of components of secretions has commonly been the end of a piece of research, rather than the beginning of it. Furthermore, the chemicals produced most copiously by insects and used in communication and defence tend to be relatively small volatile molecules with less than 20 carbon atoms. It is therefore very optimistic to believe that the information collected so far can be of a similar use in insect systematics to the information available on plant chemistry.

Insect pheromones and defensive secretions are often, and perhaps almost always, multicomponent and multifunctional (Silverstein and Young, 1976; Howse *et al.*, 1977). Only in relatively recent years has this been realized. This means that there is a danger in using the presence or absence of arbitrarily selected major components as systematic criteria: such criteria tend to lose their force when subsequent research shows a substance to be present as a minor component rather than absent altogether in a given species. The ratios of major components in the sexual attractant pheromones may be characters of sibling species of Lepidoptera (see Roelofs and Cardé, 1977) acting as species isolating mechanisms. However, studies of the variation in ratios of the quantities of pheromonal components in geographically distinct populations of the same species have rarely been carried out.

Volatile secretions used in alarm and defence are very common among social insects. In ants, defensive secretions are largely confined to the poison gland and dispensed through a sting, or as a spray or volatile droplet when a sting is absent. In termites, a gland with a large reservoir in the head capsule of soldiers, the frontal gland, is used generally for production of a defensive secretion, and the alarm function appears to be a secondary and subsidiary one. Some species have, alternatively or additionally, glands with reservoirs extending into the abdomen and opening in the region of the labium. No volatile secretions have been reported from such glands, which contain substances such as mucopolysaccharides, quinones and proteins (Moore, 1974), and insufficient work has been done on them to be of any systematic value.

The ants as a whole produce a considerable variety of volatile organic compounds. As a rule these are contained in a number of exocrine glands which are distributed throughout the body (Fig. 1). Not all of these glands occur in all species; the mandibular, poison

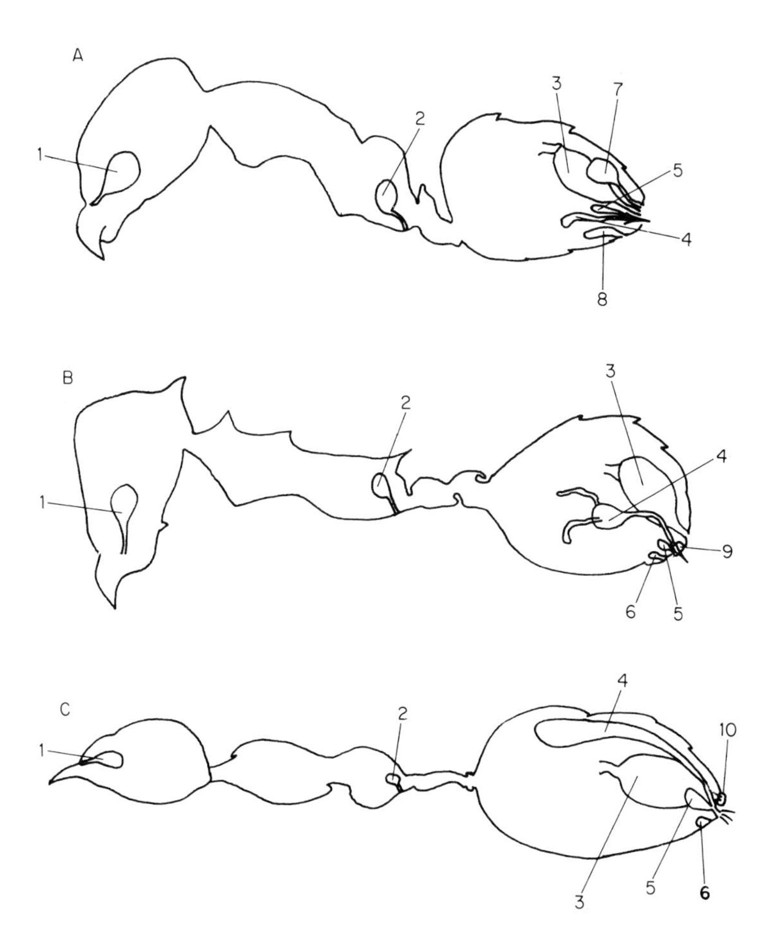

Fig. 1. Diagrams of some exocrine glands of worker ants in three sub-families. A. Dolichoderinae (*Iridomyrmex*). B. Myrmicinae (*Atta*). C. Formicinae (*Oecophylla*). 1: Mandibular gland. 2: Metapleural gland. 3: Rectal sac. 4: Poison gland. 5: Dufour's gland. 6: Sternal gland. 7: Anal gland. 8: Pavan's gland. 9: Sting sheath gland. 10: Rectal gland.

and Dufour's glands are probably ubiquitous, but others may either be absent in a few species, as is the metapleural gland (Brown, 1968), or restricted to a particular group, as is the anal gland in the Dolichoderinae. Attempts to establish chemotaxonomic principles based on analyses of whole ants would result in complex chromatograms, the composition of which would be affected drastically by the preparation of the sample, particularly by the release of venom and alarm-releasing secretions during handling, a problem not encountered in plants! The majority of studies have therefore been carried out on particular glands, mainly those associated with aggressive and defensive behaviour. The chemistry of the venom glands has been extensively reviewed by Blum and Hermann (1978), and no attempt at a comprehensive treatment will be made here.

From the point of view of the taxonomist, the more species that can be analysed the more weight can be given to any particular approach, be it chemical or otherwise. In ants, the ease with which samples can be prepared may be critical in determining the amount of data available. From Fig. 1 it is apparent that the mandibular glands are likely to be the only source of volatile materials in the head of a species to be analysed. On the other hand, the gaster contains a number of glands which vary from one group to another and would have to be dissected individually, and the thorax usually produces little volatile material. The head therefore appears to be the most suitable part of the body for chemotaxonomy. Moreover, Dufour's gland, which is usually the major source of volatile materials in the gaster (except in the Dolichoderinae, where the anal gland predominates), shows little diversity in its contents through the whole family. Although attempts have been made to distinguish poorly separated species on the basis of the contents of this gland, the key differences are usually trace components separated only by the most exacting analyses (e.g. Bergström and Löfqvist, 1972).

Comparison of the cephalic volatiles of worker ants has been approached in several ways. The family Formicidae consists of at least 12 000 species (Wilson, 1971), and is accordingly divided into a number of ill-defined subfamilies. Wilson (1971) considers the subfamily Myrmeciinae to have evolved ant-like characteristics completely independently of the subfamily Ponerinae, whereas Taylor (1978) places the primitive ponerines as ancestors of the myrmeciines. Chemical studies of the more primitive subfamilies might help to shed light on the inter-relationships. The phylogeny of smaller groups, such as tribes, has also been investigated chemically. Finally, as mentioned above for Dufour's gland, chemical data can be used to separate species which are difficult to distinguish on morphological grounds. However, the usefulness of such data depends upon the level of variation within a species as compared with that between species.

DEFENSIVE SECRETIONS IN TERMITES

This topic has been reviewed by Moore (1974), Prestwich (1979), Evans *et al.* (1978) and others. Termites resist predation, which in most species comes mainly from ants, by a variety of adaptations, including nest fortifications. The soldiers are the main defensive elements and different species have characteristic head morphologies that form the basis of most systematic studies. Some species have large biting or snapping mandibles. Some others have a well-developed cephalic gland, the frontal gland, which produces a secretion that has toxic or repellent effects against at least some potential predators. In the simplest form, the secretion runs down the front of the head on to the mandibles and so can be applied to a surface which the insect bites. In some genera of the subfamily Nasutitermitinae the opening of the gland is at the tip of a nozzle, and in the extreme case of the so-called *nasute* soldiers the mandibles are vestigial or absent and the secretion is sprayed forward as a jet, which rapidly dries to a sticky solution. Ants are usually entangled in the secretion, at least temporarily, and may be killed by its toxic action.

The chemistry of termite defensive secretions is very varied, and the reasons for this may be threefold: (a) specificity of toxic or repellent action against different spectra of predators may be involved; (b) the physico-chemical properties must be adapted to the mode of application; (c) it must be relatively non-toxic to the soldiers themselves. The first factor is supposition and can only be substantiated by extensive experimental work. Some evidence for selective toxicity or repellency is found in the work of Longhurst, Briner and others (unpublished), who studied neighbouring populations of the termite *Macrotermes subhyalinus* around Mokwa, Nigeria. These are subject to raids by the ponerine ant *Megaponera foetens.* Termites from around Mokwa were resistant to predation, while those from riverain areas of Zugurma and Rabba, within a radius of 10 km, were heavily preyed upon by the ants. This can be correlated with the relatively high levels of a C_{16} diene that are found in the Mokwa populations. The same compound was present in relatively low levels in soldiers from the other areas.

The chemistry of the secretions of nasute soldiers has been studied extensively by Czech and American workers (see, for example, Vrkoč *et al.*, 1977; Prestwich, 1979). All the secretions so far investigated consist of mixtures of monoterpenes and diterpenes, sometimes with minor amounts of sesquiterpenes or hydrocarbons. The monoterpenes act as solvents for the resinous and sticky diterpenes. The toxic role of the different compounds is not yet understood; there is evidence that monoterpenes are insecticidal in their own right (Hrdý *et al.*, 1977; Howse, 1975), as are some diterpenes (Hrdý *et al.*, 1977) but Eisner *et al.* (quoted in Prestwich, 1979) believe that the diterpenes provide a slow release mechanism for the monoterpenes, so prolonging the toxic action.

Prestwich *et al.* (quoted in Prestwich, 1979) investigated the variation in soldier frontal gland secretions of the nasute species of *Trinervitermes* from East Africa. They found striking differences between major and minor workers but negligible variation within a mound or within a population from a limited area. Important differences were found, however, between populations of the same species from different areas. One population of *T. gratiosus* lacked monoterpenes present in two other populations and contained a new major diterpene in the secretion of the major soldiers.

While further studies remain to be done to establish the degree of interspecific variability that may exist generally, it has already been established that considerable interspecific diversity exists among the nasutes, following the work of Prestwich and his co-workers (see Prestwich, 1979) and of Czech workers (e.g. Vrkoč *et al.*, 1977), that is especially marked in the diterpene components. These are products found only in the frontal gland and not in other parts of the body or in workers. If, as this implies, they are synthesized by the glandular tissue, it is difficult to suggest why so much biosynthetic energy should be devoted to this purpose if their role is purely mechanical. It is equally difficult to see why marked species diversity of diterpenes should exist, extending to compounds with novel tricyclic skeletons. It may be that there are fine balances to be achieved in producing persistent selectively toxic secretions that can be stored and used with minimum danger to the species. With no knowledge of the differing detailed behavioural strategies used by nasute species in defence or of the nature and susceptibilities of their predators, the design features of the defensive secretions can be discussed only speculatively, and therefore their use as aids to systematics can also be only speculative.

ANT SECRETIONS USED IN ALARM AND DEFENCE

1. Relationships Between Subfamilies – the Ponerinae

The ponerines are a group which at the same time are known to produce a wide range of mandibular gland compounds and are important in the construction of phylogenetic relationships within the ants. In a recent reappraisal of the ancestry of the Formicidae, Taylor (1978) places primitive poneroids as ancestors of the subfamilies Myrmeciinae and Pseudomyrmeciinae, and primitive ponerines as ancestors of the army ant subfamilies Dorylinae, Ecitoninae and Leptanillinae, the large and advanced subfamily Myrmicinae and the advanced Ponerinae. Unfortunately no reports of mandibular gland chemicals have been published for the primitive ponerines, particularly the tribe Amblyoponini, which is usually placed near to the ancestral poneroids. All the available literature relates to the "advanced" tribes Ponerini and Odontomachini and is summarized in Fig. 2. Four species of the Odontomachini produce alkyl pyrazines (Wheeler and Blum, 1973; Longhurst *et al.,* 1978) as do three species from three genera of Ponerini (Duffield *et al.,* 1976; Longhurst *et al.,* 1978). Also in the Ponerini, two species with large workers produce alkyl sulphides (Crewe and Fletcher, 1974; Longhurst *et al.,* 1979) while methyl salicylates are found in two species (Duffield and Blum, 1975; Longhurst, 1978). One of the latter, *Bothroponera soror,* also contains a range of straight-chain alcohols and ketones similar or identical to those found in *Tetramorium termitobium* (Longhurst *et al.,* 1980) and in the Dufour's gland of many formicine species (Blum and Hermann, 1978). Workers of *Neoponera villosa* contain 4-methyl-3-heptanone (Duffield and Blum, 1973), a compound found commonly in several myrmicine

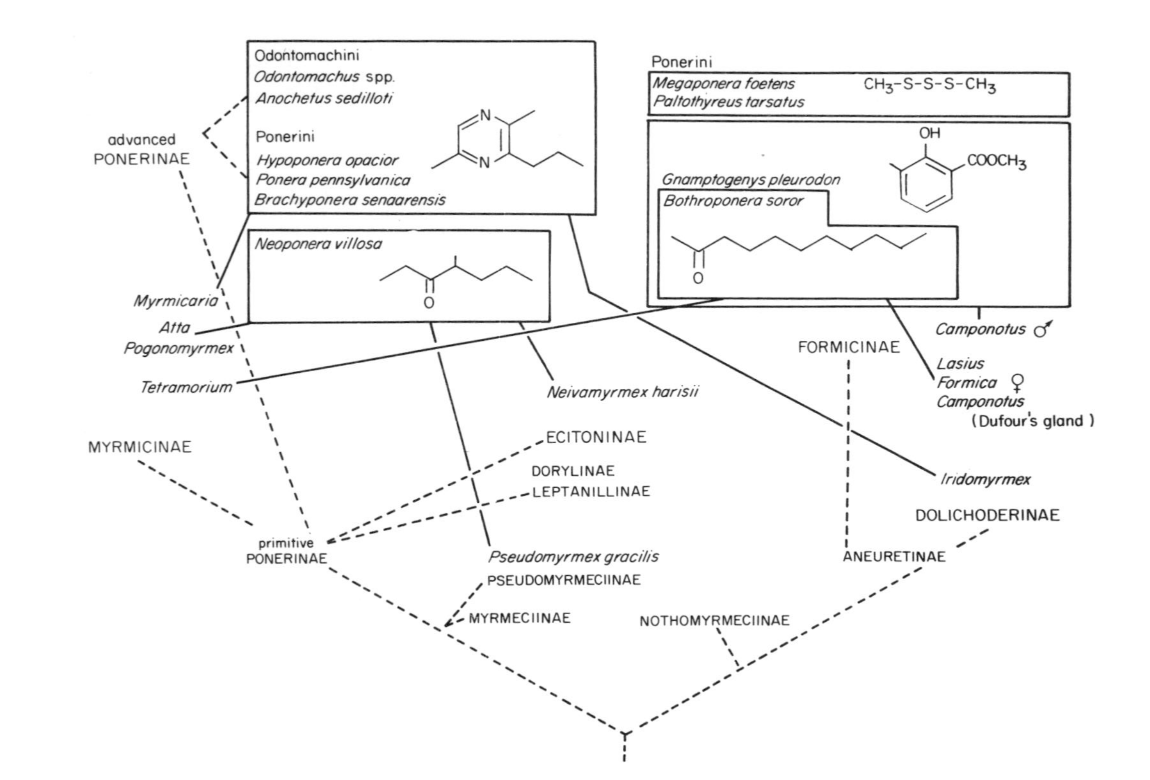

Fig. 2. Representative compounds isolated from heads of ponerine ants, their occurrence in genera from other subfamilies (solid lines) and a proposed phylogenetic scheme (dotted lines) based on other characters (Taylor, 1978). References in the text.

genera, also in one of the few army ants to have been analysed, *Neivamyrmex harisii,* and in a pseudomyrmeciine *Pseudomyrmex gracilis* (Blum and Hermann, 1978). This last compound has not yet been detected in any member of the formicoid complex of subfamilies (principally Formicinae and Dolichoderinae) but is apparently ubiquitous within the poneroid complex as defined by Taylor (1978). On the other hand, methyl salicylates are found in the Ponerini and in males of the formicoid genus *Camponotus* (Brand *et al.*, 1973).

When first isolated from *Odontomachus,* the alkyl pyrazines had not previously been identified from ants (Wheeler and Blum, 1973), but subsequently compounds of this type have been detected in *Iridomyrmex* (Cavill and Houghton, 1974) and in *Myrmicaria* (Longhurst, unpublished). They cannot therefore be considered as typical even of the Ponerinae. Within this subfamily they cross well-defined boundaries between tribes, and within the Ponerini occur in genera with poorly developed sociality (*Hypoponera* and *Ponera*) and comparatively advanced sociality (*Brachyponera*). Only the alkyl sulphides are apparently confined to the ponerines.

It appears that within this subfamily, it is not yet possible to trace phylogenetic relationships using the mandibular gland chemicals alone. The compounds identified so far fall into five different structural types with, presumably, different biosynthetic routes. The situation is apparently far more complex than in the Formicinae, where relationships can be traced between genera and subgenera based on chemical communication systems (Wilson and Regnier, 1971). As the functions of the ponerine compounds become known in terms of communication and the maintenance of sociality, so the basis for this unusual diversity in exocrine chemistry may become apparent.

2. Relationships within a Tribe – the Fungus-growing Ants

The tribe Attini of the subfamily Myrmicinae consists of a group of neotropical species which hold the tending of fungus gardens in common. They range from a primitive species, *Cyphomyrmex rimosus,* in which the fungus is cultured on insect faeces and is kept separated from the brood, to the genera *Atta* and *Acromyrmex,* in which a number of large fungus gardens are maintained in each colony and the brood is kept in chambers within the fungus (Weber, 1958). The genus *Trachymyrmex* is considered to be transitional between *Cyphomyrmex,* with monomorphic workers, and *Acromyrmex* and *Atta* whose workers exhibit considerable polymorphism. In a chemical study of the mandibular glands in these genera, Crewe and Blum (1972) isolated four compounds, 3-octanone, 3-octanol, 4-methyl-3-heptanone and 4-methyl-3-heptanol, which they considered to be of phylogenetic importance. However, our own studies of *Acromyrmex* and *Atta* and those of Riley *et al.* (1974) and Schildknecht (1976) of *Atta* have indicated that the mandibular gland secretions are far more complex than this. In particular, seventeen compounds were identified from *Atta sexdens rubropilosa,* including such diverse chemical groups as monoterpenes, branched and straight-chain aliphatics, an aromatic and a lactone (Schildknecht, 1976). Workers of *Acromyrmex octospinosus* produce at least fifteen compounds in their mandibular glands (Bradshaw, Baker and Howse, unpublished). Even *Cyphomyrmex rimosus* apparently produces at least four compounds (data in Crewe and Blum, 1972). It may therefore be premature to consider phylogenetic relationships within this group until more compounds have been identified from a wider range of species.

Given the diversity of chemicals from *Atta,* it is possible that specificity of gland secretions in this group of species has been achieved by changes in the proportions of constituents held in common and in their minor structural modification, rather than by adoption of new biochemical pathways. The significance of such minor and potentially reversible changes may have more relevance to behaviour and ecology than to the reconstruction of phylogeny.

3. Variation within Species

(*a*) *Weaver ants* (*Oecophylla*). Chemical communication in ants of the genus *Oecophylla* is highly complex. The genus is usually divided into two extant species, *Oecophylla longinoda* from Africa and *O. smaragdina* from Asia and Australasia (Wheeler, 1922; Cole and Jones, 1948). *O. longinoda* is divided into a number of varieties based mainly upon colour, and *O. smaragdina* has two subspecies, *virescens* and *subnitida,* in the south-eastern part of its range, of which there are several varieties.

Chemical analysis of exocrine glands from *O. longinoda* collected in West Africa showed a large number of volatile compounds; over thirty in the mandibular glands of major workers alone (Bradshaw *et al.*, 1979a). A solid-sampling gas chromatography technique was used to analyse the cephalic volatiles of individual workers from nests collected from a number of localities in Nigeria, Ghana and Sierra Leone (Fig. 3). The colonies were maintained in the laboratory on identical diets for at least one month before analysis, to minimize differences due to nutritional factors. Although a limited number of individuals were used, it was possible to distinguish two groups, depending on whether six- or eight-carbon aldehydes predominated (Fig. 4). Two colonies, collected at Ibadan and Benin City and typical of each group respectively, differed to at least $P < 0.005$ levels for

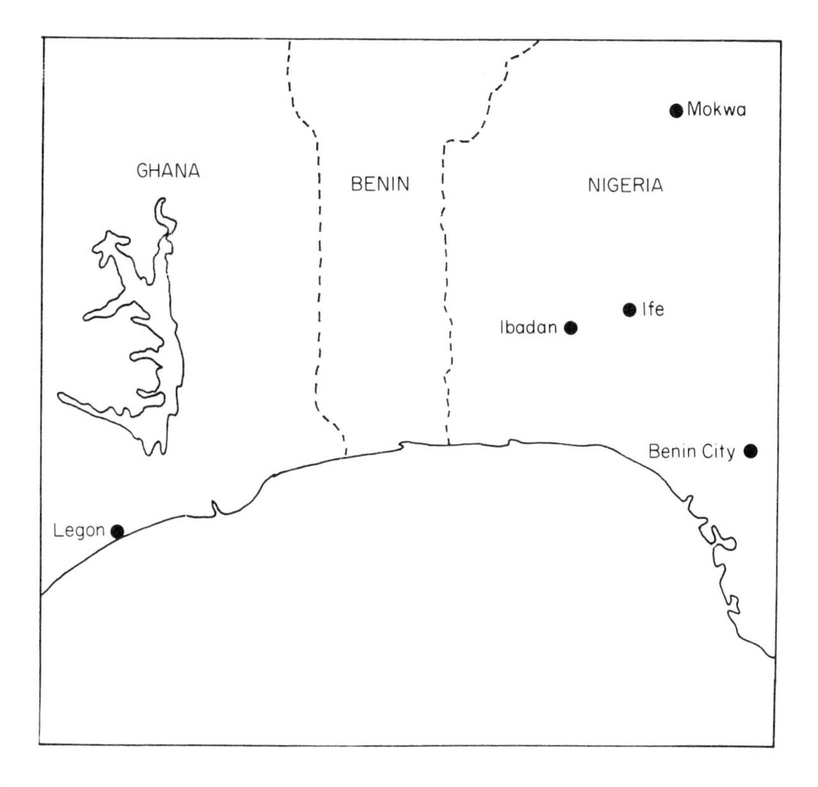

Fig. 3. Locations in Nigeria and Ghana from which weaver ant colonies were collected and analysed (see Fig. 4).

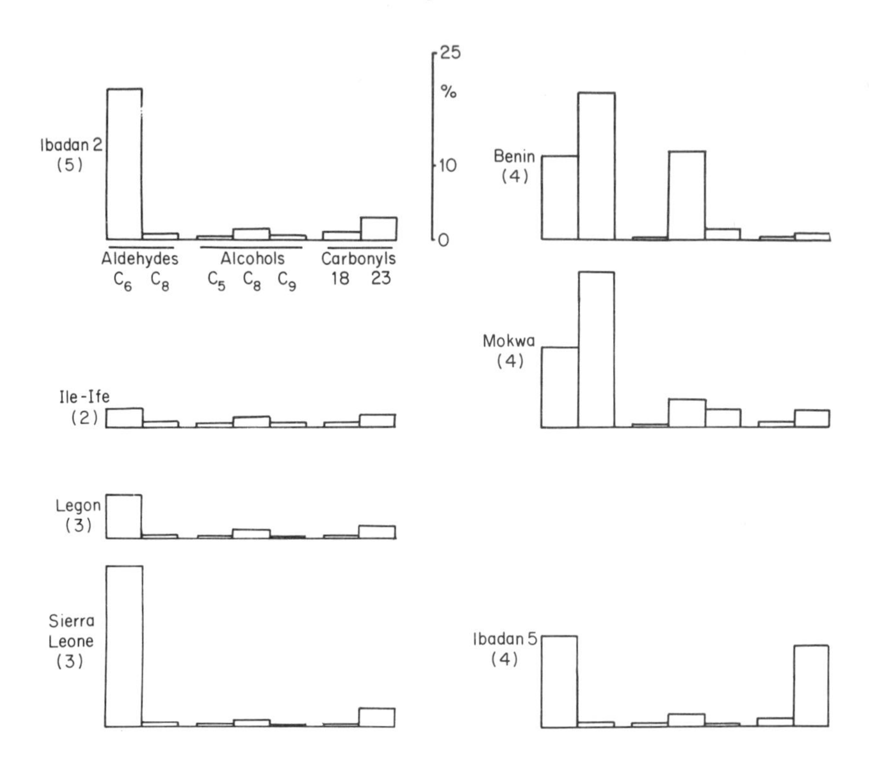

Fig. 4. Quantities (% of major component, 1-hexanol) of selected mandibular gland components in major workers of *Oecophylla longinoda*. Seven colonies are shown, collected at six locations in West Africa (see Fig. 3). The compounds are (left to right) hexanal, octanal, 1-pentanol, 1-octanol, 1-nonanol, 3-undecanone (component 18) and 2-butyl-2-octenal (component 23). Numbers of analyses of individual ants from each colony are indicated in brackets. Details of analysis in Bradshaw *et al.* (1979a, b).

each of nine well-resolved peaks on the gas chromatographs (Bradshaw *et al.*, 1979b). Even within a comparatively small area variations were found; one of five colonies collected at Ibadan had consistently large amounts of component 23, 2-butyl-2-octenal and in two colonies collected at Ile-Ife, about 70 km to the east, the ratio of components 23 and 22 was reversed compared to the four "typical" colonies from Ibadan.

In some of the Nigerian colonies it was possible to correlate these chemical differences with altered behaviour. In a colony where six-carbon aldehydes predominated, a biting response was released by

2-butyl-2-octenal alone; in an "eight-carbon colony" collected at Mokwa, the addition of an equal amount of 1-hexanol was required to elicit the full response and the same was also found for the aberrant Ibadan colony. Thus even within Nigeria, there appears to be a number of chemically distinct varieties of *O. longinoda*, two of which are sympatric. However, the natural distribution of these varieties cannot be inferred directly from Fig. 3 since small nests may be accidentally transported over long distances in transplanted ornamental trees (Longhurst, personal communication).

The chemical characterization of *O. longinoda* as a species is further complicated by differences between the four castes: females, males, major workers and minor workers. Minor workers produce some mandibular gland compounds in common with majors, and some of their own; males produce a series of acids in this gland which are not found in workers (Bradshaw *et al.*, 1979b). Significant variations between members of a caste were not detected in the individuals sampled, but in at least one species, *Myrmica rubra*, the mandibular gland contents are known to vary with the age of the worker (Cammaerts-Tricot, 1974). Considerable caution is therefore needed in selecting samples of ants for chemosystematic studies.

Analysis of a single colony of *O. smaragdina* (typical) from Sri Lanka has indicated that some components are held in common between the mandibular glands of major workers of this species and *O. longinoda* from West Africa, but qualitative differences are apparent between the two species (Bradshaw, unpublished). The situation is further complicated by the division of the major worker caste into two chemically distinct groups (Fig. 5). Thus in this species even greater care is required in selecting individuals for analysis in order that spurious differences between areas do not appear due to uneven sampling of the two groups. In contrast with the major workers, very little difference has been found between minor workers from different areas of West Africa (Bradshaw *et al.*, 1979b) or from the two species.

(*b*) *Myrmicaria.* In the West African predatory ant, *Myrmicaria eumenoides*, the poison gland secretion has a dual function: produced as a droplet retained at the tip of the abdomen it serves for communication of alarm and coordination of associated behaviour, but forcibly ejected in greater quantities as a spray it acts as a

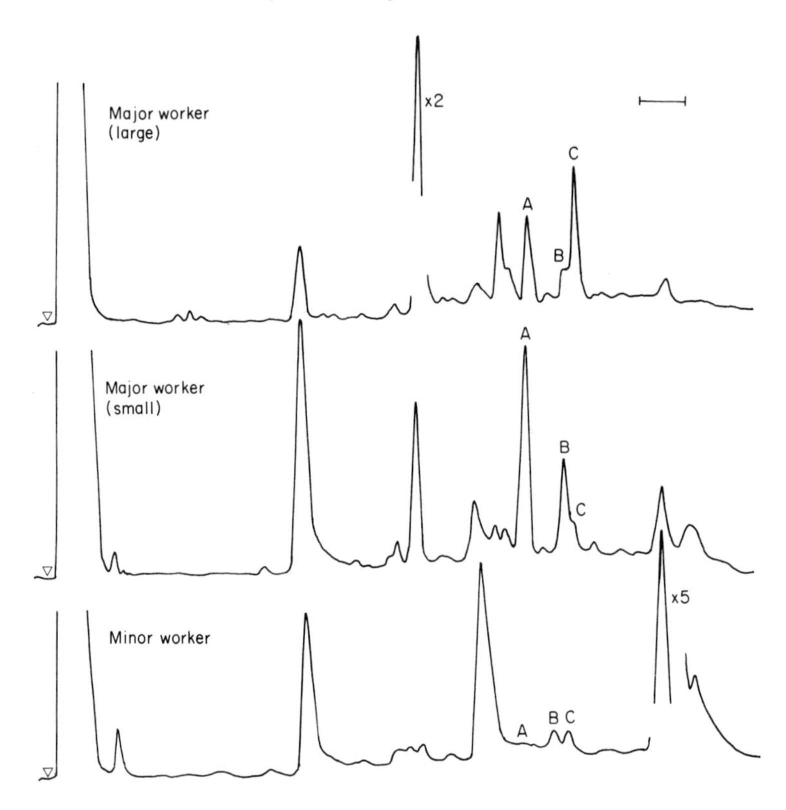

Fig. 5. Gas chromatograph traces of heads from workers of *Oecophylla smaragdina* from one colony collected in Sri Lanka. Note particularly the relative proportions of peaks A, B and C. Column: PPGA 10%, N_2 carrier at 40 ml min^{-1}, oven temperature 80°C for 5 min, 4°C min^{-1} increase to 168°C, injection point arrowed. Figures preceded by "x" are attenuation increases.

defensive secretion and induces "panic alarm" in nest-mates. Work of Longhurst and Bolwell (unpublished) in our laboratory has shown that the main constituents are the monoterpenes α- and β-pinene, limonene, myrcene, sabinene, terpinolene and α-phellandrene. The first five attract workers directly to the source if they are presented in the quantities equivalent to a droplet from one ant. Limonene, however, becomes repellent at close range and elicits circling movements 1–2 cm from the source, and α-pinene and myrcene are repellent at source (see Howse *et al.*, 1977). Quantities equivalent to five droplets will kill the ants that produce them if they spread to the body cuticle, and must therefore be rapidly ejected. At the quantity in five droplets, α-pinene, β-pinene, myrcene and limonene elicit rapid direct approach to the source and then rapid erratic retreat in which the alarm is communicated to other ants which run around rapidly in an alert fashion.

Gas chromatography was carried out on a number of individual workers from each of eight colonies of *M. eumenoides* collected within a one mile radius near Mokwa, Nigeria. Differences in the percentage titre of the major components, limonene and β-pinene, showed very little variation among individuals of the same colony (Fig. 6). Intercolony differences, on the other hand, were sometimes quite marked, so that the major component in some was limonene,

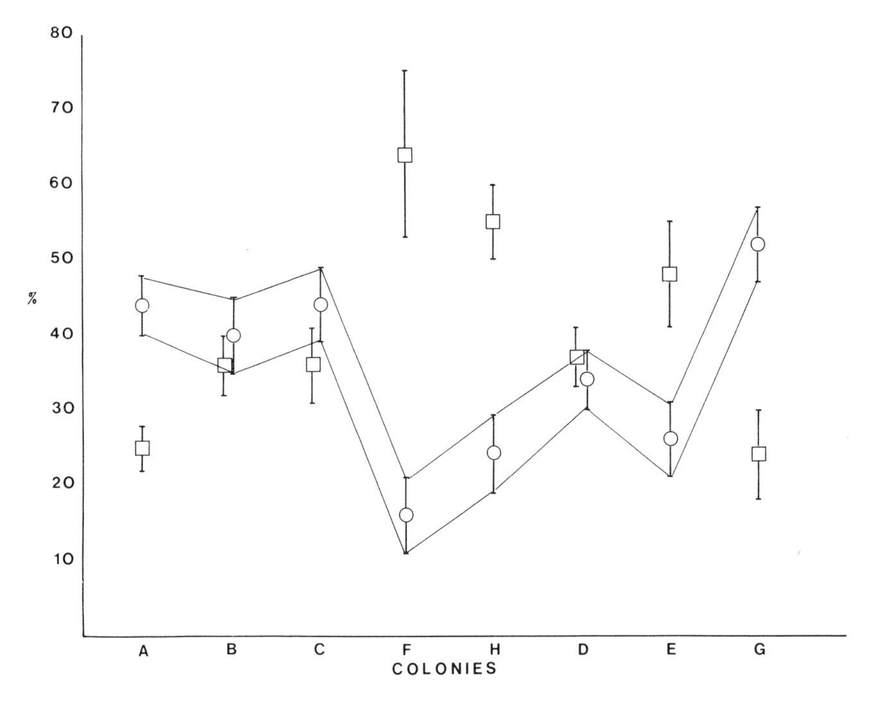

Fig. 6. Intracolony and intercolony variations in the proportions of two major components, limonene and β-pinene, in poison gland secretions of workers of eight colonies of *Myrmicaria eumenoides* from the Mokwa region, Nigeria (see text). The percentages with standard deviations are those of total poison gland volatiles.

and in others β-pinene. On this basis, ants from different colonies can sometimes, but not always, be distinguished, providing that a number from each colony are examined. Samples of the eight colonies taken together show that limonene is the major component for the species. Samples of the related species, *M. striata*, from Ibadan in Nigeria had the same seven major constitutents, with limonene also the most abundant, but myrcene is the second largest component in *M. striata*, while in *M. eumenoides* it is β-pinene. Data from Brand *et al.*, (1974) suggest that sabinene is the major component for *M. natalensis* from S. Africa, and our analyses of two colonies of *M. brunnea* from India show that terpinolene, β-pinene and limonene are the main components, in that order of importance.

Multivariate analysis (single-link method) of the data confirms the integrity and division of these four species on the basis of six selected major components. We can therefore infer that colony distinctiveness may sometimes be shown by the major component, but only if a number of samples are taken from the same colony. Species differences can not be confirmed by examination of the major component in insects from one or a very few colonies, but analyses of variations in the titres of five or more components can provide useful taxonomic information providing that intracolony variation is taken into account.

CONCLUSIONS

1. Procedures in Insect Chemical Taxonomy

Although there have been a number of attempts to use the presence or absence of major components of exocrine glands in systematic studies on social insects (e.g. Blum, 1973; Crewe and Blum, 1972), the work we have outlined here suggests that such an approach can be misleading. Before chemical criteria are used in social insect systematics, we suggest that the following procedures are observed.

(1) Criteria of presence or absence should not be used unless minor components of secretions, present at levels down to 1 ng per insect, have been examined.

(2) Functional homology of secretions, or of components of secretions, should not be assumed (e.g. as alarm pheromones,

defence secretions etc.) unless bioassays have been done with pure compounds in the quantities normally employed by the insects.

(3) Comparisons should be based on equivalent castes, especially where worker polymorphism exists, as glandular secretions may differ widely according to caste and, possibly, with age.

(4) A measure of intracolony variation should be taken before comparing different colonies on the basis of multicomponent secretions. More information can be obtained if ratios of several components are compared.

(5) Intercolony variation, over as wide an area as possible, should be determined before closely related species can be justifiably compared on chemical criteria alone. Such variation can be assessed only when the extent of intracolony variation is known.

(6) The effects of diet on biosynthesis of volatile secretions in social insects are not generally known. They constitute a further possible source of variation which it may be possible to control in laboratory colonies.

The use of extracts of a large number of insects from a given colony allows for intracolony variation, but has a number of disadvantages compared with measurements on a number of separate individuals. Most importantly, the extent of intracolony variation will be concealed. The greater the intracolony variation the more it will approach possible interspecies differences and tend to invalidate them as systematic criteria. Additionally, extracts of large numbers will include insects of different ages and may include different castes or subcastes.

2. Ecological and Behavioural Problems

Ultimately, chemical information will be of greater value in insects generally when the ecological significance of exocrine secretions is understood and their role and method of usage can be evaluated in adaptive terms and correlated with morphological features.

In *Oecophylla* and *Myrmicaria* it is tempting to relate the low intracolony variation and the relatively much larger intercolony variation in pheromone components with "colony odour" that controls recognition of nest-mates and discrimination against those of other colonies or species. Ants may be able to habituate, during a development period, to the blend of pheromones produced by nest-mates, and behavioural isolation will then be achieved if colonies of the same or other species have a different blend. Indeed, we can expect strong "alarm" between two different *Myrmicaria* species in which a component of the poison gland secretion of one may be produced in the quantity normally used in a defensive spray of the other. Evidence for habituation to odours may be found in the early work of Fielde (1904) who developed techniques of mixing colonies of the same and normally inimicable species. Jaisson (1975) has also shown that ants of the genus *Formica* imprint on the odour of cocoons during a period immediately following eclosion, and subsequently reject others. If exposed to cocoons of another species, they subsequently reject those of their own species when given a choice.

ACKNOWLEDGEMENTS

We are grateful to the following colleagues for discussions and permissions to quote unpublished work: Professor R. Baker, Dr C. Longhurst and Dr D. A. Evans. Professor U. Maschwitz kindly supplied speciemens of *Oecophylla smaragdina* for some of the work described here.

REFERENCES

Bergström, G. and Löfqvist, J. (1972). Similarities between the Dufour gland secretions of the ants *Camponotus ligniperda* (Latr.) and *Camponotus herculeanus* (L)(Hym.). *Ent. scand.* 3, 225–238.

Blum, M. S. (1973). Comparative exocrinology of the Formicidae. *Proc. VII Int. Congr. IUSSI (London)*, pp. 23–40.

Blum, M. S. and Hermann, H. R. (1978). Venoms and Venom Apparatuses of the Formicidae. *In* "Arthropod Venoms" (S. Bettini, Ed.), pp. 801–869. Springer-Verlag, Berlin.

Bradshaw, J. W. S., Baker, R. and Howse, P. E. (1979a). Multicomponent alarm pheromones in the mandibular glands of major workers of the African weaver ant, *Oecophylla longinoda. Physiol. Ent.* 4, 15–25.

Bradshaw, J. W. S., Baker, R., Howse, P. E. and Higgs, M. D. (1979b). Caste and colony variations in the chemical composition of the cephalic secretions of the African weaver ant, *Oecophylla longinoda. Physiol. Ent.* 4, 27–38.

Brand, J. M., Duffield, R. M., MacConnell, J. G., Blum, M. S. and Fales, H. M. (1973). Caste-specific compounds in male carpenter ants. *Science* **179**, 388–389.

Brand, J. M., Blum, M. S., Lloyd, H. A., and Fletcher, D. J. C. (1974). Monoterpene hydrocarbons in the poison gland secretion of the ant *Myrmicaria natalensis* (Hymenoptera: Formicidae). *Ann ent. Soc. Am.* **67**, 525–526.

Brown, W. L. (1968). An hypothesis concerning the function of the metapleural gland in ants. *Am. Nat.* **102**, 188–191.

Cammaerts-Tricot, M. C. (1974). Production and perception of attractive pheromones by differently aged workers of *Myrmica rubra* (Hymenoptera: Formicidae). *Insectes soc.* **21**, 235–247.

Cavill, G. W. K. and Houghton, E. (1974). Volatile constituents of the Argentine ant, *Iridomyrmex humilis. J. Insect Physiol.* **20**, 2049–2059.

Cole, A. C. and Jones, J. W. (1948). A study of the weaver ant, *Oecophylla smaragdina* (Fab.). *Am. Midl. Nat.* **39**, 641–651.

Crewe, R. M. and Blum, M. S. (1972). Alarm pheromones of the Attini: their phylogenetic significance. *J. Insect Physiol.* **18**, 31–42.

Crewe, R. M. and Fletcher, D. J. C. (1974). Ponerine ant secretions: the mandibular gland secretion of *Paltothyreus tarsatus. J. ent. Soc. Sth. Afr.* **37**, 291–298.

Duffield, R. M. and Blum, M. S. (1973). 4-Methyl-3-heptanone: identification and function in *Neoponera villosa. Ann. ent. Soc. Am.* **66**, 1357.

Duffield, R. M. and Blum, M. S. (1975). Methyl 6-methylsalicylate: identification and function in a ponerine ant (*Gnamptogenys pleurodon*). *Experientia* **31**, 466.

Duffield, R. M., Blum, M. S. and Wheeler, J. W. (1976). Alkylpyrazine alarm pheromones in primitive ants with small colonial units. *Comp. Biochem. Physiol. B* **54**, 439–440.

Evans, D. A., Baker, R. and Howse, P. E. (1978). The chemical ecology of termite defence behaviour. *In* "Chemical Ecology: Odour Communication in Animals" (F. J. Ritter, ed.). Elsevier, N. Holland.

Fielde, A. M. (1904). Power of recognition among ants. *Biol. Bull.* **7**, 227–250.

Howse. P. E. (1975). Chemical defenses of ants, termites and other insects: Some outstanding questions. *In* "Pheromones and Defensive Secretions in Social Insects", pp. 23–38. IUSSI, Dijon, France.

Howse, P. E., Baker, R. and Evans, D. A. (1977). Multifunctional secretions in ants. *Proc. VIII Int. Congr. IUSSI (Wageningen)*, pp. 44–45.

Hrdý I., Křeček, J. and Vrkoč, J. (1977). Biological activity of soldiers secretions in the termites: *Nasutitermes ripertii, N. costalis,* and *Prorhinotermes simplex. Proc. VIII Int. Congr. IUSSI (Wageningen)*, pp. 303–304.

Jaisson, P. (1975). L'impregnation dans l'ontogenese des comportements de soin aux cocons chez la jeune fourmi rouse (*Formica polyctena* Forst.). *Behaviour, Leiden* **52**, 1–37.

Longhurst, C. (1978). Behavioural, chemical and ecological interactions between West African ants and termites. Ph.D. thesis, University of Southampton.

Longhurst, C., Baker, R., Howse, P. E. and Speed, W. (1978). Alkylpyrazines in ponerine ants: their presence in three genera, and caste specific behavioural responses to them in *Odontomachus troglodytes*. *J. Insect Physiol.* **24**, 833–837.

Longhurst, C., Baker, R. and Howse, P. E. (1979). Termite predation by *Megaponera foetens* (Fab.) (Hymenoptera: Formicidae): Coordination of raids by chemicals. *J. chem. Ecol.* **5**, 703–719.

Longhurst, C., Baker, R. and Howse, P. E. (1980). A comparative analysis of mandibular gland secretions in the ant tribe Tetramoriini. *Insect Biochem.* **10**, 107–112.

Moore, B. P., (1974). Pheromones in termite societies. *In* "Pheromones" (M. C. Birch, ed.), pp. 250–265. Elsevier, New York.

Prestwich, G. D. (1979). Chemical defense by termite soldiers. *J. chem. Ecol.* **5**, 459–480.

Riley, R. G., Silverstein, R. M. and Moser, J. C. (1974). Isolation, identification, synthesis, and biological activity of the volatile compounds from the heads of *Atta* ants. *J. Insect Physiol.* **20**, 1629–1637.

Roelofs, W. L., and Cardé, R. T., (1977). Responses of Lepidoptera to synthetic sex pheromone chemicals and their analogues. *A. rev. Ent.* **22**, 377–406.

Schildknecht, H. (1976). Chemical ecology – a chapter of modern natural products chemistry. *Angew. Chem. Int. Ed. Engl.* **15**, 214–222.

Silverstein, R. M., and Young, J. C. (1976). Insects generally use multicomponent pheromones. *In* "Pest management with insect sex attractants" (Beroza, ed.) *Am. chem. Soc. Symp.* **23**, 1–29.

Taylor, R. W. (1978). *Nothomyrmecia macrops*: a living-fossil ant rediscovered. *Science* **201**, 979–985.

Vrkoč, J., Budesinsky, M., Křeček, J. and Hrdý, I. (1977). Diterpenes from secretions of *Nasutitermes* soldiers. *Proc. VIII Int. Congr. IUSSI (Wageningen)*, 320–321.

Weber, N. A. (1958). Evolution in fungus-growing ants. *Proc. 10th Int. Congr. Entomol., 1956* **2**, 459–473.

Wheeler, J. W. and Blum, M. S. (1973). Alkyl pyrazine alarm pheromones in ponerine ants. *Science* **182**, 501–503.

Wheeler, W. M. (1922). Ants of the Belgian Congo. *Bull. Am. Mus. nat. Hist.* **45**, 1–269.

Wilson, E. O. (1971). "The Insect Societies". Belknap, Cambridge, Massachusetts.

Wilson, E. O. and Regnier, F. E. (1971). The evolution of the alarm-defense system in the formicine ants. *Am. Nat.* **105**, 279–289.

5 | Automatic Amino Acid Analysis Data Handling

B. V. CHARLWOOD

and

E. A. BELL

Department of Plant Sciences, University of London King's College, 68 Half Moon Lane, London SE24 9JF, England

INTRODUCTION

This paper describes an approach towards the completely automated handling of data accumulated from a research programme aimed at analysing the pools of free protein and non-protein amino acids in seeds of the Leguminosae. The term non-protein amino acid will be used in this paper to specify all those amino acids, imino acids and amino acid amides that are *not usually* found as constituents of protein. Most of these non-protein amino acids can be considered to be secondary metabolites in the accepted sense of the word, although some of them (for example ornithine and homoserine) may also be involved in primary metabolic pathways.

The distribution of non-protein amino acids and their taxonomic significance has been reviewed by Bell and Fowden (1964), by Bell (1976) and by Bell *et al.* (1978), and it is generally accepted that a study of this group of compounds can provide valuable information

---

Systematics Association Special Volume No. 16, "Chemosystematics: Principles and Practice", edited by F. A. Bisby, J. G. Vaughan and C. A. Wright, 1980, pp. 91–102, Academic Press, London and New York.

concerning the relationships within and between different genera, tribes and families of plants.

At the present time over 240 non-protein amino acids are known to occur in plants, and the spectrum of structures is very wide. Some members of this group are simple homologues of protein amino acids, for example azetidine-2-carboxylic acid (I, a component of many species of the Liliaceae) and pipecolic acid (III, which occurs regularly in legume species) are the lower and higher homologues respectively of proline (II). Others of the group are analogues of their protein amino acid counterparts – for example indospicine (IV, a toxic amino acid found in some *Indigofera* species) and canavanine (VI, which is characteristic of the subfamily Papilionoideae of the Leguminosae) are both analogues of the basic amino acid arginine (V). A large number of protein amino acid derivatives also occur in plants; 3-cyanoalanine (VII, a toxin from *Vicia* species) and 4-methyleneglutamic acid (VIII) are examples of this group. Other non-protein amino acids have complex and sometimes surprising structures which are seemingly unrelated to any of the protein amino acids. It is within this last group that one sees the vast range of structural diversity that is associated with secondary metabolites of, say, the acetate-malonate or mevalonate pathways. Some examples of this group are lathyrine (IX, found in certain

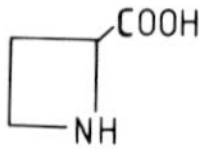

COOH; NH (I) — COOH; N H (II) — COOH; N H (III)

I 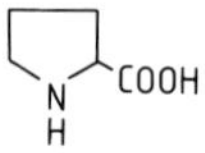 II 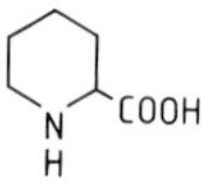 III

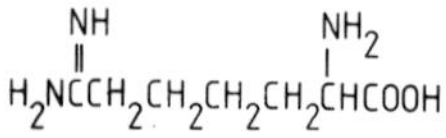

$H_2NC(=NH)CH_2CH_2CH_2CH_2CH(NH_2)COOH$ (IV)

$H_2NC(=NH)NHCH_2CH_2CH_2CH(NH_2)COOH$ (V)

$H_2NC(=NH)NHOCH_2CH_2CH(NH_2)COOH$ (VI)

IV 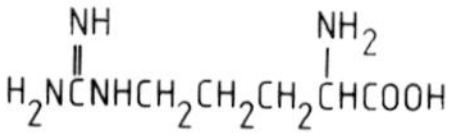 V 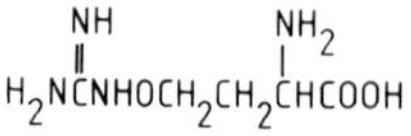 VI

$NCCH_2CH(NH_2)COOH$ (VII)

$HOOCC(=CH_2)CH_2CH(NH_2)COOH$ (VIII)

H_2N–pyrimidinyl–$CH_2CH(NH_2)COOH$ (IX)

O=, N H, O — $NCH_2CH(NH_2)COOH$ (X)

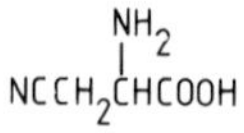

VII 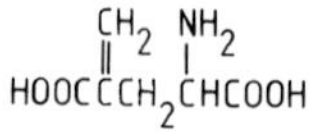 VIII 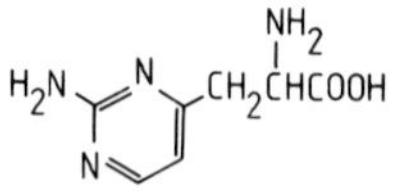 IX 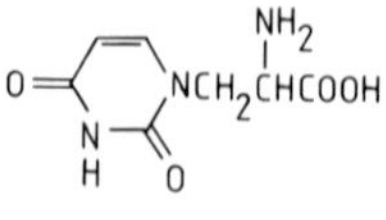 X

Lathyrus species) and willardine (X, from *Acacia willardiana*). In higher plants non-protein amino acids normally occur either in the free state, or as simple condensation derivatives such as the 4-glutamyl, acetyl or oxalyl derivatives. In micro-organisms, on the other hand, the non-protein acids often occur in the bound form as bacterial cell wall components or as constituents of small antibiotic-type polypeptides.

THE ANALYSIS OF FREE AMINO ACIDS

It is the wide range of structural types found among the non-protein amino acids that makes it difficult to establish a single standard technique for the simultaneous determination of all possible amino acids in any particular species. Much of the early work in this field was carried out using two-dimensional paper chromatography, and this technique does allow the compositions of large numbers of seed samples to be surveyed in a qualitative manner very quickly and cheaply. One of the useful features of the method is its reproducibility, non-protein amino acids being identifiable from their characteristic positions on the chromatograms, their "spots" often appearing as additions to a basic pattern of spots contributed by the free "protein" amino acids.

Amino acids are usually visualized on paper chromatograms by treatment with ninhydrin followed by heating to 110°C for 10 min. Whereas most protein amino acids produce a characteristic blue colour during the heating stage, many non-protein amino acids give other colours which may be highly characteristic. Thus *N*-ethyl and *N*-hydroxymethyl derivatives of asparagine (like asparagine itself) all react brown; some unsaturated acids (such as 4-methyleneglutamic acid) give spots that are yellow or brown; pipecolic acid produces a vivid purple colour, whilst its 4-hydroxy derivative is green with ninhydrin. The most characteristic colour of all is perhaps the brilliant red produced by lathyrine.

In order to make a general survey of the non-protein amino acids of the Leguminosae (*c.* 17 000 species) it was necessary to use an analytical technique that could be totally automated. Furthermore, in order that comparisons could be made between the amino acid profiles of all seeds analysed at any time during the survey, it was necessary that a single analytical protocol was developed and retained throughout the whole research programme. The analysis of amino acids by ion exchange chromatography using a commercial amino acid analyser can be performed automatically, although until recently the resolution of non-protein amino acids has been far from perfect. The advent of a new generation of high speed (i.e. high pressure) amino acid analysers, coupled with the use of a five buffer lithium elution system, has led to a large improvement in resolution with no increase in the time taken for a complete analysis. Our present system employs an LKB Model 4400 analyser which is equipped with a 4.6 mm diameter, 30 cm long stainless steel high pressure column packed with very fine resin (Durrum DC6A). The resolution of non-protein amino acids has been optimized using the Durrum Pico-Buffer System IV and the total analysis time has been reduced to 4 hours. With suitable sample preparation (involving the cleaning of the crude ethanol/water seed extracts for every trace of protein, lipid and carbohydrate), the analyses are highly reproducible, the sensitivity is in the region of 1–5 nmol for any one amino acid and the resin will sustain about 200 analyses before requiring extensive cleaning.

Transfer of the analogue data from the chart recorder output of the analyser into a digital form for direct input into a computer can be carried out in several ways:

(1) using an on-line dedicated integrator which integrates each peak as it is detected on the analyser and supplies the integrated data to the computer at the end of one, or several analyses;

(2) using a data logger system which is essentially an analogue to digital converter and a tape recorder to store the digital information for further manipulation;

(3) using an on-line mini-computer as an intelligent data logger.

We have found that a dedicated integrator is not ideal for use with an amino acid analyser in that the parameters to be used for the mathematical integration of peaks during the analysis have to be chosen before the analysis is carried out. Although this procedure would be perfectly satisfactory for the analysis of samples which were similar in their composition (i.e. for protein hydrolysates or physiological fluids) it is certainly not suitable for the integration of spectra containing protein and non-protein amino acids which inevitably overlap to varying degrees. More sophisticated integrators that are able to store spectra and calculate them off-line are more expensive and offer little advantage over a data logger system and a separate computer. The on-line dedicated mini-computer is now quite a feasible alternative to the data logger since the cost of the former has plummeted in recent years.

The data logger system that we employ is the Digitronix Data 100 which consists of an eight channel logger to record data from the analyser in Binary Coded Decimal and a replay unit which reads the data tape, converts it into standard ASCII and outputs it at the RS 232 interface level. One of the most useful features of the data logger system is the ability to calculate data using any computer facility that is available.

Having obtained a digital record of an analysis, the first problem to be solved before this data can be evaluated automatically is one of peak assignment. It is chastening to admit that automatic amino acid analysis has not reached the same degrees of sophistication as has gas liquid chromatography. Peak shapes, particularly of non-protein amino acids, are not entirely reproducible, and peaks do not become gradually broader as analysis proceeds – a sharpening of peaks can occur at each buffer and temperature change. Naturally, a slight pH change of any one of the 5 eluting buffers, or of the loading buffer, can have a "knock-on" effect on the absolute retention times of peaks throughout the spectrum. Absolute retention times are also extremely susceptible to the slightest change in flow rate, which is in turn governed to some extent by column bed compression, and these factors do tend to fluctuate very slightly from run to run. It is thus necessary to employ relative retention times as assignment criteria and we use alanine, phenylalanine and arginine as standards since these amino acids are fairly universal in their occurrence in plant tissue and cover a wide range of the spectrum. The problem of peak assignment is thus reduced in the first stage to one of computer recognition of one or more of these three amino acid peaks. In commercial analysers the method normally employed is to set up an absolute retention time window, which is fairly wide, for the standard acid and to assign the largest peak within this window to that standard. Again this method is probably satisfactory for spectra containing nothing but protein amino acids but is less useful when one is likely to have a high concentration of a non-protein amino acid whose peak is very close to the standard. We have thus utilized two further criteria of amino acid identity drawn from observations on paper chromatography and high voltage paper electrophoresis. The first of these criteria is that of the colour of the particular amino acid-ninhydrin complex. The second criterion is concerned with the rate at which the final colour and intensity of the ninhydrin complex is attained. Some amino acids reach their final intensities almost as soon as they are heated with ninhydrin reagent, others take upwards of 10 min. Furthermore, some non-protein amino acids initially produce characteristic colours but these may gradually revert to the normal blue colour on prolonged heating. In order to measure the colour of the amino acid-ninhydrin complex and its rate of formation for each amino acid in the spectrum we have modified an LKB 4101 amino acid analyser as shown in Fig. 1. In the unmodified analyser the effluent from the ion exchange column is mixed with ninhydrin reagent and passed through a heated coil for 15 min. After this time the reaction mixture is passed through two flow cells, one that measures the absorption at 570 nm and the other at 440 nm (for

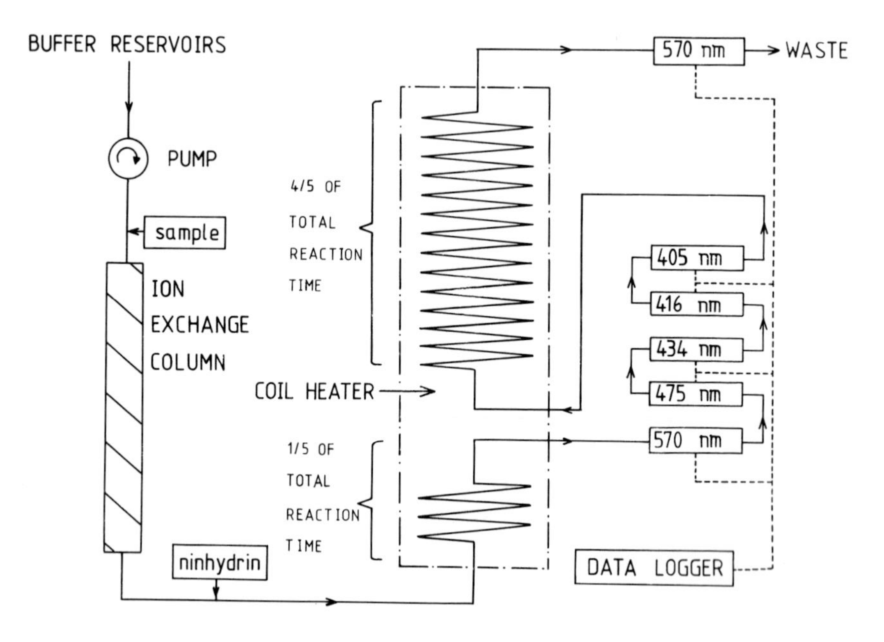

Fig. 1. Flow chart of modified amino acid analyser.

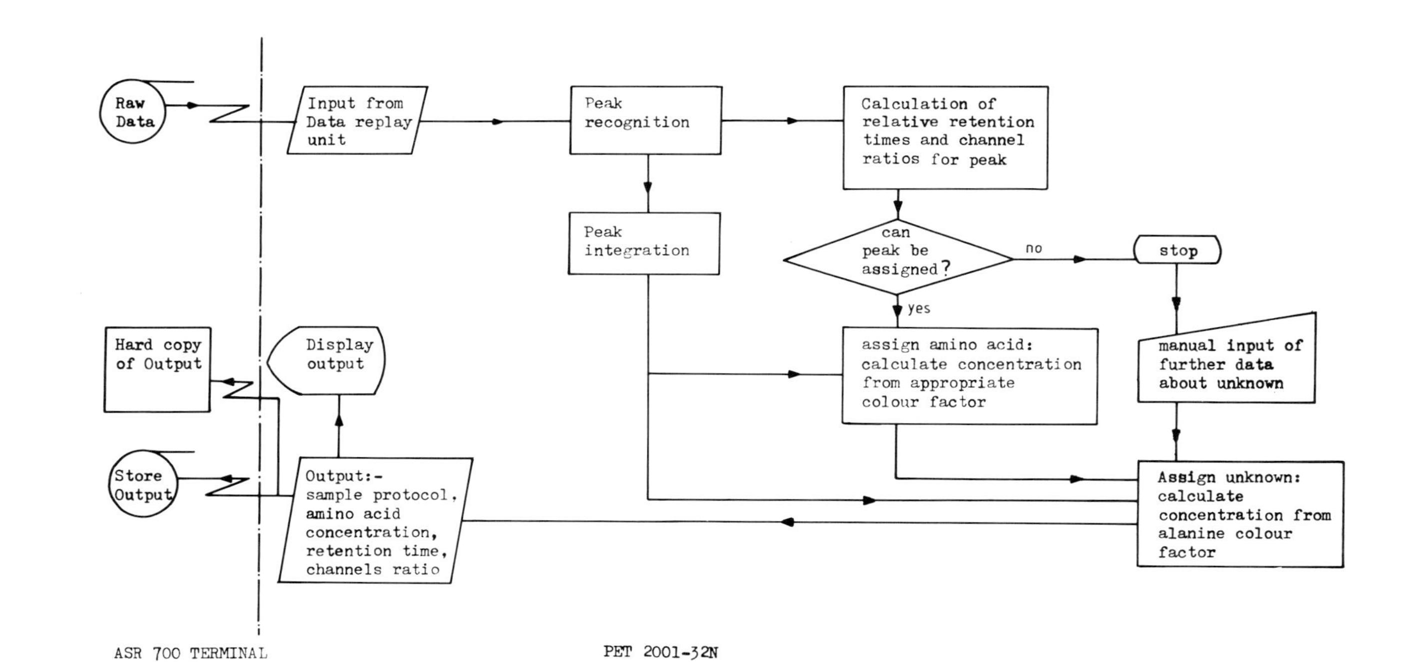

Fig. 2. Flow chart of data reduction and storage programs.

proline). Our modification involves carrying out the heating of the reaction mixture in two stages using two separate coils. After one fifth of the total reaction time has passed the mixture is taken from the first coil and its absorptions at 570, 475, 434, 416 and 405 nm are measured in five flow cells. The reaction mixture is then reheated in a second coil for the balance of the reaction time and its absorption at 570 nm remeasured in a further flow cell. The output from all six flow cells are recorded by the data logger such that complete colour and rate of development data are available for peak correlation. Examples of the information available using this technique are available in an earlier publication (Charlwood and Bell, 1977).

DATA PROCESSING

The suite of computer programmes that are used in the automated handling of the complete data output from the modified analyser are depicted in Figs. 2 and 3. All programmes are written in BASIC in a form suitable for the PET 2001-32N mini-computer, and require the availability of a separate tape cassette read/write facility and printer (Texas Instruments ASR Silent 700 Terminal fulfils these functions) and a Graphics Plotter (Hewlett Packard Model 7202A). The first programme set (Fig. 2) takes raw data from the data logger replay unit and carries out peak detection, integration and assignment and sets up a permanent data bank of the calculated values on tape cassettes. The raw data from each of the six channels of the analyser are transferred into a memory array and the 570 nm channel output is scanned for changes in the transient slope as each point is accessed. The value of the change in slope is compared with values of various parameters which can be set by the operator after the complete spectrum has been assessed visually for noise, separation and complexity. By this means points representing the start of a peak, the inflection points, the peak maxima and peak terminate or overlap, are obtained. Each peak or peak complex is integrated as it is detected using a modified form of Simpson's rule to determine the fractional increase in area of the peak as each new point is accessed: appropriate baseline corrections are made after this integration. For peaks which overlap by more than 20% the leading or trailing edge of each peak is mirrored into the overlapping

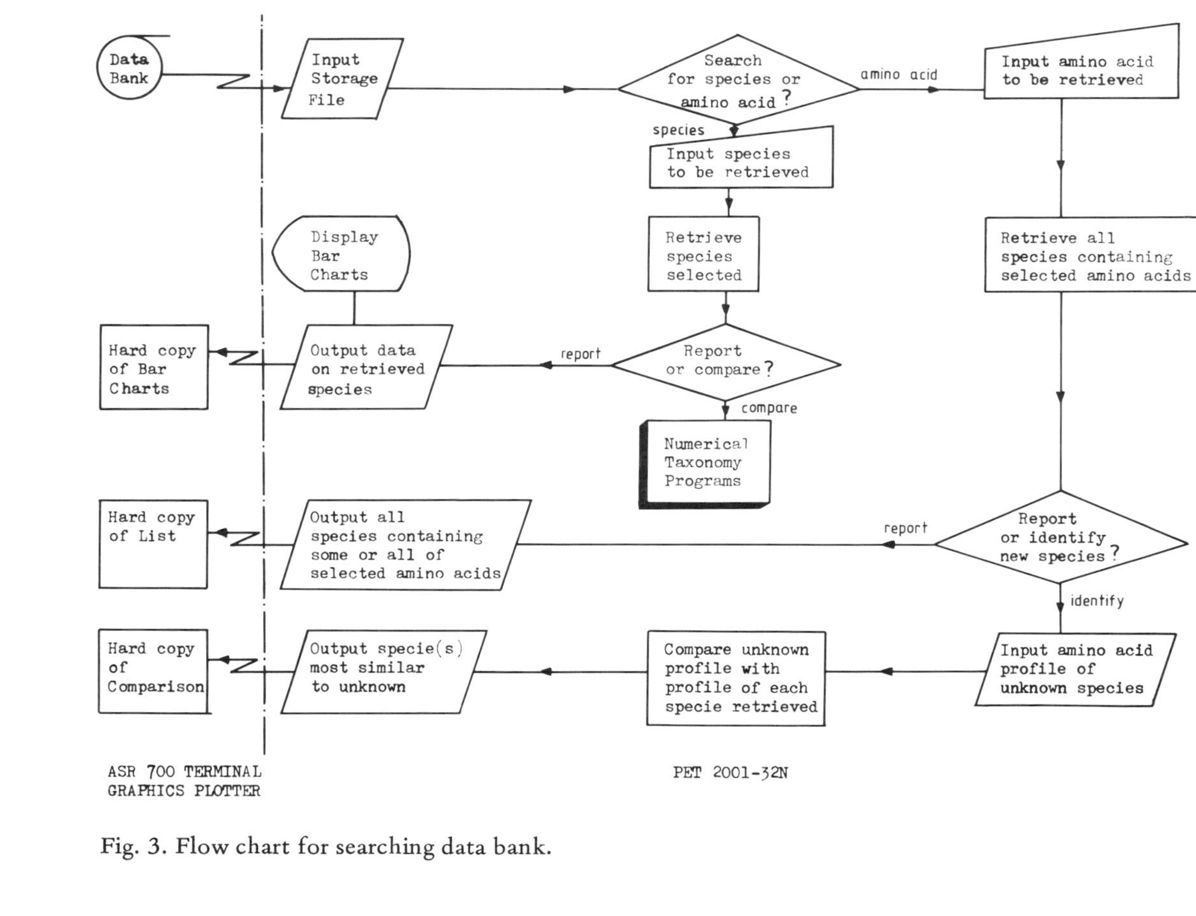

Fig. 3. Flow chart for searching data bank.

edge. When all peaks have been integrated their absolute retention times are calculated by reference to the position of maximum absorption in the 570 nm channel, and from this timing count the maximum absorptions for this amino acid in all other channels can be retrieved from the input array. The output from this part of the programme consists of a list of the peaks detected together with the area of each peak in arbitrary terms, the absolute retention time for each peak, the development time ratio and the wavelengths ratios. The next stage is to identify alanine, phenylalanine and/or arginine from this list. The values of absolute retention time, development time ratio and wavelength ratios for each standard are retrieved from the latest calibration analysis (this being updated every 10 analyses), and comparison of this data with the data for each peak in the newly calculated spectrum enables the standard amino acids to be assigned.

Having ascribed the standard peaks to the correct amino acids, the relative retention times for the rest of the peaks in the spectrum can be calculated. By recourse to a permanent memory file, the peaks are then compared one by one with respect to relative retention time, development time ratio and wavelength ratios with the stored values for non-protein amino acid reference standards. This process leads either to a positive identification of each peak or a request for further information about a peak that does not satisfy any of the criteria. If such a peak is completely unidentifiable then an identification code is input manually and the data for this unknown are returned to the memory for future reference. Once a peak has been recognized as a particular amino acid, the Beer-Lambert's Law colour factor is retrieved from memory and the true concentration of the amino acid is calculated from the area integration. The calculated information for the complete amino acid profile can be obtained as a list for record purposes and it is also sent for store on a cassette tape as part of the permanent data bank. On this tape the numerical information is stored in a 50 line string (each line can be 40 characters in length), the first item in each line being the amino acid name followed by its relative retention time and its concentration. The sample protocol is stored in a separate string with the species name as string identifier – this string is also used for a self-generating index to each tape file. Access of species name or amino acid components of a species are thus

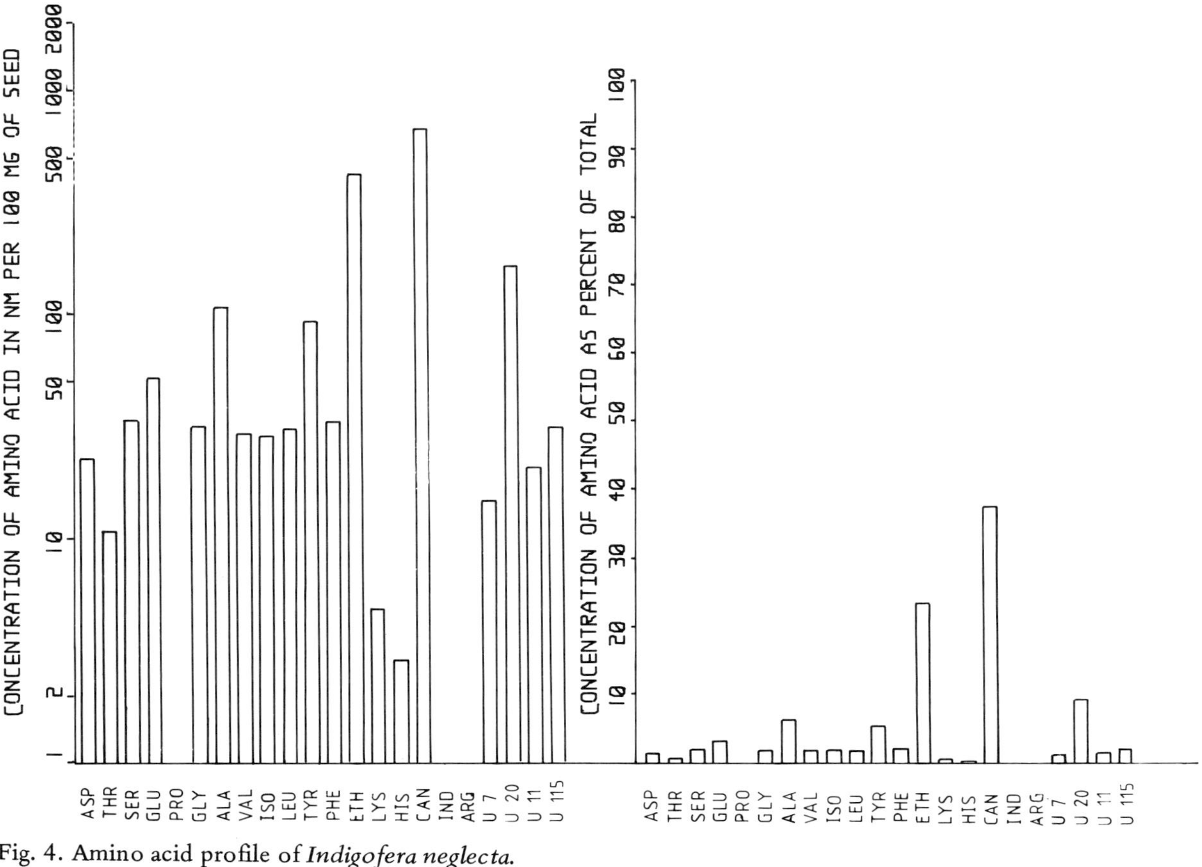

Fig. 4. Amino acid profile of *Indigofera neglecta*.

readily accessible on search. The data bank will be made available to taxonomists with an interest in the chemical characters of Legume seeds.

The second programme set (Fig. 3) enables the rapid searching of the data bank in a number of ways. Thus a search may be made for one or more amino acids in all species analysed, or a complete profile for any species analysed may be retrieved and presented in a bar chart form on a VDU or as hard copy (Fig. 4). The speed with which a bar chart may be displayed on the PET screen is particularly useful for the visual comparison of the amino acid profiles of a number of species which can be displayed two at a time in any order on a split screen.

ACKNOWLEDGEMENTS

We would like to thank LKB Biochrom Ltd, Science Park, Cambridge, England, for the kind loan of a model 4400 Automatic Amino Acid Analyser.

REFERENCES

Bell, E. A. and Fowden, L. (1964). Studies on amino acid distribution and their possible value in plant classification. *In* "Taxonomic Biochemistry and Serology" (C. A. Leone, ed.), pp. 203–223. Ronald Press, New York.

Bell, E. A. (1976). Uncommon amino acids in plants. *FEBS Letters* **64**, 29–35.

Bell, E. A., Lackey, J. A. and Polhill, R. M. (1978). Systematic Significance of Canavanine in the Papilionoideae (Faboideae). *Biochem. Syst. Ecol.* **6**, 201–212.

Charlwood, B. V. and Bell, E. A. (1977). Qualitative and quantitative analysis of common and uncommon amino acids in plant extracts. *J. Chromat.* **135**, 377–384.

6 Enzymes as a Taxonomic Tool: a Botanist's View

H. HURKA

Botanisches Institut und Botanischer Garten der Universität, Schlossgarten 3, D 44 Münster, Germany

INTRODUCTION

In 1966 Harris, Hubby and Lewontin introduced gel electrophoresis and isozyme assays into population genetics. Since then, electrophoresis has been intensively used by population geneticists and has become equally important in biosystematics and evolutionary biology. The main reasons for the rapid incorporation of electrophoretic technique as an experimental tool are its following advantages: (1) monomorphic and polymorphic loci can be identified which had been impossible beforehand; (2) large numbers of single-gene characters can be measured in the same individual; (3) most banding patterns are inherited codominantly. The enormous amount of heterallelism in natural populations was not primarily revealed by electrophoretic studies, as is sometimes stated. But since the incorporation of isozyme assays in population analyses it was possible to estimate the proportions of heterallelism among loci of which the sampled enzymes may be considered as representative.

Electrophoretic surveys so far seem to justify the assumption that most species populations of plants and animals contain remarkable stores of genic variation. The functional and evolutionary

Systematics Association Special Volume No. 16, "Chemosystematics: Principles and Practice", edited by F. A. Bisby, J. G. Vaughan and C. A. Wright, 1980, pp. 103–121, Academic Press, London and New York.

significance of this variation and the reasons for its maintenance in natural populations, however, are still under discussion. One hypothesis suggests that the greater part of the variation is maintained by some sort of balancing selection. The alternative hypothesis suggests that much of the variation in populations is neutral to selection and persists for that reason (cf. Le Cam *et al.*, 1972; Lewontin, 1973 and 1974; Kimura and Ohta, 1974). This highly debated question has occupied population geneticists in the last decade and many, if not most, studies of enzyme polymorphism address themselves to this controversy. Although this problem will not be discussed in this paper, every systematist using enzymes as a taxonomic tool must always keep it in mind, for it might influence the interpretation of his data.

This paper does not aim at presenting a review of electrophoretic studies in plant populations. It rather concentrates on some field of plant systematics where progress could be achieved by using enzymes, and it also points to problems originating from shortcomings of techniques and sampling strategy. As our own field of research is the biosystematics of the genus *Capsella*, I occasionally would like to quote from results of these investigations.

PATTERNS OF VARIABILITY IN PLANT POPULATIONS

Until now electrophoretic studies of enzyme variation mainly concentrate on animals and not so much on plants. Yet the majority of data suggests that many plant species populations contain remarkable stores of genetic variation, comparable with the situation in animals (for review see Allard and Kahler, 1972; Allard *et al.*, 1975; Brown, 1978). Some results, however, contrast sharply with this picture. In *Xanthium strumarium* populations in Australia, for instance, there is very little variation at the 13 tested enzyme loci within races but considerable interracial differentiation at several loci (Moran and Marshall, 1978). Remarkable within-species isozyme uniformity is also reported for some species of the genus *Glycine* (Broué *et al.*, 1977). In a study of six species of *Amaranthus* most populations are monomorphic for almost all the nine enzyme "phenotypes" scored (Hauptli and Jain, 1978). However, in most reports of apparently no or very little variation among enzyme loci there remains the question of sample size and of the enzyme systems scored. One often gets the impression that more accessions, collected properly and in sufficient numbers, are needed. In some cases the existence of a clinal variation pattern could be proved (e.g. for acid phosphatase in *Picea abies*, Bergmann, 1978), and sometimes a more patchy discontinuous mosaic pattern is reported (in *Silene maritima,* Baker *et al.*, 1975). But different enzyme loci may display different variation patterns within the same species. In *Capsella bursa-pastoris*, for instance, some electromorphs are apparently constant over the whole area from Scandinavia to Switzerland, whereas others vary proportionally to increasing adversity of climatic conditions, and others again are characterized by an irregular and discontinuous variation pattern. This situation was not only found for different enzyme systems but also for different loci within one and the same ensyme system (Bosbach, 1978; Bosbach and Hurka, in press).

Numerous papers tried to relate the levels of enzyme polymorphisms and heterozygosity to environmental heterogeneity, especially to the amplitude of environmental variation and to niche breadth. All the hitherto proposed genotype–environment interactions have not so far been demonstrated unequivocally, though considerable evidence has accumulated indicating that genetic polymorphisms may be related to environmental heterogeneity (cf. Hedrick *et al.*, 1976). For several reasons one should be very careful to relate isozyme polymorphisms to environmental heterogeneity. There is a high probability that relationships exist between levels of enzyme polymorphism and (1) population sizes which, unfortunately, are seldom reported in the literature; (2) the length of time that has passed since the population was established in the natural site, and since it has evolved; (3) the breeding system. There is also the danger of simple misinterpretation which is sometimes not taken into account. Reports of electrophoretic surveys are not always extensive or intensive enough to guarantee that the difference does not result from a small sample of electromorphs, a restricted geographical representation or a difficulty in techniques.

I am sure that the want of knowledge of the biology of the investigated organism often prevents satisfactory explanations. In *Capsella*, a study of seed ecology could help to explain a rather complicated variation pattern of isozyme banding patterns. A strong relationship between habitat conditions and genotypic composition is obvious in this species. Thus in highly disturbed sites (newly broken or freshly filled up soils) populations are built up with many different genotypes whereas in less disturbed sites only few genotypes are found, which form a very mosaic-like variation pattern (Bosbach and Hurka, in press). This variation pattern can be explained by the properties of the soil seed bank. In highly disturbed soils *Capsella bursa-pastoris* is one of the first plants to emerge, providing that the seeds had been stored in the soils beforehand. The shepherd's purse is able to build up large populations before succession begins, but successional displacement is rapid. The soil as a seed bank is a long-known phenomenon, the significance of which has recently been stressed, especially by Harper (1977). The analysis of formation and structure of the *Capsella* seed bank provides the key for an understanding of the observed variation pattern. Experiments showed (Haase, unpublished) that by the activity of earth worms *Capsella* seeds are continuously transported into the soil at random. In the course of time many generations can be deposited there where they are not influenced by selection. It has been demonstrated that after a period of 35–40 years of storage in the soil, *Capsella* seeds are still viable (Salisbury, 1964). Given time the stored genotypic variability in undisturbed soils eventually exceeds that which is estimated by taking samples of the actual population. This is because strong selection will operate on the plants when interspecies competition is becoming more and more rigorous in the course of successional change. If the soils containing a seed bank are broken, many seeds comprising an array of different genotypes are brought to the surface, and a highly polymorphic population emerges because germination and growth take place quickly thus reducing interspecies competition in the early stages of colonizing bare soils.

Thus highly disturbed soils may carry populations that are much more genetically polymorphic than those in less disturbed soils. These circumstances are obviously not directly correlated with the amplitude of environmental variation nor with the predictability of trophic resources nor with any other of the components so far discussed in literature. Differences between populations growing on disturbed and less disturbed sites are also reflected in characters other than enzyme banding patterns (Hurka and Benneweg, 1979).

BREEDING SYSTEMS

As emphasized by Wright (1951 and 1978) and others, environmental heterogeneity is not a prerequisite for gene frequency variance between subpopulations. Gene frequency heterogeneity among subpopulations may arise as a by-product of gene flow restriction, which is especially striking in plants (Levin and Kerster, 1974; Schaal, 1975). Gene flow restrictions can be imposed by the mating system, which in the case of assortative mating may take the form of self-pollination or pollination among neighbouring plants.

There exists an extensive literature on the impact of breeding systems on genotypic distributions within a population and on the long-term evolutionary potential of plant species (for review see Baker, 1955; Stebbins, 1957; Allard, 1975; Jain, 1976). For theoretical reasons a loss of variability is expected to occur when high self-pollination is introduced into a population. The "classic" theory of the influence of selfing on population structure summarized by Stebbins (1957) assumed that predominantly self-pollinated populations are built up by one or two, and occasionally by some more homozygous biotypes each of which are thought to be represented in numerous individuals. Heterozygotes that are sometimes found are thought to be the outcome of sporadic outcrossing. But it is well established that highly inbreeding populations can exhibit a high degree of heterozygosity (Allard *et al.*, 1968).

In order to analyse the influence of breeding systems on population variability experimentally, one has to quantify the effective breeding system and to estimate the genetic variability. With regard to the latter, the main experimental designs are progeny tests and the measuring of segregation of morphological characters and, secondly, the marking of genotypes by allozymes which gives the opportunity of recognizing homozygous and heterozygous loci provided that the genetics of the electromorphs has been established.

The quantification of the actual breeding system in a given population is a much more difficult task not free from fallacy. The difficulty is that the breeding system has to be determined without relying on estimates of genetic variability, otherwise one easily gets oneself into a vicious circle. For instance, heterozygotes are not necessarily the outcome of high outcrossing. There are several mechanisms which can account for the presence of a higher proportion of heterozygotes even in heavy inbreeding populations (Allard *et al.*, 1968). Nevertheless, by using enzyme assays a more detailed insight into the relation between breeding system and population structure has been gained, and the picture has become very puzzling.

In *Leavenworthia*, electrophoretic banding patterns conform to a simple model regarding the effect of autogamy (Solbrig, 1972), whereas in *Limnanthes*, inbreeder and outbreeder had similar genetic structures judged by electrophoretic assays (Arroyo, 1975), but this statement could not be corroborated by using morphological traits as genetic markers (Jain, 1978). *Avena barbata* outcrosses at a rate about three times higher on average than *Avena fatua*, but *Avena fatua* is much more variable genetically than *Avena barbata* (Allard *et al.*, 1975). Comparisons of within-population variability using coefficients of variation and allozyme polymorphisms for seven allozyme loci showed that the sexual lines in *Panicum maximum* on average have more variation than the asexual ones, but within-population variability in the asexual group is such that a few even exceed the sexual group in heterogeneity (Usberti and Jain, 1978). As was mentioned for *Limnanthes*, difficulties may arise when comparing arguments based on morphological evidence with allozyme data. Measuring segregation of morphological characters in progeny tests performed with *Capsella bursa-pastoris* reveals an unexpectedly high degree of heterozygosity (Hurka and Wöhrmann, 1977), whereas allozyme loci are not in favour with these findings (Bosbach, 1978). From morphological studies of barley (Jain and Allard, 1960) and lima beans (Harding and Tucker, 1964) evidence was provided for wide differences in the outcrossing rate for different genotypes. When allelic isozymes were used in estimating mating system parameters, no differences in the outcrossing rate between different genotypes could be detected in corn (Brown and Allard, 1970) and in *Avena barbata* (Marshall and Allard, 1970), but could be in *Lolium multiflorum* (Allard *et al.*, 1977). These discrepancies suggest two things: (1) outcrossing rates often depend upon genotypes for morphological variants. However, this is not the case if the markers are enzyme loci, and (2) the sample of enzyme markers studied is still too small or biassed by technique and/or function of enzymes. The above cited example of *Capsella* populations growing on highly disturbed and less disturbed soils provides an example where estimates of the influence of the breeding system on population structure could easily lead to misinterpretations.

Despite the puzzled and sometimes contradictory picture often brought about by isozyme data themselves, there is no doubt that isozyme data contributed a lot towards a more profound understanding of breeding systems and their impact on the genetic structure of populations.

(1) The influence of breeding systems on population variability cannot be described in absolute terms, but it varies with the intensity of other factors determining gene frequencies.

(2) The mating systems in different populations of one species may change significantly, even over short distances (Hamrick and Allard, 1972, for *Avena barbata*; Phillips and Brown, 1977, for *Eucalyptus pauciflora*).

(3) It seems to be established that different genotypes do not outcross to the same extent and hence there is genetic variability for mating systems within one and the same population. Genotypic differences in outcrossing rates can lead to stable polymorphisms (Wöhrmann and Lange, 1970).

(4) It seems to be evident that the amount of outcrossing is correlated with environmental factors and in many species is a genetic adaptation to the environment. This implies that the mating system itself is under selection and that it is subject to evolutionary change, which was found in *Hordeum vulgare* (Kahler *et al.*, 1975).

The usual procedure found in botanical textbooks to dichotomize plant species into "selfers" and "outcrossers" seems not to be justified because a wide spectrum of mating systems is often represented within a single species. Comparisons among populations and species of different levels of outcrossing reveal no clear relationship between mating systems and the amount of genetic variability.

SPECIATION

A crucial step of evolution is the speciation process. Analyses of variation within species supports evidence that the coming into being of a new species is a relatively rare event. It is an important task for the evolutionist to define the conditions that are most favourable for speciation. The classic theory points out that in outcrossing populations, speciation first requires spatial isolation and then divirgent selection pressures that establish new barriers of reproductive isolation. There are several arguments in favour of this geographical theory of speciation, both in animals and in plants (Mayr, 1963). But ever since the speciation process became a main issue in evolutionary biology, alternative methods of speciation have been proposed. The principal ones are embraced under the term "quantum speciation" (discussed by Grant (1971) and Carson (1975)) and called "saltational speciation" by Ayala (1975).

By introducing electrophoretic data it has become possible to analyse the speciation hypotheses more quantitatively. The data available so far seem to indicate that the acquisition of reproductive isolation in some *Drosophila* species does not necessarily require changes in a large proportion of the genome (Ayala, 1975). This situation seems to hold true for diploid annual plants, too (Gottlieb, 1973a and 1974, for *Clarkia*; Gottlieb, 1973b, for *Stephanomeria*). It is concluded that "all studies of genetic differentiation demonstrate that shortly after their origin species are still limited genetic versions of their progenitors or possess very few or no unique alleles insofar as their genomes have been assayed" (Gottlieb, 1976). Genetic differentiation between species is thought to be time-dependent and not accelerated by the species formation process itself.

As interesting and worth-while as it may be to compare enzyme data between derivative and progenitor species, the danger of over-interpretation and premature generalizations should not be overlooked. It is not evident up to now whether that fraction of the total genome marked by allozymes plays a decisive role in the course of species formation. In addition, I refer to the major unresolved problems which arise when estimating genetic variation detected by electrophoretic techniques. One of those problems is that electrophoresis detects only a fraction of all amino acid substitutions in proteins; it is not known exactly how much additional variation exists. The other unresolved problem emanates from the fact that only genes that qualitatively affect the structure of soluble proteins are examined by electrophoretic methods. Attention has been directed at the intriguing possibility that most significant evolutionary changes may not involve changes in the structural gene loci themselves, but rather changes in gene regulation, whatever this may stand for (Britten and Davidson, 1969; Dobzhansky *et al.*, 1977).

One would be ill-advised to take for granted arguments about the genetics of species formation based on enzyme assays. Some biologists even insist on the conclusion that isozyme data cannot even help in understanding the speciation process (Bush, 1975), and Lewontin (1974) writes, "We know virtually nothing about the genetic changes that occur in species formation".

PHYLOGENY

The degree of similarity of the enzymic pattern can be used as a marker of the degree of the phylogenetic relationships among the species. Basically, this procedure is the same as with other characters (i.e. morphological, cytological and physiological characters) employed in evolutionary studies. The advantages of using isozymes markers are (1) their less complicated genetic structure; (2) they seem to be less modifiable; (3) banding patterns are additive. It is thought that they exhibit a conservative nature of evolution and are more direct estimates of gene homology than morphology. Electrophoretic techniques have been adapted to problems such as the identification of closely related lines (Singh *et al.*, 1973, for oat varieties, Almgard and Landegren, 1974, for barley cultivars; Shechter and De Wet, 1975, for cultivated Sorghum races; Stegemann and Loeschcke, 1976, for potato varieties) and to the study of phylogenetic and evolutionary relationships (Scogin, 1973, in *Lupinus*; Rick and Fobes, 1975, in *Lycopersicon*; Broué *et al.*, 1977, in *Glycine*; Esen and Scora, 1977, in *Citrus*). I would like to comment on some problems connected with the use of enzyme markers for unravelling phylogenetic relationships.

1. Appropriate enzymes

It is not possible to decide *a priori* which enzyme system can serve as a genome marker in taxonomic studies. For instance, isozyme systems that have been used successfully in genetic analyses of other taxa were not successfully applied to *Galeopsis* (Priddle Houts and Hillebrand, 1976). This is especially true for malate dehydrogenase. Peroxidase isozyme patterns did not show much variation in different species and amphidiploids of *Brassica* (Yadava *et al.*, 1979), whereas peroxidase data agree well with differences between *Aegi-*

lops species (Symeonidis *et al.*, 1979). Lactate dehydrogenase varies considerably in *Picea glauca* (Tsay and Taylor, 1978), but is monomorph in *Capsella bursa-pastoris* from Scandinavia to Switzerland (Bosbach, 1978).

In general, the number of enzymes suitable to serve as genetic markers in the study of phylogeny has been until now quite limited, even in taxa which have been investigated for a longer time and rather intensively, like polyploid wheats (cf. Jaaska, 1978). There is an urgent need to search for suitable enzyme markers of genomes whenever a relevant study is undertaken.

2. *Additivity of Banding Patterns*

Some workers give the impression that it seems to be a fact that electrophoretic banding patterns are additive in hybrids. This additivity can partly be expected by the codominant inheritance of allozymes which, however, may not be true for silent alleles (null alleles) which are claimed to be recessive in some cases (Allard and Kahler, 1972). Strict additivity is reported for tobacco (Sheen, 1972), *Phaseolus* (Garber, 1974) and *Tragopogon* (Roose and Gottlieb, 1976). On the other hand, some species-specific bands were found to be missing in the amphidiploids *Brassica carinata* and *Brassica juncea* (Yadava *et al.*, 1979). The non-additive type of banding pattern is also reported for *Avena* (Murray *et al.*, 1970) and *Gossypium* (Cherry *et al.*, 1971). Sing and Brewer (1969) and Brewer *et al.* (1969) have shown that the average number of enzyme morphs in hexaploid wheat isozymes was not proportional to the ploidal level for the 12 enzyme systems studied, and there was almost no additivity of isozyme pattern. The reports of relationships between isozymes and polyploidy within the Triticinae, however, are still debated. Cubadda *et al.* (1975), for instance, demonstrated that several isozymes in wheats are clearly increasing from tetraploid to hexaploid level, and they also demonstrated their additivity (cf. also Hart, 1975 and Jaaska, 1978).

These different findings in regard to the additivity of enzymatic banding patterns are reflected in judgements about the usefulness of enzymes as genomic markers. They are not thought to be promising by some authors, and high resolving power is given to them by others.

When the diploid parents of a tetraploid species differ in their allelic composition at genes which code for different subunits of multimeric enzymes, novel heteromers can be produced which never occur in the parental species. Unlike heterozygosity at the diploid level, these phenotypes do not segregate. They are fixed heterozygotes. High proportions of novel heteromeric enzymes in tetraploids were observed in *Tragopogon* (Roose and Gottlieb, 1976). It is sometimes concluded that the enzyme multiplicity observed in polyploids extends the range of possible environments and that this may explain the ecological success.

The varying success in marking genomes by isozymes is certainly a result of differences in both techniques and material used (species-dependent and tissue-dependent).

3. *Lack of Data on Intraspecific Variability*

When tracing the history of a species and its phylogenetic relationships by means of enzymes it is not sufficient to rely on more or less accidental accessions as is sometimes the case. Extensive efforts to uncover geographic variability in the enzyme profile are unfortunately lacking. There is always the problem of genetic difference between two related species as compared to the genetic difference between two populations of the same species. There is strong demand for thorough biosystematic investigations throughout the species area, and in the case of polyploids also throughout the range of their putative parental species. For instance, investigations including different geographic areas and different *Aegilops* species could help to unravel the history and origin of hexaploid wheats in West Europe in more detail (Jaaska, 1978).

4. *How Convincing are Enzyme Studies?*

Different lines of investigations sometimes suggest different conclusions. For instance, because of the geographic distribution and the fertility of synthesized amphidiploids, the parents of *Nicotiana tabacum* seem to be *Nicotiana sylvestris* and *N. otophora*, whereas based on isozyme variation and other characters, the parental species are thought to be *Nicotiana sylvestris* and *N. tomentosiformis* (cf. Pickersgill, 1977). To cite another example: for *Brassica carinata*, isozyme similarities suggest *Brassica nigra* and *B. campestris* as parental species, whereas morphological and cytological results suggest that instead of *Brassica campestris, B. oleracea* is the other parent (Yadava *et al.*, 1979).

Similarity indices based on electrophoretic bands are due to alterations as the resolving power of the involved techniques is improved. They are also dependent on the number of enzymes assayed and on their variability, that means on the number of individuals scored and on their provenances within the species area. The main pitfalls in using enzymes as a taxonomic tool, therefore, seem to be sample error and technique.

In the last section of this Chapter I would like to point to some technical difficulties.

REMARKS ON TECHNIQUES

When measuring genetic variations by means of electrophoretic mobility data several questions arise.

When enzymes were first used to estimate levels of genetic variations, these markers were presented as a random sample of the genome. Since then several observations suggest, for a variety of reasons, that groups of enzymes differ in their levels of genetic variation. As an explanation for this variation, Gillespie and Kojima (1968) suggested that the degree of polymorphism in enzymes varies according to function. The expressions group I and group II enzymes were coined, and redefined by Gillespie and Langley in 1974 according to the number of physiological substrates generated by the enzymes. In contrast to group I enzymes, group II enzymes are thought to be almost universally variable in populations. Although this grouping is helpful, it is by no means clear-cut. Difficulties arise for special enzymes as to which group they should be placed in, and some enzymes which seem to be correctly placed exhibit strong polymorphism in one species and nearly none in another (cf. lactate dehydrogenase cited above). If groups of enzymes differ in their levels of genetic variation, it is not likely that electrophoretically detected variation will give an accurate estimation of the total genetic variation of the population.

The method of electrophoretic separation of proteins does not detect all of the allelic substitution at a locus. It is estimated that only as little as one quarter of all substitutions is detected. Apart from these implications thermal denaturation studies have revealed large numbers of variants not detected by other methods (for literature cf. Throckmorton, 1977 and Johnson, 1977a, b). However, the question remains as to how much of the detectable variation is allelic.

Cryptic variation of a very different sort was reported soon after the thermal stability studies were first published. By varying the pore size of electrophoretic gels many new variants can be detected. The reason for this additional variation is that protein, in migrating through a gel, interacts with the fibres of the gel. Its rate of migration is thus not only a function of its charge, but also of its size and asymmetry, and of the pore size of the gel. The major problems caused by the "hidden heterogeneity among electrophoretic alleles" (Johnson, 1977b) are the following. What is the genetic basis of this variation? Are these cryptic alleles functionally different? How much enzyme variation is there?

It seems that the amount of variation at the molecular level is much greater than that detectable by electrophoresis. The genes detected by electrophoresis, like those detected by ordinary observation, tend thus to be composites.

In the light of the alterations of electrophoretic banding patterns brought about by changing technical details, serious doubts arise when electrophoretic data gathered in different laboratories are compared. Some workers characterize their electromorphs by "mobility data", that is measurement of migration distance per time. One would prefer internal standards and characterize the band by R_f values, as these increase the chance of detecting congruence of data of different working groups. To make comparisons possible one would also prefer slab gels to tube techniques especially when coordinating closely adjacent bands. (The electrophoresis apparatus Panta-Phor developed by Stegemann turned out to be a very useful device, cf. Stegemann, 1972 and 1979.)

As Payne and Fairbrothers (1973) and others more recently pointed out, the selection of different enzyme systems and different electrophoretic approaches could lead to different interpretations regarding systematic relationships. In addition, it stands to reason that comparing banding patterns from different tissues or from different developmental stages, and ignoring possible changes brought about by growing conditions, as is for instance known for alcohol dehydrogenase (Crawford and McManmon, 1968), could also lead to serious misinterpretations. I wonder whether these aspects are always adequately taken into account when the plant material is cultivated and harvested.

CONCLUSIONS

The incorporation of enzyme studies in systematics and evolutionary biology has often brought about a better understanding of important issues, and electrophoresis will certainly see intensive use in the future. Plants with their variety of breeding systems, their array of speciation processes, their different ploidy levels and their easy handling in population experiments, both in nature and in artificial cultivation, will especially gain further importance in comparative studies. But the enthusiasm with which attention is sometimes given to isozyme investigations, does certainly overrate the possibilities of the method itself. In handling and interpreting isozyme data we need more critical interpretation, more knowledge about the ecology of a species, more character weighting and improved techniques.

ACKNOWLEDGEMENTS

Investigations on the biosystematics of the genus *Capsella* are supported by the Deutsche Forschungsgemeinschaft (DFG). I thank Miss Cornelia Papke for correcting the English text.

REFERENCES

Allard, R. W. (1975). The mating system and microevolution. *Genetics* **79**, 115–126.

Allard, R. W. and Kahler, A. L. (1972). Patterns of molecular variation in plant populations. *In* "Darwinian, Neo-Darwinian, and Non-Darwinian evolution" (L. M. Le Cam, J. Neyman and E. L. Scott, eds), pp. 237–254. Proc. Sixth Berkeley Symp. Mathem. Statist. Probab. Vol. V. University of California Press, Berkeley and Los Angeles.

Allard, R. W., Jain, S. K. and Workman, P. L. (1968). The genetics of inbreeding populations. *Adv. Genet.* **14**, 55–131.

Allard, R. W., Kahler, A. L. and Clegg, M. T. (1975). Isozymes in plant population genetics. *In* "Isozymes. Vol. IV: Genetics and Evolution" (C. L. Markert, ed.), pp. 261–272. Academic Press, New York and London.

Allard, R. W., Kahler, A. L. and Clegg, M. T. (1977). Estimation of mating cycle components of selection in plants. *In* "Measuring Selection in Natural Populations" (F. B. Christiansen and T. M. Fenchel, eds), pp. 1–19. Springer, Berlin.

Almgard, G. and Landegren, U. (1974). Isoenzymatic variation used for the identification of barley cultivars. *Z. Pfl Zücht.* **72**, 63–73.

Arroyo, de, M. T. K. (1975). Electrophoretic studies of genetic variation in natural populations of allogamous *Limnanthes alba* and autogamous *Lim nanthes floccosa* (Limnanthaceae). *Heredity* **35**, 153–164.

Ayala, F. J. (1975). Genetic differentiation during the speciation process. *Evol. Biol.* **8**, 1–78.

Baker, H. G. (1955). Self-compatibility and establishment after "long distance" dispersal. *Evolution* **9**, 347–348.

Baker, J., Maynard Smith, J. and Strobeck, C. (1975). Genetic polymorphism in the bladder campion, *Silene maritima. Biochem. Gen* **13**, 393–410.

Bergmann, F. (1978). The allelic distribution at an acid phosphatase locus in Norway spruce (*Picea abies*) along similar climatic gradients. *Theor. Appl. Gen.* **52**, 57–64.

Bosbach, K. (1978). Enzympolymorphismus in natürlichen Populationen des Hirtentäschelkrautes. Ein Beitrag zur Biosystematik der Gattung *Capsella* (Brassicaceae). Ph. D. Thesis, University of Münster.

Bosbach, K. and Hurka, H. (in press). Biosystematic studies on *Capsella bursa-pastoris* (Brassicaceae). Enzyme polymorphism in natural populations. *Pl. Syst. Evol.*

Brewer, G. J., Sing, C. F. and Sears, E. R. (1969). Studies of isozyme patterns in nullisomic–tetrasomic combinations of hexaploid wheat. *Proc. Nat. Acad. Sci. U.S.* **64**, 1224–1229.

Britten, R. J. and Davidson, E. H. (1969). Gene regulation for higher cells: a theory. *Science* **165**, 349–357.

Broué, P., Marshall, D. R. and Müller, W. J. (1977). Biosystematics of subgenus *Glycine* (Verdc.): Isoenzymatic data. *Aust. J. Bot.* **25**, 555–566.

Brown, A. H. D. (1978). Isozymes, plant population genetic structure and genetic conservation. *Theor. Appl. Gen.* **52**, 145–157.

Brown, A. H. D. and Allard, R. W. (1970). Estimation of the mating system in open-pollinated maize populations using isozyme polymorphisms. *Genetics* **66**, 133–145.

Bush, G. L. (1975). Modes of animal speciation. *Ann. Rev. Ecol. Syst.* **6**, 339–364.

Carson, H. L. (1975). The genetics of speciation at the diploid level. *Am. Nat.* **109**, 83–92.

Cherry, J. P., Kattermann, F. R. H. and Endrizzi, J. E. (1971). A comparative study of seed proteins of allopolyploids of *Gossypium* by gel electrophoresis. *Can. J. Genet. Cytol.* **13**, 155–158.

Crawford, R. M. M. and McManmon, M. (1968). Inductive responses of alcohol and malic dehydrogenases in relation to flooding tolerance in roots. *J. Exp. Bot.* **19**, 435–441.

Cubadda, R., Bozzini, A. and Quadrucci, E. (1975). Genetic control of esterases in common wheat. *Theor. Appl. Gen.* **45**, 290–293.

Dobzhansky, T., Ayala, F. J., Stebbins, G. L. and Valentine, J. W. (1977). "Evolution". Freeman and Comp, San Francisco.

Esen, A. and Scora, R. W. (1977). Amylase polymorphism in *Citrus* and some related genera. *Am. J. Bot.* **64**, 305–309.

Garber, E. (1974). Enzymes as taxonomic and genetic tools in *Phaseolus* and *Aspergillus. Israel. J. Med. Sci.* **10**, 268–277.

Gillespie, J. H. and Kojima, K. (1968). The degree of polymorphisms in enzymes involved in energy production compared to that in nonspecific enzymes in two *Drosophila ananossae* populations. *Proc. Nat. Acad. Sci. U.S.* **61**, 582–585.

Gillespie, J. H. and Langley, C. H. (1974). A general model to account for enzyme variation in natural populations. *Genetics* **76**, 837–884.

Gottlieb, L. D. (1973a). Enzyme differentiation and phylogeny in *Clarkia franciscana, C. rubicunda* and *C. amoena. Evolution* **27**, 205–214.

Gottlieb, L. D. (1973b). Genetic differentiation, sympatric speciation and the origin of a diploid species of *Stephanomeria. Am. J. Bot.* **60**, 545–553.

Gottlieb, L. D. (1974). Genetic confirmation of the origin of *Clarkia lingulata. Evolution* **28**, 244–250.

Gottlieb, L. D. (1976). Biochemical consequences of speciation in plants. *In* "Molecular Evolution" (F. J. Ayala, ed.), pp. 123–140. Sinauer Association, Sunderland, Massachusetts.

Grant, V. (1971). "Plant speciation". Columbia University Press, New York.

Hamrick, J. L. and Allard, R. W. (1972). Microgeographical variation in allozyme frequencies in *Avena barbata. Proc. Nat. Acad. Sci. U.S.* **69**, 2100–2104.

Harding, J. and Tucker, C. L. (1964). Quantitative studies on mating systems. I. Evidence for the non-randomness of outcrossing. *Heredity* **19**, 369–381.

Harper, J. L. (1977). "Population Biology of Plants". Academic Press, London and New York.

Harris, H. (1966). Enzyme–polymorphisms in man. *Proc. R. Soc.*, Ser. B, **164**, 298–310.

Hart, G. (1975). Glutamate oxalacetate transanimase isozymes of *Triticum*: evidence for multiple systems of triplicate structural genes in hexaploid wheat. *In* "Isozymes. Vol. III: Developmental Biology" (C. L. Markert, ed.). Academic Press, New York and London.

Hauptli, H. and Jain, S. K. (1978). Biosystematics and agronomic potential of some weedy and cultivated *Amaranthus. Theor. Appl. Gen.* **52**, 177–185.

Hedrick, P. W., Ginevan, M. E. and Ewing, E. P. (1976). Genetic polymorphism in heterogeneous environments. *Ann. Rev. Ecol. Syst.* **7**, 1–32.

Hubby, J. L. and Lewontin, R. C. (1966). A molecular approach to the study of genic heterozygosity in natural populations. I. The number of alleles at different loci in *Drosophila pseudoobscura. Genetics* **54**, 577–594.

Hurka, H. and Benneweg, M. (1979). Patterns of seed size variation in populations of the common weed *Capsella bursa-pastoris* (Brassicaceae). *Biol. Zbl.* 98, 699–709.

Hurka, H. and Wöhrmann, K. (1977). Analyse der genetischen Variabilität natürlicher Populationen von *Capsella bursa-pastoris* (Brassicaceae). *Bot. Jahrb. Syst.* 98, 120–132.

Jaaska, V. (1978). NADP-dependent aromatic alcohol dehydrogenase in polyploid wheats and their diploid relatives. On the origin and phylogeny of diploid wheats. *Theor. Appl. Gen.* 53, 209–217.

Jain, S. K. (1976). The evolution of inbreeding in plants. *Ann. Rev. Ecol. Syst.* 7, 469–495.

Jain, S. K. (1978). Breeding system in *Limnanthes alba*: several alternative measures. *Am. J. Bot.* 65, 272–275.

Jain, S. K. and Allard, R. W. (1960). Population studies in predominantly self-pollinated species. I. Evidence for heterozygote advantage in a closed population of barley. *Proc. Nat. Acad. Sci. U.S.* 46, 1371–1377.

Johnson, G. B. (1977a). Assessing electrophoretic similarity: The problem of hidden heterogeneity. *Ann. Rev. Ecol. Syst.* 8, 309–328.

Johnson, G. B. (1977b). Hidden heterogeneity among electrophoretic alleles. *In* "Measuring Selection in Natural Populations" (F. B. Christiansen and T. M. Fenchel, eds), pp. 223–244. Springer, Berlin.

Kahler, A. L., Clegg, M. T. and Allard, R. W. (1975). Evolutionary changes in the mating system of an experimental population of barley (*Hordeum vulgare* L.). *Proc. Nat. Acad. Sci. U.S.* 72, 943–946.

Kimura, M. and Ohta, T. (1974). On some principles governing molecular evolution. *Proc. Nat. Acad. Sci. U.S.* 71, 2848–2852.

Le Cam, L. M., Neyman, J. and Scott, E. L., (eds) (1972). "Darwinian, Neo-Darwinian, and Non-Darwinian Evolution". Proc. Sixth Berkeley Symp. Mathem. Statist. Probab., Vol. V. University of California Press, Berkeley and Los Angeles.

Levin, D. A. and Kerster, H. W. (1974). Gene flow in seed plants. *Evol. Biol.* 7, 139–220.

Levin, D. A. and Schaal, B. A. (1970). Reticulate evolution in Phlox as seen through protein electrophoresis. *Am. J. Bot.* 57, 977–987.

Lewontin, R. C. (1973). Population genetics. *Ann. Rev. Gen.* 7, 1–17.

Lewontin, R. C. (1974). "The Genetic Basis of Evolutionary Change". Columbia University Press, New York.

Marshall, D. R. and Allard, R. W. (1970). Maintenance of isozyme polymorphisms in natural populations of *Avena barbata*. *Genetics* 66, 393–399.

Mayr, E. (1963). "Animal Species and Evolution". Belknap Press, Cambridge, Massachusetts.

Moran, G. F. and Marshall, D. R. (1978). Allozyme uniformity within and variation between races of the colonizing species *Xanthium strumarium* L. (Noogoora Burr). *Aust. J. Biol. Sci.* 31, 283–291.

Murray, B. E., Craig, J. L. and Rajhathy, T. (1970). A protein electrophoretic study of three amphidiploids and eight species of *Avena*. *Can. J. Genet. Cytol.* 12, 651–655.

Payne, R. C. and Fairbrothers, D. E. (1973). Disc electrophoretic study of pollen proteins from natural populations of *Betula populifolia* in New Jersey. *Am. J. Bot.* **60**, 182–189.

Phillips, M. A. and Brown, A. H. D. (1977). Mating system and hybridity in *Eucalyptus pauciflora*. *Aust. J. Biol. Sci.* **30**, 337–344.

Pickersgill, B. (1977). Taxonomy and the origin and evolution of cultivated plants in the New World. *Nature* **268**, 591–595.

Priddle Houts, K. and Hillebrand, G. R. (1976). An electrophoretic and serological investigation of seed proteins in *Galeopsis tetrahit* L. (Labiatae) and its putative parental species. *Am. J. Bot.* **63**, 156–165.

Rick, C. M. and Fobes, J. F. (1975). Allozymes of Galapagos tomatoes: polymorphism, geographic distribution, and affinities. *Evolution* **29**, 443–457.

Roose, M. L. and Gottlieb, L. D. (1976). Genetic and biochemical consequences of polyploidy in *Tragopogon*. *Evolution* **30**, 818–830.

Salisbury, E. (1964). "Weeds and Aliens". 2nd edition. Collins, London.

Schaal, B. (1975). Population structure and local differentiation in *Liatris cylindracea*. *Am. Nat.* **109**, 511–528.

Scogin, R. (1973). Leucine aminopeptidase polymorphism in the genus *Lupinus* (Leguminosae). *Bot. Gaz.* **134**, 73–76.

Shechter, Y. and De Wet, J. M. J. (1975). Comparative electrophoresis and isozyme analysis of seed proteins from cultivated races of *Sorghum*. *Am. J. Bot.* **62**, 254–261.

Sheen, S. (1972). Isozymic evidence bearing on the origin of *Nicotiana tabacum*. *Evolution* **26**, 143–154.

Sing, C. F. and Brewer, G. J. (1969). Isozymes of a polyploid series of wheat. *Genetics* **61**, 391–398.

Singh, R. S., Jain, S. K. and Qualset, C. O. (1973). Protein electrophoresis as an aid to oat variety identification. *Euphytica* **22**, 98–105.

Solbrig, O. T. (1972). Breeding system and genetic variation in *Leavenworthia*. *Evolution* **26**, 155–160.

Stebbins, G. L. (1957). Self-fertilization and population variability in the higher plants. *Am. Nat.* **91**, 337–354.

Stegemann, H. (1972). Apparatur zur thermokonstanten Elektrophorese oder Fokussierung und ihre Zusatzteile. *Z. analyt. Chem.* **261**, 388–391.

Stegemann, H. (1979). Gel-Elektrophorese und Fokussieren in Platten. Arbeitsvorschrift für das Inst. f. Biochemie, Biol. Bundesanstalt, Braunschweig, F. R. G. Revised edition from 1968.

Stegemann, H. and Loeschcke, V. (1976). Identification by electrophoretic spectra. "Index of European Potato Varieties". National registers, appraisal of characteristics, genetic data. Mitteilungen Biol. Bundesanstalt f. Land- und Forstwiss. Heft 168. Parey, Berlin.

Symeonidis, L., Karataglis, S. and Tsekos, J. (1979). Electrophoretic variation in esterase and peroxidases of native Greek diploid *Aegilops* species (*Ae. caudata* and *Ae. comosa*, Poaceae). *Pl. Syst. Evol.* **131**, 1–15.

Throckmorton, L. H. (1977). *Drosophila* systematics and biochemical evolution. *Ann. Rev. Ecol. Syst.* **8**, 235–254.

Tsay, R. C. and Taylor, E. P. (1978). Isoenzyme complexes as indicators of genetic diversity in white spruce, *Picea glauca*, in southern Ontario and the Yukon Territory. Formic, glutamic, and lactic dehydrogenases and cationic peroxidases. *Can. J. Bot.* **56**, 80–90.

Usberti, J. A. and Jain, S. K. (1978). Variation in *Panicum maximum*: a comparison of sexual and asexual populations. *Bot. Gaz.* **139**, 112–116.

Wöhrmann, K. and Lange, P. (1970). Untersuchungen zur Wechselwirkung von Selektion und Selbstungsrate auf das genetische Gleichgewicht unter besonderer Berücksichtigung tetraploider Populationen. *Theor. Appl. Gen.* **40**, 289–296.

Wright, S. (1951). The genetical structure of populations. *Ann. Eugen.* **15**, 323–354.

Wright, S. (1978). "Evolution and the Genetics of Populations". Vol. 4. "Variability within and among Natural Populations". University of Chicago Press, Chicago.

Yadava, J. S., Chowdhury, J. B., Kakar, S. N. and Nainawatee, H. S. (1979). Comparative electrophoretic studies of proteins and enzymes of some *Brassica* species. *Theor. Appl. Gen.* **54**, 89–91.

7 | Enzymes as a Taxonomic Tool: a Zoologist's View

D. ROLLINSON

Department of Zoology, British Museum (Natural History), London SW7 5BD, England

INTRODUCTION

In the concluding chapter of "Origin of Species", Darwin (1859) foresaw that, with the acceptance of the ideas that he had put forward, there might be a considerable revolution in natural history. He also suggested that "Systematists will only have to decide (not that this will be easy) whether any form be sufficiently constant and distinct from other forms, to be capable of definition, and if definable, whether the differences be sufficiently important to deserve a specific name." One hundred and twenty years later, although the revolution has subsided a little, the decision is still not easy.

Taxonomists have traditionally identified and classified organisms primarily by the consideration of their morphological characters and, in many instances, this is still the only feasible approach. The discrete morphological species was not found to be incompatible with the biological species concept because it was apparent that reproductive isolation allows morphological divergence to take

---

Systematics Association Special Volume No. 16, "Chemosystematics: Principles and Practice", edited by F. A. Bisby, J. G. Vaughan and C. A. Wright, 1980, pp. 123–146, Academic Press, London and New York.

place. Animal species do, however, show a range of variation for most of their phenotypic characteristics and it is now well established that this variation often has both a genetic and an environmental component. When the morphological diversity within a group is such that a broad phenotypic overlap exists between related groups or when morphological differences between reproductively isolated groups are very small, problems of recognition arise. Furthermore, due to the difficulties associated with measuring the relationship between phenotypic and genotypic variation, it has not been possible to quantify satisfactorily the degree of genetic variation occurring both within and between animal populations.

For the taxonomist fortunate enough to be dealing with living material, additional tools to aid the identification and classification of animals have been provided by the recent introduction and application of the methods of molecular genetics, particularly enzyme electrophoresis. Earlier biological studies on the electrophoretic mobility of enzymes were not concerned with taxonomic problems but concentrated on such aspects as ontogeny and tissue distribution. In 1965, Shaw drew together information concerning sixteen enzymes in twenty species of organisms, from groups as diverse as protozoa and mammals, that indicated some variation in electrophoretic mobility. It was not long before the potential of the technique was realized and, in 1966, the first detailed estimates of heterozygosity and allelic distributions in natural populations were published independently by Harris for ten enzymes in man and by Hubby and Lewontin for eight enzymes in *Drosophila pseudoobscura*. Considerable interest was generated by the possibility of determining how much genetic differentiation accompanies the speciation process and a limited number of organisms, particularly fruit flies, mice and men, have since been the subject of intensive investigations (see Lewontin, 1974; Ayala, 1975 and Avise, 1976).

The extent to which biochemical characters have been used in animal taxonomy varies greatly in different groups (see Wright, 1974). The formidable task of reviewing the protein variation recorded in natural populations of animals has been admirably attempted elsewhere (Powell, 1975) and it is not my intention to provide an updating. Other authors have, in part, evaluated electrophoresis as a taxonomic method (Richardson *et al.*, 1973 and Avise, 1974) and it is with questions regarding the use, interpretation and value to the taxonomist of enzyme electrophoresis that this paper will be concerned.

WHY ENZYMES?

Enzymes are potential reservoirs of chemical variation and, as such, are an obvious target for taxonomic investigations. Primarily, the information about variation, that is mutation, is stored in the nucleic acids. Because the sequence of nucleotides that make up a gene is translated with a high degree of accuracy into a sequence of amino acids that make up a polypeptide chain, most of the mutations occurring in the genetic material are reflected by the amino acid composition of the resulting protein. Whitfield *et al.* (1966) determined that of every 24 random mutations involving base substitutions, 17 are missense mutations that pass on to the proteins in the form of altered amino acids. Fortunately, the technique of electrophoresis provides a simple way by which certain amino acid alterations can be detected. In synthesis beyond the enzyme level, most of the qualitative information which can be used for taxonomic purposes is lost. Usually, what passes on to the product of an enzyme reaction, if anything, is quantitative information due to the different catalytic efficiencies of the enzyme variants. Enzymes can be considered, therefore, to represent a particularly useful stage, not too far removed from the source of genetic variation, at which there is still an abundance of qualitative information which can be easily monitored.

ELECTROPHORESIS

The amino acids, of which a polypeptide is composed, possess different charges. Out of the 20 common amino acids, 16 can be considered neutral, two, arginine and lysine, positive and two, aspartic and glutamic acids, negative (Lewontin, 1974). The net charge of a polypeptide depends upon the balance of the charges of the constituent amino acids, the folding of the molecule and the hydrogen ion concentration. The hydrogen ion concentration at which a protein is effectively neutral, that is when the positive and negative charges of the molecule balance out, is known as the isoelectric point (*pI* value). If a change in the amino acid composition takes place, then it is possible that this will be reflected by a change in the overall charge of the molecule. It is primarily difference in charge, at a given pH, which can be detected by electrophoresis, as proteins will migrate through an electric field at different rates according to their net charge.

Obviously, only certain amino acid changes will alter the net charge of a molecule. Estimates of the likelihood of changes taking place which can be detected by electrophoresis vary a little. Shaw (1965) calculated that a single substitution in the nucleotide will produce a change in net charge of the polypeptide in 27.56% of the cases. Fitch (1966) estimated that of the possible missense mutations, 40% alter the net molecular charge of the polypeptide. Whereas, Henning and Yanovsky (1963) reported that seven out of nine induced single amino acid substitutions in the enzyme tryptophane synthetase were detectable as electrophoretic variants. The important point here is that electrophoresis only detects certain amino acid changes, while a large amount of variation remains undetected.

Samples for electrophoresis are usually crude water-soluble extracts from tissues of individual organisms, or, when size is a limiting factor, a whole organism or group of organisms. Proteins are separated in gels, usually starch or acrylamide; certain gels act as molecular sieves and will separate molecules, to some extent, by their size and shape as well as net charge. Appropriate stains are applied to localize specific enzymes and analysis can be made of the position, intensity and number of fractions which give enzyme activity. Considerations of various electrophoretic techniques can be found in Smith (1968), Latner and Skillen (1968) and Brewer (1970); recent advances in techniques of electrophoresis have been briefly reviewed by Hopkinson (1979). Staining methods are now available for the detection of more than one hundred different enzymes (Harris and Hopkinson, 1976), and the taxonomist is unlikely to be short of enzyme characters to investigate.

The enzyme patterns (or zymograms) obtained after electrophoresis often appear very complex and necessitate a combination of genetic and biochemical methods if they are to be fully elucidated. Almost every enzyme exhibits multiple forms; the occurrence of multiple loci, multiple alleles at a single locus and post-translational modification of enzyme molecules, all of which might contribute to the banding pattern in the gel, obviously demands careful interpretation of the enzyme fractions. Members of a species will, in general, possess the same multiple loci which will be seen as a common overall banding pattern; it is multiple allelism which is of particular benefit to the taxonomist.

At the enzyme production level, there appears to be little or no dominance and each allele functions to make its product. Therefore, if an organism is heterozygous for particular alleles, this can be detected when the products of the alleles are distinguishable by electrophoresis. If the enzyme is monomeric, the heterozygote can be recognized by the possession of two different forms of the enzyme each corresponding to one of the homozygotes. If the enzyme is dimeric then an additional intermediate or hybrid band will also be found. At any single locus, it is possible for two diploid individuals from a population to be genetically identical (both homozygous or heterozygous for the same alleles) or 50% alike (one homozygous and one heterozygous) or totally different (both homozygous or heterozygous for different alleles). Little taxonomic information can be gleaned from the fact that two individuals from a population bear no resemblance to one another, what is required is an examination of many different enzyme loci in a large number of individuals. Limitations to this approach are usually imposed by the number of organisms available rather than by the technique, which is capable of dealing rapidly with many samples.

TERMINOLOGY AND THE NEED FOR STANDARDIZATION

Different proteins with similar enzyme activity are commonly referred to as isoenzymes (isozymes). Such a broad definition, however, seemed unsatisfactory to many biochemists who felt that the term should be restricted to those enzyme forms arising from the genetic control of primary protein structure. The IUPAC-IUB Commission on Biochemical Nomenclature (1971) supported this view and recommended:

(1) The term "multiple forms of the enzyme" should be used as a broad term covering all proteins possessing the same enzyme activity and occurring naturally in a single species.

(2) The term "isoenzyme" or "isozyme" should apply to those multiple forms of enzymes arising from genetically determined differences in primary structure, and not to those derived by modification of the same primary sequence.

This terminology excludes such phenomena as enzymes conjugated with other chemical groups, different enzymes derived from one polypeptide chain by hydrolytic cleavage and polymers of a single sub-unit.

In addition to isoenzymes, many other terms have been used to describe the bands of enzyme activity identified after electrophoresis, including fractions, variants, types, phenotypes, zones, electromorphs, allelomorphs and patterns; the study of enzymes would profit by the introduction of a more limited and precise terminology. One useful term which has been widely accepted is that of Prakash *et al.* (1969) who proposed the term allozymes for the protein products of a single genetic locus which differ in electrophoretic mobility and which segregate according to Mendelian laws.

The ease with which the electrophoretic mobilities of enzymes can be misinterpreted or altered might well give the taxonomist cause for concern. Many examples can be found in the literature which question the interpretation of fractions identified as isoenzymes after electrophoresis. Chen and Sutton (1967) explained the complexity of the homozygous bovine transferrin patterns by the number of sialic residues per molecule and stressed that there was no need to consider heterogeneity of the polypeptide portion of the molecule. Only rarely, however, will the taxonomist be able to analyse further the enzyme fractions separated by electrophoresis, and confidence in using certain active fractions for taxonomic purposes can only be gained by repeated sampling to demonstrate that the enzyme patterns obtained are constant. The technique of isoelectric focusing (IEF) facilitates the separation of more active fractions than do conventional methods of electrophoresis. For example, Kaloustian *et al.* (1974) analysed glucose 6-phosphate dehydrogenase (G6PDH) of human erythrocytes by IEF and detected six active bands, whereas only one major and minor fraction were revealed by starch-gel electrophoresis. They proposed that the different forms of G6PDH were not genetic in origin but the result of post-transcriptional changes. Nevertheless, the repeatability and constancy of the enzyme patterns obtained by IEF are such that it is rapidly becoming an important technique for enzyme analysis.

Standardization of protein extraction and storage procedures is essential, because various chemical and physical conditions may alter the charge or chemical properties of a protein. Variations of isoenzyme and protein patterns have been reported after freezing and thawing of samples (Smith, 1968), after ultrasonic treatment (Dubbs, 1966), after the use of Triton X-100 (Momen *et al.*, 1975) and after lyophilization, refrigeration and incubation of extracts (Ruff *et al.*, 1971). Similarly, the conditions of electrophoresis, particularly the pH of the buffer (Gill, 1978a) are likely to affect migration rates and should be accurately recorded.

If an enzyme character is to be of diagnostic value in more than one laboratory, it is necessary to stipulate the way in which the electrophoretic mobility of an enzyme is to be measured and recorded. A number of methods are commonly employed for measuring the position (in the gel) of the active fractions:

(1) Distance from the origin
(2) Distance compared to the corresponding allozymes of a named strain
(3) Distance compared to the commonest allozyme
(4) Distance compared to the mobility of a standard marker
(5) Direct comparison between forms
(6) Isoelectric point.

Whichever method of measurement is utilized, and this usually depends upon the technique of electrophoresis, the purpose of study and the animal under investigation, it is desirable that the measurements of enzyme mobility can be easily repeated by other workers; unfortunately, this is often not the case. Furthermore, the nomenclature chosen for the bands of enzyme activity should be adaptable and able to incorporate any new enzyme forms that are detected by subsequent investigations.

FACTORS AFFECTING ENZYME EXPRESSION

The metabolism of an animal may vary both quantitatively and qualitatively at different times during the life cycle and with different environmental conditions. For the most part, isoenzyme expression appears to be remarkably constant and, as such, is of particular taxonomic value. Morphologically distinct stages occurring in the life of an organism can often be linked together and identified as one, due to the constancy of the isoenzyme expression. However, it is necessary, especially as electrophoretic studies usually depend on large samples of animals of mixed age and sex collected at different times from their natural habitat, to be aware of the possible existence of isoenzymes which might be induced or repressed by changes in the environment or which might reflect a particular state of the animal under study.

Alterations in the isoenzyme composition of certain animals have been correlated with changes in temperature; the role of isoenzymes in the process of temperature acclimatization in fish has been well documented (e.g. Hochachka, 1965; Baldwin and Hochachka, 1970; Yamawaki and Tsukuda, 1979). Shaklee *et al.* (1977) emphasized the importance of understanding the genetics of the isoenzyme systems and the need to use large samples to expose polymorphisms in such studies. Out of 26 systems studied by Shaklee *et al.* (1977) in the green sunfish, however, a few of the esterases in liver and eye did exhibit significant changes in isoenzyme distribution during acclimatization. Differences in the isoelectric point of fructose 1-6-diphosphatase in the skeletal muscle of the Alaskan king crab, between warm and cold acclimatized animals, were observed by Behrisch (1975). The results suggested that instead of an entire new isoenzyme being formed, it may only have been a small portion of the enzyme molecule that was altered during thermal acclimatization. Similarly, the pyruvate kinase from hibernating squirrels had a *pI* value of 5.7 (Behrisch, 1974), whereas that observed in the non-hibernating squirrel was *pI* 5.3 (Behrisch and Johnson, 1974). An ingenious approach to the study of temperature-induced isoenzyme variants was carried out by Marcus (1977) who monitored the isoenzyme composition of the tube feet of individual sea urchins, *Arabacia punctulata*, exposed to different temperature regimes. While three enzyme systems were found to be monomorphic and no change in the band migration was observed, alteration of the esterase pattern occurred and the change was reversible.

An environmental influence on esterase isoenzymes was also implicated by Flowerdew and Crisp (1976) who argued that the seasonal variation of certain esterase allozymes in the barnacle, *Balanus balanoides*, was probably not due to selection operating on a balanced polymorphism but was dependent on some environmental agent, possibly temperature or day-length. Oxford (1975) found that the land snail, *Cepaea nemoralis*, often contained three or four heavily staining esterase zones when analysed straight from the field; however, after three months under laboratory conditions, the snails were found to possess only two such zones. Interestingly, the ingestion of nettle induced extra esterase zones in laboratory reared snails. Different isoenzyme patterns of lactate dehydrogenase in the hepatopancreas of *Cepaea nemoralis* were observed by Gill (1978b) at different times of the year and variation was also detected between aestivating and feeding snails. Furthermore, similar differences occurred in the isoenzyme patterns and intensities of acid phosphatases, alkaline phosphatase and α-glycerophosphate dehydrogenase (Gill, 1978c).

The precise nature of environmental influences on enzyme expression may often be rather obscure. For instance, the form of lactate dehydrogenase in the liver of the creek chub, *Semotilus atromaculatus*, was influenced by photoperiod only when the fish was kept at an acclimatization temperature of 5° C (Kent and Hart, 1976). Variations in expression of acetylcholinesterases in the parasitic nematode of rats, *Nippostrongylus brasiliensis*, were related to the effects of immunity (Edwards *et al.*, 1971). The increasing number of reports relating differences in isoenzyme expression by similar genotypes in different environments makes it particularly necessary for the taxonomist to establish the stability of his enzyme characters.

The taxonomist must also consider whether other factors, such as age, sex, type of tissue used for analysis, or the physiological or disease state of the animal will influence the enzyme separations. Erlanger and Gershon (1970) studied enzymes in the nematode *Turbatrix aceti* as a function of age. Acrylamide gel electrophoresis of malate dehydrogenase revealed three isoenzymes, one of which disappeared completely in older animals (days 25 and 27). Several changes in the pattern of activity were also found in acid phosphatases of nematodes of various ages. Ross (1976) examined enzymes of the parasitic blood flukes, *Schistosoma* spp., by isoelectric focusing and drew attention to differences occurring between male and female worms. The tissue specificity of isoenzymes is now well documented and care must be taken in the choice of tissue for analysis.

Many observations have been published in relation to the usefulness of isoenzyme studies in the diagnosis of human ailments (e.g. Latner and Skillen, 1968), but little is known of isoenzyme changes in other diseased animals. Changes in activities of lactate dehydrogenase isoenzymes in the gut and fat body of the insects *Barathra brassicae* and *Galleria mellonella* infected by the microsporidian *Nosema plodiae* were observed by Kucera and Weiser (1975). When whole animals or crude tissue homogenates are used for enzyme analyses, it is possible that parasitic organisms within the tissues could contribute to the final enzyme pattern. Wright *et al.* (1979) detected a number of parasites in the freshwater snail, *Bulinus senegalensis*, during a routine analysis of enzymes in the snail digestive gland by isoelectric focusing. The differences in the *pI* values of the bands of enzyme activity of the snail compared to those of the parasite allowed each to be distinguished.

ENZYME VARIATION

Selander (1976) states that populations of many animal species are polymorphic at 25–50% of their enzyme loci ($p = 0.25$–0.50), whereas individuals are, on average, heterozygous at 5–15% of their loci. An enzyme locus is usually considered polymorphic if the frequency of the commonest allozyme is equal to or less than 0.99. As noted earlier, electrophoresis underestimates the protein differences occurring between animal groups; certain mutations may not alter the charge of the resulting protein and allozymes of different amino acid constitution may have the same net charge and the same electrophoretic mobility. Evidence of more enzyme variability than previously detected has been provided by heat denaturation studies on enzymes of *Drosophila* spp. (Bernstein *et al.*, 1973; Singh *et al.*, 1974; Singh *et al.*, 1975) and by the application of isoelectric focusing techniques (e.g. Kühnl *et al.*, 1977). In three species of the *D. virilis* group, electrophoretic studies had indicated that the octanol dehydrogenase locus was monomorphic, whereas heat denaturation studies showed that the locus was polymorphic; furthermore, it was estimated that electrophoresis alone underestimated the number of alleles at this locus by a factor of 2.6 (Singh *et al.*, 1975).

Due primarily to the large amount of molecular variation revealed by electrophoresis in natural populations of animals, a controversy exists in population genetics regarding the processes responsible for maintaining high levels of genetic polymorphism. It has been suggested that most of the variation is adaptively neutral (Kimura and Ohta, 1971), whereas many workers feel that some form of balancing selection must be operating (e.g. Ayala *et al.*, 1972). Lewontin (1974) has presented a detailed consideration of the problem. It seems likely, as suggested by Bernstein *et al.* (1973), that both neutral adaptation and selective forces might be operating simultaneously in natural populations.

Not all species show the same amounts of heterozygosity. In an effort to provide an explanation for this, many attempts have been made to relate heterozygosity to environmental variability although few satisfactory correlations have yet been found (Valentine, 1976). It is also becoming apparent that certain enzymes are more likely to be polymorphic than others. Correlations have been sought between enzyme function and heterozygosity (Gillespie and Kojima, 1968; Johnson, 1974; Powell, 1975) and, more recently, between quartenary structure and heterozygosity (Ward, 1977). Few studies have attempted to relate the level of polymorphism or heterozygosity to the amount of morphological variation within a population.

It is possible to distinguish two methods of approach which have been employed in utilizing the enzyme variation found in animal populations for taxonomic purposes. The first which will be considered is perhaps the more classical approach where each enzyme is treated as a single taxonomic character, whereas the second combines the results from many enzyme loci and is a means of expressing the overall electrophoretic differences between two forms.

ENZYMES AS TAXONOMIC CHARACTERS

Enzyme loci at which two organisms are monomorphic or polymorphic for different alleles can be used to differentiate between the two forms. For taxonomic purposes, the standard practice of assessing the limits of a character in morphological studies is equally applicable to biochemical characters. It is futile to declare a particular enzyme as "species-specific" without including any reference to the range of variation within populations of the species concerned and of closely related species. The use of only one or a few enzyme characters for the solution of a taxonomic problem may also prove very misleading, especially as there is the likelihood of different proteins exhibiting identical electrophoretic mobilities. Confidence in recognition increases as both the number of enzyme loci and the number of samples examined increases.

In many cases where traditional methods of identification have proved wanting, results of electrophoretic analyses of enzymes have provided useful taxonomic characters. The value of electrophoresis for the detection of sibling species has been well demonstrated. For example, two indistinct forms of the sea cucumber, *Thyonella gemmata*, were found to be different species by Manwell and Baker (1970), the marine polychaete *Capitella capitella* is actually at least six distinct sibling species (Grassle and Grassle, 1976) and the freshwater snail *Goniobasis floridensis* can be considered to be composed of two morphologically very similar but electrophoretically different species (Chambers, 1978). In these cases, the discovery of the sibling species was the result of the detection of enzymes which appeared unique to certain groups of individuals. Ayala and Powell (1972) studied two groups of sibling species of *Drosophila*, the allozyme variation was such that, if several diagnostic loci were used, the specific status of any individual could be diagnosed with virtual certainty.

The use of enzymes as taxonomic characters is particularly well shown by many recent studies on identification of parasitic protozoa, including *Plasmodium* (Carter, 1978), *Leishmania* (Chance, 1979), *Eimeria* (Shirley and Rollinson, 1979), trypanosomes (Godfrey, 1979) and amoebae (Sargeaunt and Williams, 1978). Many of these organisms lack distinguishing morphological features and enzyme characters are now routinely being used both for specific and infraspecific identifications. The variation found between seven species of *Eimeria* which parasitize chickens was so great and variation within each species so low that species can now be identified with confidence using only a few enzyme loci (Shirley and Rollinson, 1979). The important contribution that electrophoresis can make to the elucidation of certain epidemiological problems is well illustrated by investigations on *Trypanosoma cruzi*. This protozoan is responsible for Chagas disease on the South American continent, where at least ten million people are believed to be infected. *T. cruzi* from diverse mammals and vectors are morphologically indistinguishable and although heterogeneity had been suspected, the most reliable method of identifying strain-groups has proved to be enzyme electrophoresis. Miles *et al.* (1977) examined 17 stocks of *T. cruzi* and identified two strain-groups based on the distinct mobilities of six enzymes. One strain-group (type-I) occurred in opossums living outside houses in association with a certain vector, whereas the second was found in man and domiciliary animals in association with a different vector. Further work has highlighted the role of the species of vector in the distribution of the strain-groups of *T. cruzi* (Godfrey, 1979).

Much use is made of enzymes to detect hybrids in areas where two forms overlap. If no evidence of heterozygotes for the diagnostic loci can be found, then this is taken as proof that the forms under study are reproductively isolated. For example, the mosquito *Aedes detritus* has been found to comprise a pair of sympatric sibling species in southern France (Pasteur *et al.*, 1977). Each species possessed distinct forms of glutamate-oxaloacetate transaminase (GOT), plus frequency contrasts for three other enzyme loci, and despite intermingling of both taxa in the same breeding pool, no GOT heterozygotes were found. Similarly, Britton and Thaler (1978) found four enzyme loci which could be used to identify individuals belonging to two sympatric species of mice. The absence of hybrids, particularly in localities where both types of mice were captured, confirmed the specific status of the two forms. Krepp and Smith (1974) only realized that they were dealing with two species of the periodical cicada when they were part way through their electrophoretic analyses. The evidence for two species was based entirely on the α-glycerophosphate dehydrogenase allozymes, the complete absence of heterozygotes among animals collected at the same time and place argued strongly for two sympatric gene pools. Enzyme markers are also of great help in the laboratory to demonstrate the occurrence of hybridization. For example, by utilizing differences in phosphoglucomutase and drug sensitivity as markers, Rollinson *et al.* (1979) were able to demonstrate interbreeding between a strain of the protozoan *Eimeria maxima*, a parasite of the domestic fowl, and a strain isolated from the Ceylon Jungle fowl.

When diagnostic allozymes have been found, it is possible to devise biochemical identification keys. Avise (1974) utilized *Lepomis macrochirus macrochirus* as the standard and five genetic loci as markers. By comparing the relative mobility of the common allele of the species in question to the common allele of the standard, ten species of sunfish could be distinguished. A key for the identification of murine plasmodia was given by Carter (1978) who also provided a detailed description of technique to allow for accurate repetition. Similarly, Miles (1979) devised a biochemical identification key based on four enzymes for identification of adult mosquitoes in the *Anopheles gambiae* complex. Such keys, unfortunately, do have an inherent rigidity and may not be capable of dealing with regional enzyme variation. Diagnostic allozymes may differentiate closely related taxa in one area, but may be of little use in another. Rollinson and Southgate (1979) examined freshwater snails belonging to the *Bulinus africanus* species complex from Tanzania, and discovered that certain enzyme types were diagnostic for populations of *B. nasutus* collected from coastal areas. However, the same enzymes were of little value for the recognition of *B. nasutus* from inland areas close to Lake Victoria, where the enzyme types were also common to other species.

GENETIC DIFFERENTIATION BETWEEN TAXA

A number of statistical methods have been devised using overall electrophoretic data to measure genetic similarity, or its converse, genetic difference occurring between populations. The methods most commonly used in electrophoretic studies are genetic identity, I, and genetic distance, D, proposed by Nei (1972). I values may range from 0 to 1, with 1 indicating genetic identity. When a number of genetic loci are studied, the values of genetic identity are averaged over loci. Genetic distance is defined as $D = -\log_e I$ and can be interpreted as a measure of the average number of electrophoretically detectable allelic substitutions per locus which have accumulated since two populations separated from a common ancestral one. The similarity coefficient of Rogers (1972) is also frequently used; both Nei's and Rogers' coefficients yield similar summaries of the genetic information (Avise and Smith, 1977).

An extensive study of genetic differentiation during the early stages of evolutionary divergence has been carried out on the *Drosophila willistoni* complex of fruit flies (Ayala, 1975). Five levels of divergence were recognized; local populations, subspecies, semispecies, sibling species and non-sibling species. The mean levels of genetic identity and genetic distance between these groups corresponded well with their position in the species continuum. Local populations showed the greatest identity with $I = 0.970 \pm 0.006$;

there were no significant differences between subspecies and semi-species, with $I = 0.795 \pm 0.013$ and $I = 0.798 \pm 0.026$ respectively; sibling species were easily identified $I = 0.563 \pm 0.023$, and non-sibling species, as perhaps expected, showed the least identity with $I = 0.352 \pm 0.023$. In recent years, other studies on genetic differentiation have appeared, and have been considered by Ayala (1975), Avise (1976) and Zimmerman *et al.* (1978).

It is of taxonomic interest to establish whether estimates of genetic differentiation during speciation are uniform for different groups of animals. If speciation events are accompanied by a set amount of genetic divergence, detectable by electrophoresis, it should be possible to equate taxonomic categories, e.g. subspecies, sibling species etc., to a certain degree of genetic change. Avise *et al.* (1975) state that, typically, the similarities in a group of species belonging to the same genus (congeneric) range from 0.30 to 0.80, while the mean usually lies between 0.50 and 0.60. In contrast, infraspecific populations (conspecific) have similarity values very rarely smaller than 0.80 and generally greater than 0.90, which implies that speciation may usually be accompanied by substantial genetic differentiation. Although only a limited number of studies have yet been carried out, the frequent correspondence between levels of taxonomic and allozymic divergence suggests that allozymes may provide a useful measure of genetic change but, unfortunately, they do not always do so. Some species may appear little or no more distinct than populations within a species; Carson *et al.* (1975) found that *Drosophila sentosimentum* and *D. ochrobasis*, two Hawaiian members of the *adiastola* subgroup, although differing in their chromosomes, had a similarity coefficient in some areas of 0.98. Similarly, despite the voles *Microtus breweri* and *M. pennsylvanicus* sharing 94% of their genomes, Kohn and Tamarin (1978) argued that they may not be conspecific.

Examples are available of animals which differ morphologically and yet are electrophoretically indistinguishable and, conversely, of groups which are morphologically identical and yet show large amounts of genetic variation. One of the best examples of small genetic divergence, accompanied by very considerable morphological changes is the case of man and chimpanzees (King and Wilson, 1975). The genetic distance based on electrophoretic comparisons of proteins encoded by 44 loci is very small and corresponds to the genetic distance between sibling species of fruit flies. However, the anatomical and behavioural differences between humans and chimps have led to their classification in separate families. More recently, Bruce and Ayala (1978) have reported the results of electrophoretic studies which show that humans are also very similar to seven other species of apes.

A comprehensive study was made by Johnson *et al.* (1977) on the allozyme variation occurring at 20 loci in the land snails *Partula taeniata, P. saturalis, P. olympia* and *P. mirablis* from Moorea, *P. otaheitana* from Tahiti and *P. gibba* from Saipan. The Tahitian and Moorean species, which were all outbreeders, varied at 65–85% of the loci, with an average heterozygosity per individual of 13–17%. No variation was detected in *P. gibba* which reproduces by self-fertilization. Despite the great variability within populations, there was little allozymic differentiation between species. The genetic identity between pairs of species from Moorea and Tahiti averaged 0.91 as did the genetic identity of conspecific populations; even the geographically distinct *P. gibba* had an average genetic identity of 0.75 with the species from Moorea and Tahiti. However, the Moorean *Partula* species are quite distinct from each other. They differ in size, shape, colour, morphology of the genitalia and in behaviour and ecology, they also co-exist without hybridizing. Another interesting example is provided by Kornfield and Koehn (1975) who examined genetic variability in two endemic cichlid fishes from Mexico. By current standards of fish taxonomy, the two forms were quite different and yet, despite the absence of hybrids in laboratory mating experiments, a study of 13 loci revealed no differences between them. Similar results showing morphological difference with little genetic change have been reported by Avise *et al.* (1975) for two presumed species of Californian minnows and by Turner (1974) for several species of desert pupfish. At the other extreme, although few morphological differences were apparent between populations of the flatworm *Polycelis coronata*, Nixon and Taylor (1977) reported that the mean similarity between populations was 0.51, with values ranging from 0.17 to 0.95. The range of the relationship between morphological and genetic variation within a single genus is further shown by studies on ten species of the butterfly *Speyeria* in California and Nevada by Brittnacker *et al.* (1978). After analysis of 16 loci, they found little allozyme differentiation between five of the species, and in some cases, the genetic distance between two subspecies was greater than between one of the subspecies and another species. Although there were no diagnostic loci among these species, a combination of chromosomal, physiological and morphological criteria allowed each species to be identified. The other five species studied were easily distinguished from each other using electrophoresis, including two which were virtually morphologically indistinguishable.

It seems that while many species may fall within an accepted range of genetic identity values, there are others that clearly do not and it therefore becomes increasingly hazardous to correlate general values with taxonomic levels. Genetic similarities can still be of use for assessing relationships and determining taxonomic hierarchies within groups, but the source of reference is best determined separately for each group of organisms under study. There is no reason to assume that speciation events will involve the same amount of genetic differentiation, especially as species comparisons reflect both the amount of genetic modification accompanying speciation and the changes that have accumulated afterwards. Moreover, it is apparent that both morphological divergence and reproductive isolation may take place occasionally with little or no detectable divergence at the enzyme level.

CONCLUDING REMARKS

Since the Systematics Association last met, twelve years ago, to discuss chemotaxonomy, a great deal of pertinent research has been carried out. There can now be no doubt about the impact of enzyme electrophoresis on population biology as a whole, or about the contribution that the study of enzymes has already made to the elucidation of certain taxonomic problems. The growing need to differentiate both species and infraspecific populations, especially those of economically or medically important groups, has been greatly helped by the introduction of electrophoretic methods. The considerable discriminatory power of the approach is exemplified by its use when other methods of identification have proved unsatisfactory.

The frequent correspondence between the results of electrophoretic analyses and classical approaches to the taxonomy of animals is very encouraging, especially as most electrophoretic studies are made on previously described forms. Considering that the variation found between enzymes reflects only a very small part of the genome, it is remarkable how powerful a taxonomic tool enzyme electrophoresis is. Most examples, which have so far been considered, involve variation below the level of the genus. Due to the large amount of allozyme variation found within and between species it is unlikely that genetic similarity studies will be of value above this level. Only if a reasonable number of conservative enzyme loci are found, are comparisons between genera likely to be of value in revealing systematic relationships. The contribution that the study of enzyme multiplicity (multiple loci) can make towards unravelling problems of phylogeny has been well documented by Masters and Holmes (1974).

Several advantages of electrophoresis are evident (see Avise, 1974). Electrophoretic analyses of enzymes reveal very precise information on the genetic composition of animals. Differences in the migration of enzymes reflect discrete differences in DNA, and although only a small amount of genetic variation is examined, each enzyme locus studied probably reflects a similar amount of genetic information. The genetic basis of morphological, physiological or behavioural traits is seldom known and, consequently, many characters and the conclusions drawn from their observation may involve much more of the genome than others. Furthermore, the different phenotypic characters manifested by animals when subjected to different environmental conditions may obscure similar genotypes; although isoenzyme expression may also be influenced by external factors, there will probably always be a sufficient number of stable isoenzymes for analysis and comparison. Enzyme analyses also provide an accurate assessment of gene flow between sympatric groups of animals and further use of electrophoresis will very likely reveal more sibling species.

The technique of electrophoresis is relatively quick and simple and does not require a tremendous amount of expertise; identification of bands of enzyme activity is also precise, whereas the interpretation of morphological features may often be rather subjective. For results to have significance and be of long-term taxonomic value, far greater effort must be made to standardize experimental procedures and the way in which data are recorded in order to ensure the repeatability of the findings by other workers. Too often, electrophoresis is utilized in the hope of finding quick solutions for taxonomically difficult groups, with little regard to either adequate sample sizes or sufficient enzyme loci to provide meaningful results.

Every taxonomic approach will always have advantages and limitations. Although for the purpose of this paper, electrophoresis alone has been discussed and compared to other methods, the best taxonomic approach will always be to use all the available data and to use characters from as wide a range of organizational levels as possible. Electrophoresis, as a taxonomic tool, is undoubtedly proving to be a valuable aid in the definition of many animal forms, which as Darwin (1859) predicted will continue to perplex the zoologist.

REFERENCES

Avise, J. C. (1974). Systematic value of electrophoretic data. *Syst. Zool.* **23**, 465–481.

Avise, J. C. (1976). Genetic differentiation during speciation. *In* "Molecular Evolution" (F. J. Ayala, ed.), pp. 106–122. Sinauer Associates, Massachusetts.

Avise, J. C. and Smith, M. H. (1977). Gene frequency comparisons between sunfish (Centrarchidae) populations at various stages of evolutionary divergence. *Syst. Zool.* **26**, 319–335.

Avise, J. C., Smith, J. J. and Ayala, F. J. (1975). Adaptive differentiation with little genic change between two native Californian minnows. *Evolution* **29**, 411–426.

Ayala, F. J. (1975). Genetic differentiation during the speciation process. *Evol. Biol.* **8**, 1–78.

Ayala, F. J. and Powell, J. R. (1972). Allozymes as diagnostic characters of sibling species of *Drosophila. Proc. Nat. Acad. Sci. U.S.* **69**, 1094–1096.

Ayala, F. J., Powell, J. R., Tracey, M. L., Mourao, C. A. and Peres-Salas, S. (1972). Enzyme variability in the *Drosophila willistoni* group IV. Genic variation in natural populations of *Drosophila willistoni. Genetics* **70**, 113–119.

Baldwin, J. and Hochachka, P. W. (1970). Functional significance of isoenzymes in thermal acclimatization: Acetylcholinesterase from trout brain. *Biochem. J.* **116**, 883–887.

Behrisch, H. W. (1974). Temperature and the regulation of enzyme activity in the hibernator. Isoenzymes of liver pyruvate kinase from the hibernating and non-hibernating arctic ground squirrel. *Can. J. Biochem.* **52**, 894–902.

Behrisch, H. W. (1975). Subunit structure of seasonal isoenzymes of fructose 1,6-diphosphate from muscle of Alaskan king crab, *Paralithodes cantschatia. Comp. Biochem. Physiol.* **51B**, 317–321.

Behrisch, H. W. and Johnson, C. E. (1974). Regulatory properties of pyruvate kinase from liver of the summer-active ground squirrel. *Can. J. Biochem.* **52**, 547–559.

Bernstein, S., Throckmorton, L. H. and Hubby, J. L. (1973). Still more genetic variability in natural populations. *Proc. Nat. Acad. Sci. U.S.* **70**, 3928–3931.

Brewer, G. J. (1970). "An introduction to isoenzyme techniques". Academic Press, London and New York.

Brittnacker, J. G., Sims, S. R. and Ayala, F. J. (1978). Genetic differentiation between species of the genus *Speyeria* (Lepidoptera: Nymphalidae). *Evolution* **32**, 199–210.

Britton, J. and Thaler, L. (1978). Evidence for the presence of two sympatric species of mice (Genus *Mus* L.) in Southern France based on biochemical genetics. *Biochem. Gen.* **16**, 213–225.

Bruce, E. J. and Ayala, F. J. (1978). Humans and apes are genetically very similar. *Nature* **276**, 264–265.

Carson, H. I., Johnson, W. E., Nair, P. S. and Sene, F. M. (1975). Allozymic and chromosomal similarity in two *Drosophila* species. *Proc. Nat. Acad. Sci. U.S.* **72**, 4521–4525.

Carter, R. (1978). Studies on enzyme variation in the murine malaria parasites *Plasmodium berghei, P. yoellii, P. vinkei* and *P. chabaudi* by starch gel electrophoresis. *Parasitology* **76**, 241–267.

Chambers, S. H. (1978). An electrophoretically detected sibling species of *Goniobasis floridensis* (Mesogastropoda: Pleuroceridae). *Malacologia* **17**, 157–162.

Chance, M. (1979). The identification of Leishmania. *In* "Problems in the Identification of Parasites and their Vectors". *Symposia of the British Society of Parasitology* **17**, 55–74.

Chen, S. H. and Sutton, H. E. (1967). Bovine transferrins: sialic acid and the complex phenotype. *Genetics* **56**, 425–430.

Dubbs, C. A. (1966). Ultrasonic effects on isoenzymes. *Clin. Chem.* **12**, 181–186.

Darwin, C. (1859). "On the origin of species by means of natural selection or the preservation of favoured races in the struggle for life". John Murray, London.

Edwards, A. J., Burt, J. S. and Ogilvie, B. M. (1971). The effect of immunity upon some enzymes of the parasitic nematode *Nippostrongylus brasiliensis. Parasitology* **62**, 339.

Erlanger, M. and Gershon, D. (1970). Studies on ageing in nematodes. II. Studies on the activities of several enzymes as a function of age. *Expl. Gerontol.* **5**, 13–19.

Fitch, W. M. (1966). An improved method of testing for evolutionary homology. *J. Mol. Biol.* **16**, 9–16.

Flowerdew, M. W. and Crisp, D. J. (1976). Allelic esterase isoenzymes, their variations with season, position on the shore and stage of development in the Cirripede, *Balanus balanoides*. *Mar. Biol.* **35**, 319–325.

Gill, P. D. (1978a). Survey of isoenzymes in the snail *Cepaea nemoralis* using different buffer/gel systems in polyacrylamide disc electrophoresis: validity of comparisons and effect of "nothing dehydrogenase" activity. *Biochem. Gen.* **16**, 531–540.

Gill, P. D. (1978b). Non-genetic variation in isoenzymes of lactate dehydrogenase of *Cepaea nemoralis*. *Comp. Biochem. Physiol.* **59B**, 271–276.

Gill, P. D. (1978c). Non-genetic variation in isoenzymes of acid phosphatase, alkaline phosphatase and α-glycerophosphate dehydrogenase of *Cepaea nemoralis*. *Comp. Biochem. Physiol.* **60B**, 365–368.

Gillespie, J. H. and Kojima, K. (1968). The degree of polymorphisms in enzymes involved in energy production compared to that in non-specific enzymes in two *Drosophila ananassae* populations. *Proc. Nat. Acad. Sci. U.S.* **61**, 582–585.

Godfrey, D. G. (1979). The zymodemes of trypanosomes. *In* "Problems in the Identification of Parasites and their Vectors". *Symposia of the British Society of Parasitology* **17**, 31–53.

Grassle, J. P. and Grassle, J. F. (1976). Sibling species in the marine pollution indicator *Capitella* (Polychaeta). *Science* **192**, 567–569.

Harris, H. (1966). Enzyme polymorphism in man. *Proc. R. Soc. Lond. B.* **183**, 265–284.

Harris, H. and Hopkinson, D. A. (1976). "Handbook of Enzyme Electrophoresis in Human Genetics". North Holland Publishing Co, Amsterdam.

Henning, Y. and Yanovsky, C. (1963). An electrophoretic study of mutational altered A proteins of the tryptophan synthetase of *Escherichia coli*. *J. Mol. Biol.* **6**, 16–21.

Hochachka, P. W. (1965). Isoenzymes in metabolic adaptation of a poikilotherm: subunit relationships in lactate dehydrogenase of goldfish. *Archs Biochem. Biophys.* **111**, 96–103.

Hopkinson, D. A. (1979). Introduction. *In* "Problems in the Identification of Parasites and their Vectors". *Symposia of the British Society of Parasitology* **17**, 1–6.

Hubby, J. L. and Lewontin, R. C. (1966). A molecular approach to the study of genic heterozygosity in natural populations. I. The number of alleles at different loci in *Drosophila pseudoobscura*. *Genetics* **54**, 577–594.

IUPAC/IUB (1971). Commission on Biochemical Nomenclature. The nomenclature of multiple forms of enzymes. Recommendations. *Biochemistry* **10**, 4825–4826.

Johnson, G. B. (1974). Enzyme polymorphism and metabolism. *Science* **184**, 28–37.

Johnson, M. S., Clarke, B. and Murray, J. (1977). Genetic variation and reproductive isolation. *Evolution* **31**, 116–126.

Kaloustian, V. M. D., Idriss-Daouk, S. H., Hallac, R. T. and Awdeh, Z. L. (1974). Analysis of human erythrocyte glucose 6-phosphate dehydrogenase isoenzymes by isoelectric focusing in polyacrylamide gel. *Biochem. Gen.* **12**, 51–58.

Kent, J. D. and Hart, R. G. (1976). The effect of temperature and photoperiod on isoenzyme induction in selected tissues of the creek chub, *Semotius atromaculatus. Comp. Biochem. Physiol.* **54B**, 77–80.

Kimura, M. and Ohta, T. (1971). Protein polymorphism as a phase of molecular evolution. *Nature* **229**, 467–469.

King, M. C. and Wilson, A. C. (1975). Evolution at two levels in humans and chimpanzees. *Science* **188**, 107–116.

Kohn, P. H. and Tamarin, R. H. (1978). Selection of electrophoretic loci for reproductive parameters in island and mainland voles. *Evolution* **32**, 15–28.

Kornfield, I. L. and Koehn, R. K. (1975). Genetic variation and speciation in New World cichlids. *Evolution* **29**, 427–437.

Krepp, S. R. and Smith, M. H. (1974). Genic heterozygosity in the 13-year cicada, *Magicicada. Evolution* **28**, 396–401.

Kucera, M. and Weiser, J. (1975). Lactate dehydrogenase isoenzymes in the larvae of *Barathra brassicae* and *Galleria mellonella* during microsporidian infection. *J. Invertebr. Pathol.* **25**, 109–114.

Kühnl, P., Schmidtmann, U. and Spielmann, W. (1977). Evidence for two additional common alleles at the PGM locus (Phosphoglucomutase – E. C. 2.7.5.1). *Hum. Gen.* **35**, 219–223.

Latner, A. L. and Skillen, A. W. (1968). "Isoenzymes in Biology and Medicine". Academic Press, London and New York.

Lewontin, R. C. (1974). "The Genetic Basis of Evolutionary Change". Columbia University Press, New York.

Manwell, C. and Baker, C. M. A. (1970). "Molecular Biology and the Origin of Species". University of Washington Press, Seattle.

Marcus, N. H. (1977). Temperature induced isozyme variants in individuals of the sea urchin, *Arbacia punctulata. Comp. Biochem. Physiol.* **58B**, 109–113.

Masters, C. J. and Holmes, R. S. (1974). Isoenzymes, multiple enzyme forms and phylogeny. *Adv. comp. Physiol. Biochem.* **5**, 109–195.

Miles, M. A., Toye, P. J., Oswals, S. C. and Godfrey, D. G. (1977). The identification by isoenzyme patterns of two distinct strain-groups of *Trypanosoma cruzi*, circulating independently in a rural area of Brazil. *Trans. R. Soc. trop. Med. Hyg.* **71**, 217–225.

Miles, S. J. (1979). A biochemical key to adult members of the *Anopheles gambiae* group of species. *J. med. Ent. Honolulu* **15**, 297–299.

Momen, H., Atkinson, E. M. and Homewood, C. M. (1975). An electrophoretic investigation of the malate dehydrogenase of mouse erythrocytes infected with *Plasmodium berghei. Int. J. Biochem.* **6**, 533–535.

Nei, M. (1972). Genetic distance between populations. *Am. Nat.* **106**, 283–291.

Nixon, E. S. and Taylor, R. J. (1977). Large genetic distances associated with little morphological variation in *Polycelis coronata* and *Dugesia tigrina* (Planaria). *Syst. Zool.* **26**, 152–164.

Oxford, G. S. (1975). Food induced esterase phenocopies in the snail *Cepaea nemoralis*. *Heredity* **35**, 361–370.

Pasteur, N., Rioux, J. A., Guilyard, E., Pech-Perleres, M. J. and Verdier, J. M. (1977). Existance chez *Aedes* (*Ochlerotatus*) *detritus* (Haliday, 1833) (Diptera-Culicidae) de Camargue de deux formes sympatriques et sexuellement isolées (espèces jumelles). *Annls Parasit. hum. comp.* **52**, 325–337.

Powell, J. R. (1975). Protein variation in natural populations of animals. *Evol. Biol.* **8**, 79–113.

Prakash, S., Lewontin, R. C. and Hubby, J. L. (1969). A molecular approach to the study of genic heterozygosity in natural populations. IV. Patterns of genic variation in central, marginal and isolated populations of *Drosophila pseudoobscura*. *Genetics* **61**, 841–858.

Richardson, B. J., Johnston, P. G., Clark, P. and Sharman, G. B. (1973). An evaluation of electrophoresis as a taxonomic method using comparative data from the Macropodidae (Marsupialia). *Biochem. Syst. Ecol.* **1**, 203–209.

Rogers, J. S. (1972). Measures of genetic similarity and genetic distance. *Univ. Tex. Publs.* **7213**, 145–153.

Rollinson, D., Joyner, L. P. and Norton, C. C. (1979). *Eimeria maxima*: the use of enzyme markers to detect the genetic transfer of drug resistance between lines. *Parasitology* **78**, 361–367.

Rollinson, D. and Southgate, V. R. (1979). Enzyme analyses of *Bulinus africanus* group snails (Mollusca: Planorbidae) from Tanzania. *Trans. R. Soc. trop. Med. Hyg.* **73**, 667–672.

Ross, G. C. (1976). Isoenzymes in *Schistosoma* spp.: LDH, MDH and acid phosphatases separated by isoelectric focusing in polyacrylamide gel. *Comp. Biochem. Physiol.* **55B**, 343–346.

Ruff, M. D., Werner, J. K. and Davis, G. M. (1971). Effect of extraction procedures on disc-electrophoretic patterns of *Schistosoma japonicum* proteins. *Jap. J. Parasit.* **20**, 341–358.

Sargeaunt, P. G. and Williams, J. E. (1978). The differentiation of invasive and non-invasive *Entamoeba histolytica*. *Trans. R. Soc. trop. Med. Hyg.* **72**, 519–521.

Selander, R. K. (1976). Genic variation in natural populations. *In* "Molecular Evolution" (F. J. Ayala, ed.), pp. 21–45. Sinauer Associates, Massachusetts.

Shaklee, J. B., Christiansen, J. A., Sidell, B. D., Prosser, C. L. and Whit, G. S. (1977). Molecular aspects of temperature acclimation in fish: contributions of changes in enzyme activities and isoenzyme patterns to metabolic reorganisation in the green sunfish. *J. exp. Zool.* **201**, 1–20.

Shaw, C. R. (1965). Electrophoretic variation in enzymes. *Science* **149**, 936–943.

Shirley, M. W. and Rollinson, D. (1979). Coccidia: the recognition and characterisation of populations of *Eimeria*. *In* "Problems in the Identification of Parasites and their Vectors". *Symposia of the British Society of Parasitology* **17**, 7–30.

Singh, R. S., Hubby, J. L. and Lewontin, R. C. (1974). Molecular heterosis for heat-sensitive enzyme alleles. *Proc. Nat. Acad. Sci. U.S.* **71**, 1808–1810.

Singh, R. S., Hubby, J. L. and Throckmorton, L. H. (1975). The study of genic variation by electrophoretic and heat denaturation techniques at the octanol dehydrogenase locus in members of the *Drosophila virilis* group. *Genetics* **80**, 637–650.

Smith, I. (1968). Chromatographic and electrophoretic techniques. Volume II. "Zone Electrophoresis". Heinemann, London.

Turner, B. J. (1974). Genetic divergence of Death Valley pupfish species: biochemical versus morphological evidence. *Evolution* **28**, 281–294.

Valentine, J. W. (1976). Genetic strategies of adaptation. *In* "Molecular Evolution" (F. J. Ayala, ed.), pp. 78–94. Sinauer Associates, Massachusetts.

Ward, R. D. (1977). Relationship between enzyme heterozygosity and quarternary structure. *Biochem. Gen.* **15**, 123–135.

Whitefield, H. J., Martin, R. G. and Ames, B. N. (1966). Classification of aminotransferase (C gene) mutants in the histidine operon. *J. Mol. Biol.* **21**, 335–355.

Wright, C. A. (1974). "Biochemical and immunological taxonomy of animals". Academic Press, London and New York.

Wright, C. A., Rollinson, D. and Goll, P. H. (1979). Parasites in *Bulinus senegalensis* (Mollusca: Planorbidae) and their detection. *Parasitology* **79**, 95–105.

Yamawaki, H. and Tsukuda, H. (1979). Significance of the variation in isoenzymes of liver lactate dehydrogenase with thermal acclimation in goldfish. I. Thermostability and temperature dependency. *Comp. Biochem. Physiol.* **62B**, 89–93.

Zimmerman, E. C., Kilpatrick, C. W. and Hart, B. J. (1978). The genetics of speciation in the rodent genus *Peromyscus. Evolution* **32**, 565–579.

8 | A Geneticist faced with Enzyme Variation

R. J. BERRY

Department of Genetics and Biometry, University College London, Gower Street, London WC1, England

Abstract: The possibility of identifying variation in gene products by such techniques as electrophoresis seemed at one time a logical and easy solution to taxonomic problems. This has not proved so: many enzyme variants are adaptive and respond rapidly to selective pressures, and spectra of allozymic variations are susceptible to historical events of population subdivision and colonization. Notwithstanding, they can obviously be regarded as important traits in sorting out population and speciation events, but much more work needs doing on the pathways between gene product and morphological character.

Some of the problems and contexts of the use of allozymic variants are illustrated by a description of the relationships between races of house mice on different islands in the Faroe group.

Since the distinctions between taxa are by definition inherited, it follows that systematists must be interested in the nature and extent of genetical differences between groups. Indeed, older publications often imply that classifications are only imperfect because of the impossibility of directly investigating the genome. It is therefore not surprising that a considerable amount of energy has been invested in methods which profess to produce data reflecting "objective" measures between intra- and interspecific groups. The 1967 Symposium of the Systematics Association was devoted to reviewing such methods and their results (Hawkes, 1968) and at least one

---

Systematics Association Special Volume No. 16, "Chemosystematics: Principles and Practice", edited by F. A. Bisby, J. G. Vaughan and C. A. Wright, 1980, pp. 147–166, Academic Press, London and New York.

major later book comprehensively summarized the achievements of chemical and immunological approaches to taxonomy (Wright, 1974).

Electrophoresis, immunophoresis, isoelectric focusing and protein sequencing now make it possible to study inherited variation in primary gene products, in many cases on population samples. There can be no doubt that they have made it feasible to dissect speciation processes, but they have likewise introduced complications into systematics. The problem is that speciation is not a direct correlate of genetical divergence. The glib phrases of those who sought a "natural" phylogenetic classification through genic evolution have been found wanting; enzyme variation is a useful – possibly a necessary – tool in the systematist's kit, but it is not the answer to all his difficulties.

It is revealing to skim through the 1967 Symposium for the aspirations and hopes of twelve years ago. All the techniques now used routinely are described there but, perhaps significantly, two papers published in 1966, which applied electrophoresis to population samples and produced a revolution in population genetics, are not even mentioned. In retrospect, the subservience of population geneticists to mathematical theorists which was broken in 1966 seems almost culpable, yet other disciplines (not excluding systematics) need occasional blasts of facts to be able to advance their own understanding. From the vantage point of 1979 it can be seen that no amount of data about protein and enzyme variation is going to solve taxonomic problems at a stroke, and that protagonists of phylogenesis as the El Dorado of systematists have had their bluff called.

THE HISTORY OF MOLECULAR TAXONOMY

We can recognize four steps in the growth of understanding of the nature and significance of protein variation in populations. Apart from these, many lesser but painful lessons have had to be learnt – changes in isozyme expression with age, sexual activity, following death, etc. The usual response of electrophoresis workers is to assume an electrophoretic pattern represents a segregating locus if different gel phenotypes segregate in a binomial (or polynomial) fashion, but this may be dangerous as the statistical tests used are inefficient. The only safe way to determine the inheritance of an electrophoretic phenotype is to carry out family studies as for any other putatively inherited trait. These have been carried out for most allozymic loci in man, mouse and several *Drosophila* species, but beyond these the inheritance of enzyme variants is merely assumed. I have accepted in this chapter that the variants described by different authors are all inherited, but it must be born in mind that this is no more than a charitable assumption.

With that reservation, the history of understanding inherited variations in proteins can be divided into several stages.

(1) The 1967 Symposium represented a stage when the really significant techniques for studying variation in enzymes and proteins had already been developed. The most important advances in the 1970s have been the use of restriction enzymes to compare proteins in different populations or species and the development of immunogenetic methods to examine the products of particular genes. But in general the last decade has seen the accumulation, analysis and interpretation of data collected by established techniques.

(2) In 1966 Lewontin and Hubby scored an assumed randomly chosen group of 18 enzymes in population samples of *Drosophila pseudoobscura* for allozymic variation. They found that seven of these loci were polymorphic, and each fly was on average heterozygous at 11.5% of its loci. Harris looked at ten loci in man and discovered that three out of the ten were polymorphic and a mean of 9.9% were heterozygous. Both these studies involved the direct application of a technique described by several authors in the 1967 Symposium (Hawkes, 1968), but neither of them was mentioned at that meeting. They recall the remark of the President of the Linnean Society in 1858 after Darwin and Wallace had given their original papers on natural selection, that nothing of note happened in that year. It is now accepted that these two papers signalled a major re-thinking for population biology.

The paradox revealed by the application of electrophoresis to population samples was that populations possess "too much" variation: orthodox theory placed an upper limit on the amount of variation a population could support before its genetic load became too heavy or the cost of natural selection too great, and the population became extinct (Muller, 1950; Haldane, 1957). With few exceptions, virtually every species of animal and plant examined has proved to be theoretically inviable.

The first response of the theoreticians to this embarrassment was to suggest that electrophoretically-detected protein variation had no effect on the fitness of its carriers, i.e. was neutral (King and Jukes, 1969; Clarke, 1970; Richmond, 1970). This convinced few biologists: the arguments adduced in support were largely statistical and ignored the enormous problem that Darwinian adaptation and evidence of natural selection are everywhere apparent. The protagonists of neutralism were quickly challenged on the grounds that it is the phenotype rather than individual genes that are subject to selection, and many alleles may function as modifiers or complementary to other loci rather than as major genes in their own right (Sved *et al.*, 1967; King, 1967; Milkman, 1967; Smith, 1968; Cook, 1971). However, one limited aspect of the neutrality model has been shown to be largely correct, and that is that certain alternatives based in the third position of codons are indeed adaptively equivalent (Salser and Isaacson, 1976; Milkman, 1978).

(3) Confidence in classical mechanisms was only restored when work on enzyme variants showed that many of them – perhaps most – changed the properties of the proteins in which they were substituted, and were thus potentially subject to selection. Studies have been carried out on biochemical properties *in vitro*, on correlation of variant distribution with environmental variables (particularly temperature) and on changes of allele frequency with environmental change in time and space, in both field and laboratory (Harris, 1971; Bryant, 1974; Johnson, 1976). Although there has been statistical disagreement about details, the variances of frequencies of alleles at different loci in a group of populations are much greater than expected, indicating that all loci are not affected by the same influences, such as inbreeding (Nevo *et al.*, 1975; Berry and Peters, 1977). The firm conclusion is that allele frequencies are subject to fine adjustment by natural selection, and, although too few examples are known where enzyme properties *per se* rather than linked loci are responsible for the selective response, there can now be little argument that enzyme variants are – in the main – adaptive (Clarke, 1975; Berry, 1978, 1979*a*).

(4) Finally, current interest and speculation centre round the determinants of overall inherited variation, usually measured by the proportion of polymorphic loci (*P*) or the mean heterozygosity

per locus (H), in an assumed randomly selected range of protein coding loci. Selander and Kaufman (1973) summarized H data for 24 invertebrate and 22 vertebrate species. These showed that the mean H for invertebrates was 15.1%, but only 5.8% for vertebrates (although more recent data suggest that the invertebrate figure is largely due to high values for *Drosophila* species: Nevo, 1978). They interpreted these data as support for the hypothesis of adaptation to habitats of different "grain" (Levene, 1953; Levins, 1968). The genetical version of this is that fine-grained species (i.e. those in which an individual can exploit all or nearly all of the patches in an heterogeneous environment) should have Jack of all trades alleles at most loci, whereas selection in coarse-grained species (in which individuals spend most of their lives in one of several possible niches) will favour different alleles in different patches.

Evidence collected to test the idea that genetical variation is correlated with environmental instability in time or space seemed at first to uphold it, particularly experiments in which laboratory cultures were subjected to either genetical or environmental perturbations (Powell, 1971; Hedrick *et al.*, 1976), but specific predictions have not been fulfilled. In particular, species from the supposedly constant deep sea environment have high values of H, as do species living in trophically stable tropical areas (Somero and Soulé, 1974; Campbell *et al.*, 1975; Valentine, 1976) while burrowing animals (in an allegedly "monotonous subterranean niche": Nevo, 1976) tend to have as high heterozygosities as their non-burrowing relatives ($5.2 \pm 0.6\%$ and $5.9 \pm 0.7\%$ respectively: Selander, 1976).

Still other workers regard heterozygosity as adaptive in itself, in some unknown way (Smith *et al.*, 1975). For example, Garten (1976) found a relationship between H and aggressive behaviour in oldfield mice (*Peromyscus polionotus*), which means that the position of an animal in the social structure of the population may be related to its variation. To complicate the situation still further, H is sensitive to bottle-necks in population history, and can be very variable within the same species (Bonnell and Selander, 1974; Berry and Peters, 1977; Berry, 1979b).

Notwithstanding, there seems to be some correlation between heterozygosity and way of life. Ayala and Valentine (1974, 1978) have suggested that predictability or dependability of food resources may be the key. Their argument is that in situations where resource (mainly food) abundances are highly predictable, the best genetical strategy is to produce a variety of offspring able to exploit different foods, thus reducing sib and intra-population competition. In unpredictable circumstances, a generalist and uniform genome is supposed to be better suited for coping with adversity. On this interpretation deep-sea invertebrates and tropical *Drosophila* species are expected to be composed of segregating groups of trophic specialists capable of finding the foods for which they are adapted. Valentine and Ayala (1976) cite differences between krill (*Euphausia*) species in support of their hypothesis: *E. superba* lives in the circumpolar sea, is said to have a high trophic seasonality and possesses a heterozygosity of 5.7%; *E. distinguenda* is an eastern Pacific tropical species with an *H* of 21.3%; while *E. mucronata* lives in the temperate transitional water of the Peru Current and has an intermediate *H* of 14.1%.

The three main hypotheses about the factors involved in determining genetical variability (neutralism, with its dependence on time and history for the accumulation of variants; niche width; and resource predictability) involve intra- and interspecific interactions as well as straightforward autecological problems of physical hazards and food availability. The difficulty is that the convenience of lumping all the components of variation into a single measure of heterozygosity (or polymorphism) obscures the factors which contribute to the survival and fitness of individuals. Although we can be definite that the genetical makeup of an individual or population is non-random, we are still a long way from being able to understand fully its level of variation.

ENZYME VARIATION AND DIFFERENTIATION

Traditionally, taxonomically useful characters have been ones which it has been reasonable to assume were non-adaptive. The demonstration that enzymic variants are in some sense adaptive, implies that they have only a limited taxonomic value. However they have the beguiling property of more closely reflecting the genome than any other character available at present, and have attracted convincing apologists on this count (e.g. Avise, 1974, 1976). Three points are worth considering in assessing their value:

1. Time-related Changes

The possibility of determining the amino acid sequence of proteins early revealed differences in constitution between homologous proteins from different groups. Zuckerkandl and Pauling (1962) proposed that macromolecular sequences evolve at constant rates, and could therefore form an "evolutionary clock" calibrating phylogenetic trees based on protein or other evidence. A number of methods have been proposed for this exercise, as well as more sophisticated evolutionary analyses (Fitch and Farris, 1974; Moore *et al.*, 1973; Sneath *et al.*, 1975; Fitch, 1976). However the evolutionary clock has proved a very fallible time-keeper, principally revealed in the vastly different rates of change in different proteins (Table I). Biochemists have speculated about "accelerated" or "retarded" rates at different evolutionary stages and introduced concepts of "functional constraints" or "dispensability" of some parts of protein molecules (Wilson *et al.*, 1977). These ideas have proved valuable in probing the functional morphology of some well-studied molecules (notably the cytochromes, globulins and immunoglobulins), but in general nothing more needs to be assumed than that some variants are deleterious, whereas others are primarily or secondarily adaptive, i.e. that amino acid substitutions in proteins are subject to exactly the same selective constraints as any other allelic change (Levandowsky and White, 1977).

The debate about the evolutionary clock has produced evidence about the relative proportions of "selective" and "neutral" mutations. Notwithstanding, it is dangerous to make too much of this distinction, because a mutant which is neutral under many conditions experienced by its carriers may become highly adaptive when the environment changes. An example of such "preadaptation" is the availability of inherited variation conferring resistance to pesticides in a wide variety of organisms.

A rather similar argument about evolutionary divergence has been based on the connection between speciation and "genetic distance" (measuring the amount of difference between two groups in terms of the frequencies of alleles at a range of (usually) electrophoretically detected loci: Smith, 1977). Ayala (1975) has examined the distances between local populations, subspecies or "semi-species", sibling species and "good" species in the *Drosophila willistoni*

Table I. Rates of Protein Evolution (after Wilson, Carlson and White, 1977). The "unit evolutionary period" is the average time in years × 10^6 for a 1% difference in amino acid sequence to arise between two lineages.

	Unit evolutionary period
Histones: H4	400
H2	60
H1	8
Collagen	36
Albumin	3
Casein	1.4
Fibrinopeptide	1.4
Glutamate dehydrogenase	55
Lactate dehydrogenase	20
Triosephosphate isomerase	19
Carbonic anhydrase	4
Electron carriers:	
Cytochrome c	15
Cytochrome b	11
Ferredoxin	6
Hormones:	
Glucagon	43
Corticotrophin	24
Insulin	14
Prolactin	5
Growth hormone	4
Myoglobin	6
Haemoglobin α	3.7
β	3.3

complex of the American tropics and reviewed evidence from other groups (Table II). He recognized two stages in the process of geographic speciation. In the first, populations accumulate genetical differences as a result of some degree of isolation between them. This divergence may be the result of adaptation to different environments, but founding events may also play a significant role since they have the potential of changing the frequency of alleles at all segregating loci (Berry, 1975). During this stage partial or even complete reproductive isolation may develop as a by-product of genetical differences. In the second stage, selection acts to enhance reproductive isolation. This will be more intense if the originally separated populations re-establish contact. Thorpe (1979) and

Table II. Genetic Similarity at Early Stages of Evolutionary Divergence (after Ayala, 1975 and Avise, 1976).

	Local populations	*Sub- or semi-species*	*Species*	*Genera*
Invertebrates:				
Drosophila willistoni	0.97	0.79	0.47	
Drosophila obscura	0.99	0.82		
Drosophila repleta	1.00	0.88	0.78	
Vertebrates				
Lepomis (Sunfish)	0.98	0.84	0.54	
Cyprinidae (Minnows)	0.99			0.59
Taricha (Salamanders)	0.95	0.84	0.63	0.31
Sceloporus (Lizards)	0.89	0.79		
Mus (House mice)	0.95	0.77		
Dipodomys (Kangaroo rats)	0.97		0.61	0.16
Peromyscus (Deer mice)	0.95		0.65	

Thorpe *et al.* (1978) have suggested a new measure of taxonomic similarity designed for interspecies comparisons for taxonomic purposes.

The "distance" argument is concerned with genetical differences between populations in the same way as the "evolutionary clock" argument, but is less reductionist in taking into account the circumstances of the populations involved rather than the mere passage of time (Thorpe, 1979). Nevertheless, there is clearly a need to examine the properties of the variants that contribute to genetic distances, rather than regarding them merely as useful markers (Berry, 1979*a*).

2. *Biochemical-morphological Correlations*

There is considerable interest at the moment in investigating the morphological correlates of biochemically detected variation. The concern here is two-fold.

(1) The response of an enzyme variant to genetical forces (notably selection, but also gene flow, drift, etc) is relatively easy to detect, but the locus concerned may have a more important phenotypic effect for the organism than

its biochemical manifestation – for example if it affects the colour of the organism through involvement in pigment formation.

(2) Segregation of enzyme variants in fact involves segregation of linked genes as well. The linked genes may be more significant than the enzyme used as a marker.

It may be difficult to distinguish these effects. Alleles affecting haemoglobin β-chain synthesis in house mice populations change markedly in frequency in different conditions with a predictability that can only be attributed to strong selection (Berry, 1978), but mice carrying different alleles can be classified with a high degree of accuracy by traits such as organ weight and mineral composition of bones, which have nothing directly to do with haemoglobin synthesis, i.e. there are different phenotypes associated with different alleles which can most easily be regarded as the effect of linked genes (Bellamy *et al*, 1973). However the haemoglobins produced by animals with different alleles at the β-haemoglobin locus itself differ in their oxygen binding capacities, and could well affect their carrier's survival.

There is clearly need for a fuller understanding of the relation between gene and character in general. The ease of looking directly at gene products has diverted interest from this link, but now the initial euphoria induced by electrophoresis has subsided, it is to be hoped that more attention will be paid to it. So far, most studies suggest a similar pattern of distance distinction in both multifactorial (usually morphometric) and allozyme sets of data (e.g. Patton *et al.*, 1975; Mickevich and Johnson, 1976), although there seem to be enough exceptions to invalidate any general conclusions (Gorman and Kim, 1976; Rutherford, 1977; Berry *et al.*, 1978; Schnell *et al.*, 1978; Larson and Highton, 1978).

In general, the most clearcut uses of enzyme variation for taxonomy have been the description of the range of particular variants and the use of these data to back up (or contradict) traditional classification (e.g. Skibinski *et al.*, 1978; Dando *et al.*, 1979). Although the arguments used are generally valid, they cannot be made into overriding principles, because homologous enzymes are, of course, subject to similar mutants. The reasoning at this level begins to impinge upon the "evolutionary clock" dissension.

3. Genes versus Chromosomes in Speciation

A major difficulty in evaluating the importance of genic variation in systematics is uncertainty about the relative importance of gene and chromosome changes in speciation (Greenbaum and Baker, 1976). Whereas many gene differences *may* produce some reduction in hybrid viability, chromosome differences can produce a high degree of reproductive isolation instantaneously. Robertsonian variation of centromere number may produce ecological adaptation without involving any gene differences at all between ecotypes (Bantock and Cockayne, 1975). White (1978) has argued that the insistence that virtually all speciation is allopatric (as so persuasively described by Mayr) may be as misleading as were the models of classical population genetics destroyed by the application of electrophoresis to population samples. It is premature to judge the true balance between chromosomal and genic speciation, but obviously the final evaluation will be of major importance to the usefulness of enzyme variants to systematists (Bush *et al.*, 1977; Patton and Young, 1977). Data are accumulating of taxonomically distinct groups which are allozymically very similar (e.g. Turner, 1974; Avise *et al.*, 1975; Johnson *et al.*, 1977).

FAROE HOUSE MICE: A CAUTIONARY TALE

House mice from the Faroe Islands (lying between Britain and Iceland) are often cited as an example of spectacular microevolutionary divergence, and inferences from them have been used to calibrate the rate of formation of new taxa. The situation is worth enlarging, because it is relatively uncomplicated and illustrates many of the points made in preceding sections.

The first recognition of the distinctiveness of Faroe mice (from the island of Nolsøy) was by Clarke (1904) who put them into a new subspecies (*Mus musculus faeroensis*), distinguished solely by "immense size". On the same grounds, Miller (1912) promoted the race to specific rank since it "differs so conspicuously from all other known members of the (species) group". In 1942, Degerbøl examined a large number of museum specimens of mice from Faroe and created another subspecies (*M. m. mykinessiensis*) for animals from the island of Mykines, which he differentiated from the Nolsøy race by the shape of the mesopterygoid fossa on the base of the skull, which in Mykines animals tapers to a point, unlike the majority of house mice, albeit like the extinct *M. m. muralis* of St. Kilda, which lies in the Atlantic west of the Hebrides. Degerbøl judged mice from another Faroe island, Fugløy to have a "special stock of mice", but he did not give them taxonomic rank. He regarded specimens from Streymøy as "likely [to be] fairly pure representatives of the original Faroe mouse, originating through isolation for a long space of time, while the big Nolsøy mice have been developed by adaptation to the leaping life on the bird cliffs (i.e. as a kind of ecological race)".

Now the interest in the Faroe mice stems from the fact that they could not have reached the islands until brought by human colonizers, who arrived less than 1000 years ago. Huxley (1942) regarded the Faroe mouse situation as a significant example of rapid evolution. After discussing evidence that subspeciation usually takes about 5000 years, he commented, "The facts concerning rats and mice show that subspeciation may sometimes occur much more quickly. In particular the Faroe house mouse, which was introduced into the islands not much more than 250 years ago, is now so distinct that certain modern authorities have assigned full specific status to it." Matthews (1952) has argued similarly, "These island races are particularly interesting in showing the rate at which isolated wild populations can develop new genetic characteristics, for it is definitely known that their forerunners could not have been introduced into the Faroes less than 250 years or more than 1000 years ago."

In fact, the situation is even more abrupt: of the six Faroe islands where mice occur, Hestur seems to have acquired them since 1939 while both Fugløy and Mykines were recorded as being free from mice in 1800. Notwithstanding, all the island races are very distinct from each other and from other North European populations for both morphometric and allozymic traits (Berry *et al.*, 1978).

As far as enzyme variants are concerned, the Faroe mice are relatively invariable, only Sandøy having as much variation as on islands generally (an average of 4.9%: Berry and Peters, 1977); Fugløy is exceptional in showing no variation at all in 22 loci tested (Table III). Twelve of the 22 segregated in at least one of the six populations. Of particular interest is that different alleles are fixed in different populations, implying either that the populations have

Table III. Allozymic Variation in Faroe House Mice (after Berry, Jakobson and Peters, 1978).

	Mean % heterozygosity per locus	*% frequency of allele*			
		$Gpi{-}1^b$	$Mor{-}1^c$	$Es{-}2^c$	$Es{-}2^e$
Fugløy	0	100	100	100	0
Mykines	1.0	0	100	0	2.6
Streymøy	2.6	7.1	–	0	0
Nolsøy	1.2	0	0	0	1.3
Hestur	1.4	0	0	100	0
Sandøy	7.8	–	0	0	3.4

completely different origins, or that they have been founded by a very small number of individuals from segregating populations. $Es{-}2^c$ is the commonest allele at the Esterase–2 locus in Denmark, but is comparatively rare in Britain; $Es{-}2^e$ occurs in Denmark, but has only been found on the Shetland island of Yell among British populations. Narrowing of the mesopterygoid fossa also occurs in Shetland, and this suggests that the Faroe, Shetland and presumably St. Kilda populations were all derived from a common ancestral population. On historical grounds, the most likely agents for introducing mice to the islands were voyaging Vikings.

As far as inter-island differentiation is concerned, Mykines is the most distinct population, although its separation from the others is less on biochemical than on metrical data. This immediately raises a general problem about the comparative value of allozymic and morphological data. As already noted, there is often a good correlation between biochemical and morphological data, but in the Faroe data, it has already been noted that only 12 of the 22 loci tested were segregating. These were carried on only six of the 20 chromosomes and mark less than 0.1% of the total genome. Obviously an intelligently chosen set of metrical characters is more likely to be representative of the overall variation. Soulé and Yang (1974) came to a similar conclusion in a study of lizards in the Gulf of California. Lewontin (1974) has argued that data on at least 100 loci are needed to produce an adequate estimate of variation.

The genetically and morphologically most variable population came from Sandøy, one of the southern Faroe islands. The southern islands are more fertile and intensively farmed than the northern islands, and it seems plausible that the original establishment of mice in Faroe was on Sandøy, and that they spread from there to the other islands, although both Hestur and Nolsøy were probably colonized from the large neighbouring island of Streymøy (Fig. 1).

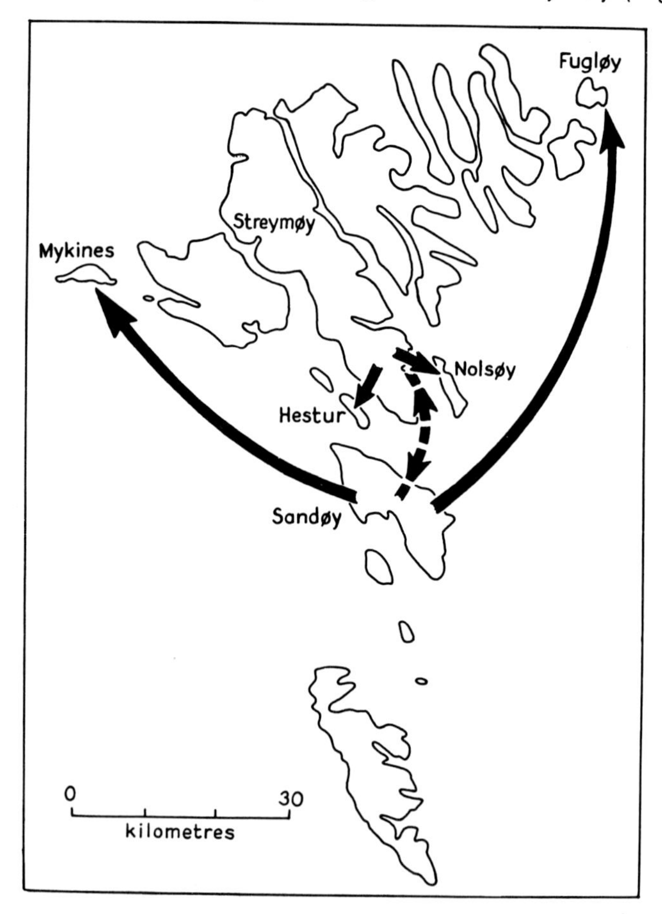

Fig. 1. Possible routes of inter-island colonization by *Mus musculus* on the Faroe group of islands (from Berry *et al.* 1978).

There is much that is speculative about the Faroe mouse story, but it does show how allozymic and morphological data can complement each other in constructing a phylogenetic sequence, and also throw light on the importance of historical bottlenecks in establishing the cause of taxonomic diversity. Other examples could be used to make similar points (e.g. Patton *et al.*, 1975; Carson and Kaneshiro, 1976).

CONCLUSION

Virtually nothing has been said about enzymic variation in plants. The main reason for this is the lack of work that has been done, together with the conviction that the same general rules that apply to animals will apply to plants (Gottleib, 1976; Harper, 1977). But a general conclusion *can* be drawn from a survey of genetical variation in enzymes from the point of view of the systematist, and it is exactly the same as that drawn by Cain (1968) in a summary paper during the 1967 Symposium, "All that the really competent taxonomist needs is to know everything about his subject and about the groups he works on. Since this is hardly possible he must go to others for help. The new taxonomy is peculiarly a field in which team work is necessary, and . . . team work by taxonomists, chemists, biochemists and anyone else with special knowledge can be stimulating, profitable and delightful." The only positive thing to add is that geneticists ought to be a necessary part of such a team, both for their own good and for their symbiosis with the heart of biology – which is not taxonomy or immunology or evolution, but the study of animals and plants as interacting individuals surrounded and subjected to an assault of environments.

REFERENCES

Avise, J. C. (1974). Systematic value of electrophoretic data. *Syst. Zool.* 23, 465–481.

Avise, J. C. (1976). Genetic differentiation during speciation. *In* "Molecular Evolution" (F. J. Ayala, ed.), pp. 106–122. Sinauer, Sunderland, Massachusetts.

Avise, J. C., Smith, J. J. and Ayala, F. J. (1975). Adaptive differentiation with little genic change between two native California minnows. *Evolution* 29, 411–426.

Ayala, F. J. (1975). Genetic differentiation during the speciation process. *In* "Evolutionary Biology" (T. Dobzhansky, M. K. Hecht and W. C. Steere, eds), Vol. 8, pp. 1–78, Plenum Press, New York.

Ayala, F. J. (1976). "Molecular Evolution". Sinauer, Sunderland, Massachusetts.

Ayala, F. J. and Valentine, J. W. (1974). Genetic variability in the cosmopolitan deep-water ophiuran *Ophiomusium lymani. Mar. Biol.* **27**, 51–57.

Ayala, F. J. and Valentine, J. W. (1978). Genetic variation and resource stability in marine invertebrates. *In* "Marine Organisms, Genetics, Ecology and Evolution" (B. Battaglia and J. A. Beardmore, eds), pp. 23–51. Plenum Press, New York.

Bantock, C. R. and Cockayne, W. C. (1975). Chromosomal polymorphism in *Nucella lapillus. Heredity* **34**, 231–245.

Bellamy, D., Berry, R. J., Jakobson, M. E., Lidicker, W. Z., Morgan, J. and Murphy, H. M. (1973). Ageing in an island population of the house mouse. *Age and Ageing* **2**, 235–250.

Berry, R. J. (1975). On the nature of genetical distance and island races of *Apodemus sylvaticus. J. Zool., Lond.* **176**, 293–296.

Berry, R. J. (1978). Genetic variation in wild house mice: where natural selection and history meet. *Am. Scient.* **66**, 52–60.

Berry, R. J. (1979a). Genetical factors in animal population dynamics. *In* "Population Dynamics" (L. R. Taylor, B. D. Turner and R. M. Anderson eds), pp. 53–80. Blackwell, Oxford.

Berry, R. J. (1979b). The Outer Hebrides: where genes and geography meet. *Proc. R. Soc. Edin.* **77B**, 21–43.

Berry, R. J., Jakobson, M. E. and Peters, J. (1978). The house mice of the Faroe Islands: a study in microdifferentiation. *J. Zool., Lond.* **185**, 73–92.

Berry, R. J. and Peters, J. (1977) Heterogeneous heterozygosities in *Mus musculus* populations. *Proc. R. Soc. B* **197**, 485–503.

Bonnell, M. L. and Selander, R. K. (1974). Elephant seals: genetic variation and near extinction. *Science, N. Y.* **184**, 908–909.

Bryant, E. H. (1974). On the adaptive significance of enzyme polymorphisms in relation to environmental variability. *Am. Nat.* **108**, 1–19.

Bush, G. L., Case, S. M., Wilson, A. C. and Patton, J. L. (1977). Rapid speciation and chromosomal evolution in mammals. *Proc. Nat. Acad. Sci. U. S.* **74**, 3942–3946.

Cain, A. J. (1968). The assessment of new types of character in taxonomy. *In* "Chemotaxonomy and Serotaxonomy" (J. G. Hawkes, ed.), pp. 229–234. Academic Press, London and New York.

Campbell, C. A., Valentine, J. W. and Ayala, F. J. (1975). High genetic variability in a population of *Tridacna maxima* from the Great Barrier Reef. *Mar. Biol.* **33**, 341–345.

Carson, H. L. and Kaneshiro, K. (1976). *Drosophila* of Hawaii: systematics and ecological genetics. *Ann. Rev. Ecol. Syst.* **7**, 311–345.

Clarke, B. C. (1970). Darwinian evolution of proteins. *Science, N. Y.* **168**, 1009–1011.

Clarke, B. C. (1975). The contribution of ecological genetics to evolutionary theory: detecting the direct effects of natural selection on particular polymorphic loci. *Genetics,* **79**, 101–113.

Clarke, W. E. (1904). On some forms of *Mus musculus,* Linn., with description of a new subspecies from the Faeroe Islands. *Proc. R. phys. Soc. Edin.* **15**, 160–167.

Cook, L. M. (1971). "Coefficients of Natural Selection". Hutchinson, London.
Dando, P. R., Southward, A. J. and Crisp, D. J. (1979). Enzyme variation in *Chthalamus stellatus* and *Chthalamus montagui* (Crustacea: Cirripedia): evidence for the presence of *C. montagui* in the Adriatic. *J. mar. Biol. Ass. U. K.* **59**, 307–320.
Degerbøl, M. (1942). Mammalia. Zoology Faroes Pt 65.
Fitch, W. M. (1976). The molecular evolution of cytochrome c in eukaryotes. *J. mol. Evol.* **8**, 13–40.
Fitch, W. M. and Farris, J. S. (1974). Evolutionary trees with minimum nucleotide replacements from amino acid sequences. *J. mol. Evol.* **3**, 263–278.
Garten, C. T. (1976). Relationships between aggressive behavior and genic heterozygosity in the oldfield mouse *Peromyscus polionotus*. *Evolution,* **30**, 59–72.
Gorman, G. C. and Kim, Y. J. (1976). *Anolis* lizards of the Eastern Caribbean: a case study in evolution. II. Genetic relationships and genetic variation of the *bimaculatus* group. *Syst. Zool.* **25**, 62–77.
Gottlieb, L. D. (1976). Biochemical consequences of speciation in plants. *In* "Molecular Evolution" (F. J. Ayala, ed.), pp. 123–140. Sinauer, Sunderland, Massachusetts.
Greenbaum, I. F. and Baker, R. J. (1976). Evolutionary relationships in *Macrotus* (Mammalia: Chiroptera): biochemical variation and karyology. *Syst. Zool.* **25**, 15–25.
Haldane, J. B. S. (1957). The cost of natural selection. *J. Genet.* **55**, 511–524.
Harper, J. L. (1977). "Population Biology of Plants". Academic Press, London and New York.
Harris, H. (1966). Enzyme polymorphisms in man. *Proc. R. Soc. B,* **164**, 298–310.
Harris, H. (1971). Protein polymorphism in man. *Can. J. Genet. Cytol.* **13**, 381–396.
Hawkes, J. G. (1968) (Ed.). "Chemotaxonomy and Serotaxonomy". Systematics Association Special Volume. Academic Press, London and New York.
Hedrick, P. W., Ginevan, M. E. and Ewing, E. P. (1976). Genetic polymorphism in heterogeneous environments. *Ann. Rev. Ecol. Syst.* **7**, 1–32.
Huxley, J. (1942). "Evolution, the Modern Synthesis". Allen and Unwin, London.
Johnson, G. B. (1976). Genetic polymorphism and enzyme function. *In* "Molecular Evolution" (F. J. Ayala, ed.), pp. 46–59. Sinauer, Sunderland, Massachusetts.
Johnson, M. S., Clarke, B. C. and Murray, J. J. (1977). Genetic variation and reproductive isolation in *Partula*. *Evolution,* **31**, 116–126.
King, J. L. (1967). Continuously distributed factors affecting fitness. *Genetics,* **55**, 483–492.
King, J. L. and Jukes, T. H. (1969). Non-Darwinian evolution. *Science, N. Y.* **164**, 788–798.
Larson, A. and Highton, R. (1978). Geographic protein variation and divergence in the salamanders of the *Plethodon welleri* group (Amphibia, Plethodontidae). *Syst. Zool.* **27**, 431–448.

Levandowsky, M. and White, B. S. (1977). Randomness, time scales and the evolution of biological communities. *In* "Evolutionary Biology" (M. K. Hecht, W. C. Steere and B. Wallace, eds.), Vol. 10, pp. 69–161. Plenum Press, New York.

Levene, H. (1953). Genetic equilibrium when more than one niche is available. *Am. Nat.* **87**, 331–333.

Levins, R. (1968). "Evolution in Changing Environments". University Press, Princeton.

Lewontin, R. C. (1974). "The Genetic Basis for Evolutionary Change" Columbia Press, New York and London.

Lewontin, R. C. and Hubby, J. L. (1966). A molecular approach to the study of genic heterozygosity in natural populations. II. Amount of variation and degree of heterozygosity in natural populations of *Drosophila pseudoobscura*. *Genetics,* **54**, 595–609.

Matthews, L. H. (1952). "British Mammals". Collins, London.

Mickevich, M. F. and Johnson, M. S. (1976). Congruence between morphological and allozyme data in evolutionary inference and character evolution. *Syst. Zool.* **25**, 260–270.

Milkman, R. D. (1967). Heterosis as a major cause of heterozygosity in nature. *Genetics,* **55**, 493–495.

Milkman, R. D. (1978). The maintenance of polymorphisms by natural selection. *In* "Marine Organisms: Genetics, Ecology, and Evolution" (B. Battaglia and J. A. Beardmore, eds.), pp. 3–22. Plenum Press, New York.

Miller, G. S. (1912). "Catalogue of the Mammals of Western Europe". British Museum (Natural History), London.

Moore, G. W., Barnabas, J. and Goodman, M. (1973). A method for constructing maximum parsimony ancestral amino acid sequences on a given network. *J. Theor. Biol.* **38**, 459–485.

Müller, H. J. (1950). Our load of mutations. *Am. J. hum. Genet.* **2**, 111–176.

Nevo, E. (1976). Genetic variation in constant environments. *Experientia,* **32**, 858.

Nevo, E. (1978). Genetic variation in natural populations: pattern and theory. *Theor. population Biol.* **13**, 121–177.

Nevo, E., Dessauer, H. C. and Chuang, K.-C. (1975). Genetic variation as a test of natural selection. *Proc. Nat. Acad. Sci. U. S.* **72**, 2145–2149.

Patton, J. L. and Yang, S. Y. (1977). Genetic variations in *Thomomys bottae* pocket gophers: macrogeographic patterns. *Evolution,* **31**, 697–720.

Patton, J. L., Yang, S. Y. and Myers, P. (1975). Genetic and morphologic divergence among introduced rat populations (*Rattus rattus*) of the Galapagos Archipelago, Ecuador. *Syst. Zool.* **24**, 296–310.

Powell, J. R. (1971). Genetic polymorphism in varied environments. *Science, N. Y.* **174**, 1035–1036.

Richmond, R. C. (1970). Non-Darwinian evolution: a critique. *Nature, Lond.* **225**, 1025–1028.

Rutherford, J. C. (1977). Geographical variation in morphological and electrophoretic characters in the holothurian *Cucumaria curata. Mar. Biol.* **43**, 165–174.

Salser, W. and Isaacson, J. S. (1976). Mutation rates in globin genes: the genetic load and Haldane's dilemma. *Progr. Nucl. Acid Res.* **19**, 205–220.

Schnell, G. D., Best, T. L. and Kennedy, M. L. (1978). Inter-specific morphologic variation in kangaroo rats (*Dipodomys*): degree of concordance with genic variation. *Syst. Zool.* **27**, 34–48.

Selander, R. K. (1976). Genic variation in natural populations. *In* "Molecular Evolution" (F. J. Ayala, ed.), pp. 21–45. Sinauer, Sunderland, Massachusetts.

Selander, R. K. and Kaufman, D. W. (1973). Genic variability and strategies of adaptation in animals. *Proc. Nat. Acad. Sci. U. S.* **70**, 1875–1877.

Skibinski, D. O. F., Beardmore, J. A. and Ahmad, M. (1978). Genetic aids to the study of closely related taxa of the genus *Mytilus*. *In* "Marine Organisms: Genetics, Ecology and Evolution" (B. Battaglia and J. A. Beardmore, eds.), pp. 469–486. Plenum Press, New York.

Smith, C. A. B. (1977). A note on genetic distance. *Ann. hum. Genet.* **40**, 463–479.

Smith, J. M. (1968). "Haldane's dilemma" and the rate of evolution. *Nature, Lond.* **219**, 1114–1116.

Smith, M. H., Garten, C. T. and Ramsey, P. R. (1975). Genic heterozygosity and population dynamics in small mammals. *In* "Isozymes. Vol. IV. Genetics and Evolution" (C. L. Markert, ed.), pp. 85–102. Academic Press, New York and London.

Sneath, P. H. A., Sackin, M. J. and Ambler, R. P. (1975). Detecting evolutionary incompatibilities from protein sequences. *Syst. Zool.* **24**, 311–332.

Somero, G. N. and Soulé, M. (1974). Genetic variation in marine fishes as a test of the niche-variation hypothesis. *Nature, Lond.* **249**, 670–672.

Soulé, M. and Yang, S. Y. (1974). Genetic variation in side-blotched lizards on islands in the Gulf of California. *Evolution,* **27**, 593–600.

Sved, J. A., Reed, T. E. and Bodmer, W. F. (1967). The number of balanced polymorphisms that can be maintained in a natural population. *Genetics,* **55**, 469–481.

Thorpe, J. P. (1979). Enzyme variation and taxonomy: the estimation of sampling errors in measurements of interspecific genetic similarity. *Biol. J. Linn. Soc.* **11**, 369–386.

Thorpe, J. P., Beardmore, J. A. and Ryland, J. S. (1978). Taxonomy, interspecific variation and genetic distance in the Phylum Bryozoa. *In* "Marine Organisms: Genetics, Ecology and Evolution (B. Battaglia and J. A. Beardmore, eds.), pp. 425–445. Plenum Press, New York.

Turner, B. J. (1974). Genetic divergence of Death Valley pupfish species: biochemical versus morphological evidence. *Evolution* **28**, 281–284.

Valentine, J. W. (1976). Genetic strategies of adaptation. *In* "Molecular Evolution" (F. J. Ayala, ed.), pp. 78–94. Sinauer, Sunderland, Massachusetts.

Valentine, J. W. and Ayala, F. J. (1976). Genetic variability in krill. *Proc. Nat. Acad. Sci. U.S.,* **73**, 658–660.

White, M. J. D. (1978). "Modes of Speciation". Freeman, San Francisco.

Wilson, A. C., Carlson, S. S. and White, T. J. (1977). Biochemical evolution. *Ann. Rev. Biochem.* **46**, 573–639.

Wright, C. A. (1974) (ed.). "Biochemical and Immunological Taxonomy of Animals". Academic Press, London and New York.

Zuckerkandl, E. and Pauling, L. (1962). In "Horizons in Biochemistry" (M. Kasha and B. Pullman eds.) pp. 189–225. Academic Press, New York.

9 | Fraction I Protein and Plant Phylogeny

J. C. GRAY

Department of Botany, Cambridge University
Cambridge, England

Abstract: Fraction I protein, the most abundant protein in the leaves of most green plants, provides excellent phenotypic markers for both the nuclear and chloroplast genomes of higher plants. The origins of genomes in amphiploid species may be determined by the analysis of the subunit composition of Fraction I proteins. Isoelectric focusing of Fraction I proteins in 8 M urea has been used to determine the origins of tobacco, wheat, oats, Brassicas, potato and cotton. Electrophoresis and immunodiffusion of Fraction I proteins have also provided information of phylogenetic interest. The advantages and limitations of these techniques are discussed.

INTRODUCTION

Hybridization plays an important role in the evolution of higher plants, and a large number of plant species, including many plants of agricultural importance, have originated through amphiploidy following interspecific hybridization. Considerable efforts, utilizing a wide range of techniques, have been made over the years to identify the parent species of many of these allopolyploid plants. However, recent studies have shown that the analysis of a single protein, Fraction I protein, the major soluble protein in the leaves of most higher plants, can provide information not only on the identity of the parent species but also on the direction of the cross

Systematics Association Special Volume No. 16, "Chemosystematics: Principles and Practice", edited by F. A. Bisby, J. G. Vaughan and C. A. Wright, 1980, pp. 167–193, Academic Press, London and New York.

giving rise to allopolyploid plants. This work has been pioneered in the laboratory of Professor S. G. Wildman in Los Angeles and the earlier work from this laboratory has been reviewed previously (Wildman *et al.*, 1975; Kung, 1976; Uchimiya *et al.*, 1977). This paper will consider the application of studies on Fraction I protein to the phylogeny of higher plants.

Fraction I protein is located exclusively in the chloroplasts of all green plants, where it may account for over 50% of the protein in the stroma. This protein catalyses the CO_2-fixation step in the Calvin cycle and is therefore also known as the enzyme ribulose 1, 5-bisphosphate carboxylase (EC.4.1.1.39). Fraction I protein has been purified to homogeneity from a wide range of plants (Kawashima and Wildman, 1970), and has been crystallized by a very simple procedure from plants in several genera of the family Solanaceae (Chan *et al.*, 1972; Sakano *et al.*, 1974b) and recently from a range of other plants by a more generally applicable technique using vapour diffusion in the presence of polyethyleneglycol solutions (Johal and Bourque, 1979). Fraction I protein is an oligomeric protein of molecular weight 550 000 composed of two different subunit types. The protein is made up of eight large subunits of molecular weight 55 000 and eight small subunits of molecular weight 12 000–15 000 (Baker *et al.*, 1975, 1977). The large subunits have been shown to contain the catalytic sites of the enzyme (Nishimura and Akazawa, 1973) but the function of the small subunits is unknown.

Examination of the nature of the subunits of Fraction I protein from interspecific hybrids in the genus *Nicotiana* indicated that the two subunit types had different modes of inheritance. The large subunits were inherited solely from the maternal parent (Chan and Wildman, 1972; Sakano *et al.*, 1974a), whereas the small subunits were inherited from both parents (Kawashima and Wildman, 1972; Sakano *et al.*, 1974a). These patterns of inheritance have also been demonstrated in interspecific hybrids of *Triticum* (Chen *et al.*, 1975a), *Avena* (Steer and Thomas, 1976), *Lycopersicon* (Uchimiya *et al.*, 1979c) and *Gossypium* (Chen and Meyer, 1979). The maternal inheritance of the large subunits suggests an extranuclear location for the genetic information specifying the primary structure of the large subunit and recently the structural gene for the large subunit has been located on chloroplast DNA in maize

(Coen *et al.*, 1977), spinach, tobacco and *Oenothera hookeri* (Bottomley and Whitfeld, 1979). Biparental inheritance of the small subunits suggests that nuclear DNA contains the genetic information for the small subunit and recently Chen and Sand (1979) have located the gene for a small subunit polypeptide on a specific chromosome in a male sterile tobacco line.

Hybridization thus results in the production of a Fraction I protein with the large subunits of the maternal parent and a mixture of small subunits from both maternal and paternal parents. The analysis of the subunit composition of Fraction I protein from interspecific hybrids is therefore able to provide information on the identity of the two parent species and the direction of the cross giving rise to the hybrid. The most extensively used technique for examining the subunit composition of Fraction I protein is isoelectric focusing in the presence of 8 M urea, but electrophoresis and serology have also provided information of phylogenetic interest. Each of these techniques is relatively simple and applicable to the analysis of large numbers of samples. These techniques will be discussed in more detail below. In addition, amino acid sequencing (Gibbons *et al.*, 1975; Iwai *et al.*, 1976; Strobaek *et al.*, 1976) and the analysis of proteolytic fragments (Kawashima *et al.*, 1976) have been used to examine the origins of *Nicotiana tabacum*, but these two techniques require fairly large amounts of pure protein and are not readily applicable to large numbers of samples.

ISOELECTRIC FOCUSING

1. *Isoelectric Focusing of Fraction I Protein from Tobacco*

The subunit composition of Fraction I protein may be determined by isoelectric focusing in the presence of 8 M urea. This technique was first applied to Fraction I protein during studies on the structure of the protein from tobacco (Kung *et al.*, 1974). Crystalline Fraction I protein was dissociated into its component subunits with 8 M urea, reduced with dithiothreitol and the cysteine residues were blocked by *S*-carboxymethylation with iodoacetic acid to prevent possible oxidation and the formation of mixed disulphides. The dissociated protein was subjected to isoelectric focusing in a 5% polyacrylamide gel slab containing 2% Ampholine pH 5–8 and

8 M urea and, on staining with bromophenol blue, gave a pattern of three large subunit polypeptides, each with a molecular weight of 55 000, in the pH 6.3 region of the gel, and two small subunit polypeptides, each with a molecular weight of 13 000, in the pH 5.5 region of the gel. This pattern was reproducible from preparation to preparation but there was some variability in the relative staining intensity of the large subunit polypeptides. These polypeptides have each been isolated and subjected to chemical analysis (Gray *et al.*, 1978). Extensive structural studies on the large subunit polypeptides failed to reveal any evidence of sequence heterogeneity between the three polypeptides and it was suggested that the three polypeptides were due to modifications to a single gene product (Gray *et al.*, 1978). One possibility considered was variation at the *N*-terminus of the polypeptide due to variable processing of the initial translation product but this appears to be eliminated by the description of a unique amino acid sequence at the *N*-terminus of the large subunit from barley (Poulsen *et al.*, 1979), which also shows three polypeptides on isoelectric focusing (Chen *et al.*, 1976a). It now seems probable that the modifications are introduced by the iodoacetic acid treatment, for it has recently been demonstrated that isoelectric focusing of Fraction I protein from tobacco and wheat, in 8 M urea under reducing conditions, but without prior modification of thiol groups, yields a single large subunit polypeptide (O'Connell and Brady, 1979) and multiple bands can be generated on treatment with iodoacetamide.

Differences between the two small subunit polypeptides of tobacco have been detected by several techniques, including fingerprinting of tryptic peptides (Gray *et al.*, 1978) and their identification as separate gene products is indicated by sequence heterogeneity in total small subunit preparations (Gibbons *et al.*, 1975; Iwai *et al.*, 1976; Strobaek *et al.*, 1976).

The isoelectric focusing pattern of *S*-carboxymethylated Fraction I protein from tobacco was completely reproducible and was obtained irrespective of the developmental age of the plant or of environmental conditions (Uchimiya *et al.*, 1977). However in old leaves of tobacco it was necessary to prevent covalent modification of Fraction I protein by phenolic materials during the extraction procedure because this introduced additional bands into the polypeptide pattern of the large subunit (Gray *et al.*, 1978). This problem has also been encountered in the extraction of Fraction I protein from potato leaves (Melchers *et al.*, 1978). In neither case was there any effect of phenolics on the pattern of the small subunit polypeptides. Polyphenol modification of the tobacco protein was prevented by the inclusion of Dowex-1, KCN and serum albumin in the extraction medium (Gray *et al.*, 1978) whereas soluble polyvinylpyrrolidone (PVP) was included as a polyphenol absorbent during the extraction of the potato protein (Melchers *et al.*, 1978). Other modifications to the structure of Fraction I protein introduced during extraction would also be expected to affect the isoelectric focusing pattern of the protein. Proteolysis of Fraction I protein during extraction from leaves of the bean, *Phaseolus vulgaris,* has been described (Gray and Kekwick, 1974) but its effect on the isoelectric focusing pattern is not known. Proteolysis of Fraction I protein is apparently not a problem during the extraction of the protein from tobacco (Gray *et al.*, 1978).

2. *Isolation of Fraction I Proteins*

The analysis of Fraction I protein by isoelectric focusing requires small amounts (20 μg) of pure protein, free from contamination by other proteins. This can be easily obtained from tobacco and other *Nicotiana* species by the very simple procedure of crystallization directly from clarified leaf extracts (Chan *et al.*, 1972; Lowe, 1978). This procedure also yields crystalline Fraction I protein from leaves of *Solanum melongena* (Sakano *et al.*, 1974b) and *Petunia hybrida* (Chen *et al.*, 1976a) but not from any plant outside the Solanaceae. Fraction I protein may also be purified by conventional salt fractionation, gel filtration and ion-exchange chromatography (Paulsen and Lane, 1966; Gray and Kekwick, 1974) but this is very time-consuming for a large number of samples. Salt fractionation followed by sucrose density gradient centrifugation (Goldthwaite and Bogorad, 1971) provides a more rapid means of purification and allows several samples to be processed at once. The protein when purified may then be crystallized by vapour diffusion against solutions of polyethyleneglycol (Johal and Bourque, 1979). Two immunochemical methods have also been developed to facilitate the isolation of Fraction I protein from leaves of any plant species. In the first, antibodies to Fraction

I protein from tobacco were immobilized on Sepharose 4 B and used to separate the Fraction I protein in crude leaf extracts from other proteins (Gray and Wildman, 1976). The retained Fraction I protein was then eluted with 8 M urea in a form ready for isoelectric focusing. This method yields pure Fraction I protein subunits from all species examined (Gray and Wildman, 1976; Chen *et al.*, 1975a; Gatenby and Cocking, 1977, 1978a, 1978b). A more useful immunoprecipitation method for the isolation of Fraction I protein from less than 0.5 g leaf material has recently been described (Uchimiya *et al.*, 1979a). Fraction I protein in crude leaf homogenates is immunoprecipitated with antibodies to purified Fraction I protein and the whole immunoprecipitate is dissociated in 8 M urea, *S*-carboxymethylated and analysed by isoelectric focusing. This method allows the analysis of Fraction I protein from large numbers of samples. The polypeptide chains of the immunoglobulins do not interfere with the separation of the subunit polypeptides of Fraction I protein, but there is an unexplained acidic shift of all three large subunit polypeptides compared to the pattern obtained from crystallized Fraction I protein. This shift in isoelectric points of the large subunit polypeptides precludes the direct comparison of results obtained with Fraction I protein isolated by the immunoprecipitation method and protein isolated by any of the other methods described above. However, results obtained with the immunoprecipitation method are internally consistent and allow comparisons between species (Uchimiya *et al.*, 1979a).

3. Isoelectric Focusing of Fraction I Proteins from Plant Species

Isoelectric focusing of Fraction I proteins from a wide range of plants has demonstrated the presence of three large subunit polypeptides in all proteins and variation in the number of small subunit polypeptides, from one to four, depending on the species analysed (Chen *et al.*, 1976a). Subsequently the Fraction I protein from a large number of plant species has been analysed in several laboratories (see Fig. 1 and Table I) and confirms the reproducibility of the polypeptide patterns obtained. For example, Fraction I protein from the tomato, *Lycopersicon esculentum*, has been analysed in three laboratories with identical results (Gatenby and Cocking, 1978c; Melchers *et al.*, 1978; Uchimiya *et al.*, 1979b).

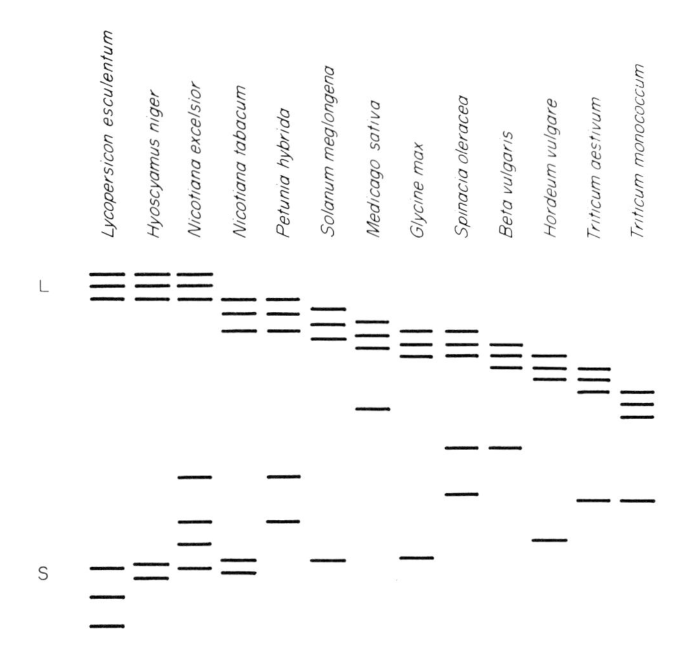

Fig. 1. Subunit polypeptide patterns of *S*-carboxymethylated Fraction I proteins obtained by isoelectric focusing in the presence of 8 M urea. Diagram adapted from Chen *et al.* (1976a) and Sakano (1975).

However, these analyses of the tomato protein highlight a problem in the interpretation of minor bands. In all three laboratories the small subunit was resolved into three polypeptides with the most acidic of these being very weakly stained. Chen *et al.* (1976a) originally showed only the two most heavily stained polypeptides in the belief that the third polypeptide was not an authentic component of tomato Fraction I protein. However, genetic analyses have now established that this minor staining polypeptide is a genuine component of the small subunit of Fraction I protein (Uchimiya *et al.*, 1979c). Similar problems have been encountered with the interpretation of the polypeptide pattern of the small subunit from potato, *Solanum tuberosum tuberosum.* Gatenby and Cocking (1978a) described a pattern of two heavily stained polypeptides and a number of minor bands. The position of the minor bands was constant and it was suggested they may be artifacts

Table I. Genera of Higher Plants from which the Subunit Polypeptide Composition of Fraction I Protein has been examined by Isoelectric Focusing

Cruciferae	*Brassica*	Gatenby and Cocking, 1978b; Uchimiya and Wildman, 1978
	Diplotaxis	Uchimiya and Wildman, 1978
	Eruca	Uchimiya and Wildman, 1978
	Raphanus	Uchimiya and Wildman, 1978
	Sinapis	Uchimiya and Wildman, 1978
Chenopodiaceae	*Beta*	Gray and Wildman, 1976
	Spinacia	Chen *et al.*, 1976a
Malvaceae	*Gossypium*	Chen and Meyer, 1979; Chen and Wildman, 1980
Leguminosae	Glycine	Chen *et al.*, 1976a
	Medicago	Chen *et al.*, 1976a
	Pisum	Ellis *et al.*, 1977; Roy *et al.*, 1978
Onagraceae	*Oenothera*	Uchimiya *et al.*, 1977
Solanaceae	*Hyoscyamus*	Sakano, 1975
	Lycopersicon	Chen *et al.*, 1976a; Gatenby and Cocking, 1978c; Melchers *et al.*, 1978, Uchimiya *et al.*, 1979b
	Nicotiana	Kung *et al.*, 1974; Sakano *et al.*, 1974a; Gray *et al.*, 1974; Sakano, 1975; Kung *et al.*, 1975a, b; Chen and Wildman, 1980
	Petunia	Chen *et al.*, 1976a; Gatenby and Cocking, 1977
	Solanum	Sakano, 1975; Chen *et al.*, 1976a; Gatenby and Cocking, 1978a, c; Melchers *et al.*, 1978
Lemnaceae	*Lemna*	Chen and Wildman, 1980
	Spirodela	Chen and Wildman, 1980
	Wolffia	Chen and Wildman, 1980
	Wolffiella	Chen and Wildman, 1980
Gramineae	*Aegilops*	Chen *et al.*, 1975a
	Avena	Steer and Thomas, 1976; Steer and Kernoghan, 1977
	Hordeum	Chen *et al.*, 1976a
	Sorghum	Uchimiya *et al.*, 1977
	Triticum	Chen *et al.*, 1975a
	Zea	Chen *et al.*, 1976c

introduced during the reaction with iodoacetic acid. However, Melchers *et al.* (1978) described a pattern of five small subunit polypeptides from potato and indicated that the number of bands could not be decreased by variation in the conditions of the

S-carboxymethylation reaction. Minor bands in the pattern of the small subunit polypeptides are most likely in plants having evolved through multiple rounds of hybridization and are thus specially likely in crop plants which have been constantly manipulated by plant breeders. Indeed, Chen and Sand (1979) have shown that a minor small subunit polypeptide has been introduced into a male sterile line of tobacco by hybridization with the Australian species, *Nicotiana debneyi*.

A further difficulty in identifying small subunit polypeptides is the wide variation in isoelectric points of these polypeptides. The first plants analysed all contained Fraction I protein whose small subunit polypeptides were more acidic than, and widely separated from, the large subunit polypeptides. However, it has recently been demonstrated that in pea, *Pisum sativum*, the two small subunit polypeptides have isoelectric points very similar to the large subunit polypeptides (Ellis *et al.*, 1977; Roy *et al.*, 1978). Further analyses have now shown that in *Oenothera, Gossypium* and *Lemna*, the isoelectric points of some of the small subunit polypeptides may be more alkaline that the large subunit polypeptides (Chen and Meyer, 1979; Chen and Wildman, 1980). The usual one-dimensional isoelectric focusing separations may not be sufficient to resolve clearly all the subunit polypeptides, but this may be achieved with a two-dimensional separation procedure, involving isoelectric focusing in 8 M urea in the first dimension followed by electrophoresis in the presence of sodium dodecylsulphate in the second dimension at right angles (Ellis *et al.*, 1977; Roy *et al.*, 1978; Chen and Meyer, 1979). The second dimension clearly separates the large and small subunit polypeptides on the basis of molecular weight.

The majority of published polypeptide patterns of Fraction I protein represent the analysis of a single, or at most a very few, seed accessions and Johnson (1976) has pointed out the possibility of variation in the polypeptide patterns. No such variation has been found in analyses of over 100 individual plants of tobacco, including over 20 different self-fertile cultivars showing considerable morphological variation (Chen *et al.*, 1975b; Uchimiya *et al.*, 1977). Similarly, analyses of Fraction I protein from numerous seed accessions of *Nicotiana glauca* (Chen *et al.*, 1975b) and *Lycopersicon esculentum* (Uchimiya *et al.*, 1979b) have failed to reveal any variation in polypeptide pattern. However, variation in the small subunit polypeptide pattern has been demonstrated in *Nicotiana suaveolens* (Uchimiya *et al.*, 1977) and in *Lycopersicon peruvianum* (Uchimiya *et al.*, 1979b). Variation in the large subunit polypeptide pattern has been encountered in only one species, *Solanum tuberosum,* where it has been shown that the subspecies *tuberosum* has a different large subunit polypeptide pattern to subspecies *andigena* (Gatenby and Cocking, 1978a). However, the taxonomic status of these potatoes is a matter of some controversy, with some workers considering the two forms to be distinct tetraploid species but with others considering the differences not to be sufficient to merit the rank of separate species. The different large subunit polypeptide patterns of these two potatoes may therefore be a useful chemical character for potato taxonomists.

The isoelectric focusing patterns of Fraction I proteins may provide additional phenotypic characters for taxonomic consideration. Most of the analyses of Fraction I proteins performed so far have included only a small number of species in any one genus, so it is not possible to generalize on the taxonomic usefulness of the subunit polypeptides as purely phenotypic characters. However, extensive analyses of Fraction I protein from species in the genera *Nicotiana* (Chen and Wildman, 1980) and *Lycopersicon* (Uchimiya *et al.,* 1979b) have shown that species regarded as closely related on morphological and cytogenetic grounds show similarities in their polypeptide compositions. The isoelectric focusing patterns of Fraction I proteins from 63 out of the 66 recognized species of *Nicotiana* have been used to generate a dendrogram based on an index of similarity (Chen *et al.,* 1976b), but it is clear that such schemes should be treated with caution because in some instances species with identical isoelectric focusing patterns can be distinguished by serological means (Gray, 1977; Gatenby, 1978) or by the analysis of tryptic peptides (Kung *et al.,* 1977).

4. Isoelectric Focusing of Fraction I Proteins from Interspecific Hybrids

Isoelectric focusing of the subunit polypeptides of Fraction I protein from reciprocal interspecific F_1 hybrids in the genus *Nicotiana* showed that the three large subunit polypeptides were inherited as a group solely from the maternal parent, whereas the small subunit polypeptides were inherited from both parents (Sakano *et al.,* 1974a). There was no indication of dominance or suppression of one of the parental small subunit types, as had been reported earlier (Kawashima and Wildman, 1972) in the analysis of the inheritance patterns by fingerprinting of tryptic peptides of the subunits of Fraction I protein. Subsequent analysis of Fraction I protein from other interspecific *Nicotiana* hybrids confirmed the maternal inheritance of the large subunit polypeptides and biparental inheritance of the small subunit polypeptides (Wildman *et al.,* 1975). This pattern has also been observed in *Gossypium* hybrids (Chen and Meyer, 1979). Maternal inheritance of the large subunit polypeptides has been observed in interspecific hybrids of *Triticum* (Chen *et al.,* 1975a) and *Avena* (Steer and Thomas, 1976), but the mode of inheritance of the small subunit could not be determined in these crosses because the single small subunit polypeptide had a similar isoelectric point in all species examined. However the biparental inheritance of the small subunit polypeptides has been observed in interspecific *Lycopersicon* hybrids (Uchimiya *et al.,* 1979c), although in this case it was not possible to determine the mode of inheritance of the large subunit polypeptides because of similarities between the species examined.

The biparental Mendelian mode of inheritance of the small subunit polypeptides has been confirmed by analysis of the F_2 generation derived from the F_1 hybrids, *Nicotiana otophora* x *N. tomentosiformis* (Chen *et al.,* 1979) and *Lycopersicon esculentum* x *L. parviforum* (Uchimiya *et al.,* 1979c). In each case the small subunit polypeptide patterns of the maternal parent, the F_1 hybrid and the paternal parent were recovered in the ratio 1:2:1. Further confirmation of the Mendelian inheritance of the small subunit polypeptides and the maternal non-Mendelian inheritance of the large subunit polypeptide pattern was obtained by backcrossing F_1 hybrids with pollen from the paternal parent species. Backcrosses of F_1 hybrids of *Nicotiana tabacum,* as the pollen parent, with *N. exigua, N. glauca, N. megalosiphon* or *N. plumbaginofolia* as maternal parents produced plants with the large subunit polypeptide pattern of the maternal parent species and the small subunit polypeptides of *N. tabacum.* These plants were all male sterile (Chen *et al.,* 1975b; Chen *et al.,* 1976c). Similar results were obtained on backcrossing the hybrid *Gossypium arboreum* × *G. anomalum* with pollen from *G. anomalum* (Chen and Meyer, 1979).

It appears likely that the different modes of inheritance of the two subunits of Fraction I protein will be a general rule in higher plants, although it is essential that this should be confirmed with each new group of plants analysed. A possible exception may be those plants which show biparental inheritance of their chloroplasts (see Gillham, 1978) and hence may show biparental inheritance of their large subunit polypeptides. An attempt to analyse the mode of inheritance of the large subunit in *Oenothera,* a genus which shows biparental chloroplast inheritance, was thwarted by the absence of any differences in Fraction I protein polypeptide pattern in 14 species analysed (Uchimiya *et al.*, 1977). However, differences between the large subunits of *Oenothera* species have been revealed by fingerprinting of tryptic peptides (Holder, 1978), so an examination of the mode of inheritance of the large subunit should be possible.

The subunit polypeptide pattern of Fraction I protein from interspecific hybrids is therefore a direct consequence of the different modes of inheritance of the large and small subunits and can be used to determine the direction of the cross giving rise to the hybrid if the identity of the parents is known. Even if the identity of the parents is not known with certainty then the polypeptide pattern of the hybrid can give a clear indication of the polypeptide patterns of the parent species and thus may help in the identification of the parent species.

Many interspecific hybrids are infertile and not capable of further sexual reproduction; however new fertile species may arise following interspecific hybridization, either by spontaneous chromosome doubling or by the fusion of unreduced gametes (Harlan and deWet, 1977). Such amphidiploid species have double the number of chromosomes of the infertile hybrid, but studies have shown that there is no corresponding change in the polypeptide pattern of Fraction I protein. This was established by an examination of the polypeptide pattern of Fraction I proteins from *Nicotiana digluta* ($2n = 72$) and the infertile hybrid, *N. glutinosa* ♀ × *N. tabacum* ♂ ($2n = 36$) from which it arose spontaneously (Clausen and Goodspeed, 1925; Kung *et al.*, 1975b). Similar results have been obtained with a polyploid series of tobacco, including haploid, diploid, triploid and tetraploid plants (Chen *et al.*, 1975b). The absence of any effects of ploidy level on the polypeptide pattern of Fraction I protein therefore allows a direct examination of the origins of amphiploid species.

5. Isoelectric Focusing of Fraction I Protein from Amphiploids

(a) *Nicotiana* species. The first species to be examined with a view to determining its origins by analysis of Fraction I protein was *Nicotiana tabacum,* the commercial tobacco plant (Gray *et al.,* 1974). This plant is not known in the wild and the determination of its origins has been the object of numerous investigations. It was believed that the diploid species *N. sylvestris* ($2n = 24$) and either *N. tomentosiformis* ($2n = 24$) or *N. otophora* ($2n = 24$) were involved in the origins of *N. tabacum* ($2n = 48$). Examination of the polypeptide patterns of Fraction I proteins of these species indicated that only *N. sylvestris* could have provided the large subunit polypeptides and hence must have been the maternal parent in the cross giving rise to *N. tabacum. N. sylvestris* also provided one of the two small subunit polypeptides of *N. tabacum,* and the other small subunit polypeptide could have been provided by *N. tomentosiformis* but not by *N. otophora.* It was concluded that tobacco originated from the hybridization of *N. sylvestris* ♀ and *N. tomentosiformis* ♂. This was supported by an analysis of Fraction I protein from the amphiploid synthesized from *N. sylvestris* x *N. tomentosiformis,* which gave a polypeptide pattern identical to *N. tabacum.* Further support for this proposed origin of *N. tabacum* has since been obtained by analysis of tryptic peptides of the subunits of Fraction I protein (Kawashima *et al.,* 1975) and by amino acid sequence analysis of the small subunit of Fraction I protein (Strobaek *et al.,* 1976; Iwai *et al.,* 1976).

This analysis of the origins of *N. tabacum* and the analysis of Fraction I protein from *N. digluta* indicate that the multiple small subunit polypeptides in Fraction I proteins of certain plants are probably a consequence of the origins of those plants by hybridization events. Thus plants with two small subunit polypeptides would have arisen by at least one round of hybridization, whereas plants with three or four small subunit polypeptides would have arisen by at least two rounds of hybridization. An alternative mechanism for the origin of multiple small subunit polypeptides by gene duplication followed by mutation without any involvements of interspecific hybridization has been suggested to account for the multiple small subunit polypeptides in the diploid species, *Lycopersicon esculentum* (Gatenby and Cocking, 1978c). However, there is as yet no experimental evidence for duplication and mutation of the genes specifying the small subunit polypeptides of Fraction I protein, whereas the origin of multiple small subunit polypeptides by hybridization is well documented (Sakano *et al.*, 1974a; Gray *et al.*, 1974; Kung *et al.*, 1975b; Chen and Meyer, 1979). This would suggest that the events giving rise to amphiploids in the genus *Lycopersicon* were very ancient and that subsequent evolution has eliminated or modified much of the chromosome duplication (Gatenby and Cocking, 1978c; Uchimiya *et al.*, 1979c).

The origin of multiple small subunit polypeptides by hybridization would appear to be the best explanation of the multiple small subunit polypeptides found in the Fraction I proteins from twenty species of *Nicotiana* restricted to Australia and certain Pacific Islands. These tetraploid species in the section *Suaveolentes* are believed to have had an amphiploid origin from ancestors of species in the present day sections *Alatae, Acuminatae* and *Noctiflorae,* and examination of the small subunit polypeptide patterns of Fraction I proteins from species in these sections reveals that four out of the six different small subunit polypeptides in the section *Suaveolentes* can be accounted for in present day species in the sections *Acuminatae, Alatae* and *Noctiflorae,* in accordance with Goodspeed's (1954) ideas on the hybrid origins of the Australian species. The Australian species all have the same pattern of large subunit polypeptides and this pattern is found elsewhere in the genus only in two species *N. noctiflora* and *N. petunioides,* in the section *Noctiflorae.* This suggests that the ancestors of these species were the maternal parents in the hybridizations which gave rise to the Australian species (Chen *et al.*, 1976b; Chen and Wildman, 1980).

Until recently it was believed that *Nicotiana* species occurred naturally only in the Americas and Australia. However a new species, *N. africana,* has recently been described from several isolated mountains in Namibia (South-west Africa). Isoelectric focusing of Fraction I protein from *N. africana* reveals a large subunit pattern similar to the Australian species and the noctifloroid species, and two small subunit polypeptides which it shares in common with species in the section *Alatae* and some Australian species (Chen and Wildman 1980). This would once again suggest that the large subunit polypeptides were provided by noctifloroid species as the maternal parent in hybridization with alatoid species to eventually produce *N. africana.*

The present-day locations of the Australian and African species and of their putative progenitor species would appear to support Goodspeed's (1954) contention that the genus reached Australia before the final break up of Gondwanaland. Similar arguments can also be made for the presence of the genus in Africa. This would put the age of the genus in excess of 75–100 million years, whereas Raven and Axelrod (1974) have expressed doubt that the Angiosperms as a phylum had evolved earlier than 120 million years ago, and their arguments would suggest that the presence of the genus *Nicotiana* in Australia and Africa was achieved by long distance dispersal within the last 10–20 million years. In an attempt to resolve this controversy, Chen and Wildman (1980) have attempted to date the age of the genera *Nicotiana* and *Gossypium* in relation to the family Lemnaceae, for which a fossil record exists, by examining mutations in the large subunit polypeptides of Fraction I protein. Their analysis suggests that the genera *Nicotiana* and *Gossypium* are at least as old as the Lemnaceae (50–75 million years) and would support their vicarious transport over land to Australia and Africa. However the validity of their arguments depends on a similar mutation rate of the large subunit of Fraction I protein in each group of plants and, as yet, there is no experimental evidence for this.

(*b*) *Wheat species.* It is well established that the polyploid wheat species have originated by interspecific hybridization (Sears, 1969). Chen *et al.* (1975a) examined the polypeptide compositions of Fraction I proteins from several *Triticum* and *Aegilops* species. All species shared a common small subunit polypeptide but there were differences between species in the isoelectric points of the large subunit polypeptides. It was established by isoelectric focusing of Fraction I proteins from reciprocal hybrids of *T. boeoticum* and *T. dicoccoides* that the pattern of large subunit polypeptides was inherited solely from the maternal parent. This allowed the direction of the crosses giving rise to the polyploid wheat species to be determined. Hexaploid wheat, *T. aestivum* (AABBDD), which is recognized on the basis of cytogenetic behaviour to have arisen by hybridization of *T. dicoccum* (AABB) and *Ae. squarrosa* (DD), could have been produced from these species only if the direction of the cross was *T. dicoccum* ♀ × *Ae. squarrosa* ♂. Similarly, the direction of the cross giving rise to the tetraploid *T. dicoccum* could be determined even though there is some controversy over the identity of the diploid donor of the B genome. *Triticum monococcum* is established as the donor of the A genome to *T. dicoccum* and the analysis of the polypeptide pattern of Fraction I protein indicated that *T. monococcum* could not have provided the genetic information for the large subunit polypeptides. This therefore suggested that the B genome donor was also the source of the large subunit polypeptides and must therefore have been the maternal parent in the cross giving rise to *T. dicoccum*. Similar analyses indicated that the G genome donor to the tetraploid *T. timopheevii* (AAGG) must have been the maternal parent in the cross with *T. monococcum* (Chen *et al.*, 1975a). These conclusions have been questioned because the plant material analysed was grown from single accessions of each of the species and possible variation in polypeptide patterns of the Fraction I protein was not examined (Johnson, 1976). However further analyses of numerous accessions of these species have failed to reveal any evidence of variation (K. Chen, personal communication).

(c) *Avena* species. Isoelectric focusing of Fraction I proteins has been used to analyse the origins of the polyploid *Avena* species (Steer and Thomas, 1976; Steer and Kernoghan, 1977). As with Fraction I protein from the *Triticum* species, all species examined contained a single small subunit polypeptide with an identical isoelectric point, but three types of large subunit pattern were found in the genus. One type of large subunit pattern was confined to the perennial tetraploid, *A. macrostachya,* which has been placed on its own in a separate section of the genus. Another type of large subunit pattern was found in the A genome diploids, as well as the A genome tetraploid *A. murphyi* and hexaploid *A. sativa,* and this pattern was distinct from that found in the C genome diploids

(Steer and Kernoghan, 1977). Steer and Thomas (1976) had previously shown the maternal inheritance of the large subunit polypeptide pattern in several reciprocal interspecific hybrids in the genus. It was therefore possible to conclude that the diploids, tetraploid and hexaploid species containing the A genome are closely related on the maternal side, and that any possible influence of the C genome diploids on the formation of the hexaploid, *A. sativa,* must have been limited to a contribution of paternal nuclear genes (Steer and Thomas, 1976; Steer and Kernoghan, 1977).

(*d*) *Brassica species.* The subunit polypeptide patterns of Fraction I proteins from *Brassica* species have been analysed in two laboratories (Gatenby and Cocking, 1978b; Uchimiya and Wildman, 1978). Gatenby and Cocking (1978b) were unable to analyse the large subunit polypeptides because of the generation of additional bands either during purification or the *S*-carboxymethylation reaction. However, their analyses of the small subunit polypeptides were identical to those of Uchimiya and Wildman (1978). Two types of large subunit polypeptide pattern and two small subunit polypeptides, which were present in different species in different ratios, were found in the genus (Uchimiya and Wildman, 1978). The origins of the amphiploid *Brassica* species were deduced from the isoelectric focusing patterns of the Fraction I proteins. The polypeptide pattern of *B. carinata* ($2n = 34$), which from cytogenetic evidence is believed to have originated from the hybridization of *B. nigra* ($2n = 16$) and *B. oleracea* ($2n = 18$), could have arisen from these species only if *B. nigra* was the maternal parent. Similarly, *B. juncea* ($2n = 36$) could have arisen from *B. nigra* and *B. campestris* ($2n = 20$) only if *B. campestris* was the maternal parent. These deduction have supposed a maternal mode of inheritance for the large subunit polypeptides, but there is, as yet, no published evidence on this point. The direction of the cross giving rise to *B. napus* ($2n = 38$) could not be determined owing to the similarity of the large subunit polypeptide in *B. napus* and its putative parents, *B. campestris* and *B. oleracea.* Densitometer tracings of the staining intensities of the small subunit polypeptides in the amphiploids and their putative parents were consistent with the origins of the amphiploids determined from cytogenetic evidence (Uchimiya and Wildman, 1978; Gatenby and Cocking, 1978b).

(*e*) *Potato.* An examination of the subunit polypeptide composition of Fraction I protein from several tuber-bearing *Solanum* species has contributed to our understanding of the origins of the modern European potato, *Solanum tuberosum tuberosum* (Gatenby and Cocking, 1978a; Uchimiya and Wildman, unpublished, cited in Gatenby and Cocking, 1978a). The subunit polypeptide compositions of Fraction I proteins from these species are inconsistent with the suggestions that *S. tuberosum tuberosum* arose by chromosome doubling of *S. stenotonum* or by hybridization of *S. sparsipilum* with *S. stenotonum* because of differences in both large and small subunit polypeptides. Differences in the large subunit polypeptide patterns also indicate that the chloroplast genome of the modern European potato was not provided by the Andean potato, *S. Tuberosum andigena,* which was the form originally introduced into Europe in the sixteenth century. The genetic information of the potato has been manipulated by plant breeders for several centuries and this has obviously contributed to the complex pattern of small subunit polypeptides (Melchers *et al.*, 1978). Further analysis of Fraction I protein from other species and synthesized hybrids may provide the information necessary to unravel the origins of the potato.

(*f*) *Gossypium species.* Chen and Wildman (1980) have analysed the subunit polypeptide patterns of Fraction I protein from 19 out of the 35 species of *Gossypium* and have found four different types of large subunit polypeptide patterns and eight different small subunit polypeptides distributed throughout the genus. Analysis of Fraction I protein from reciprocal interspecific hybrids has demonstrated maternal inheritance of the large subunit polypeptide pattern and biparental inheritance of the small subunit polypeptides (Chen and Meyer, 1979). Chen and Wildman (1980) have shown that the cultivated amphidiploid species, *G. hirsutum* and *G. barbadense,* containing the nuclear genomes A and D, have the same large subunit polypeptide pattern as the diploid species carrying the A genome, and suggest that the A genome donor was also the maternal parent in the cross giving rise to the amphiploid species. The small subunit compositions of *G. hirsutum* and *G. barbadense* are also consistent with their origins by amphiploidy from diploid species carrying the A and D genomes. The two cultivated amphidiploids both contain four small subunit polypeptides, two of which are found in *G. herbaceum,* the probable A genome donor, whereas the other two are found in *G. raimondii,* the probable D genome donor.

Isoelectric focusing of Fraction I proteins in 8 M urea appears to have considerable potential for examining the origins of genomes in amphiploid plants. The majority of the examples cited above have been concerned with the origins of crop plants, mainly because of the wealth of information available and the availability of seeds, but there is no reason why the analysis of the subunit polypeptide composition of Fraction I protein by isoelectric focusing should not become part of the armoury of the plant taxonomist.

ELECTROPHORESIS

Several studies have shown that the Fraction I proteins of closely related species may be distinguished on the basis of their electrophoretic mobility and this may be used to give information on the origins of genomes in amphiploids and hybrids. In the genera *Avena, Triticum* and *Nicotiana* it has been shown that the electrophoretic mobility of Fraction I proteins from hybrids resembled the mobility of the protein from the maternal, but not the paternal, parent species (Steer, 1975; Reichenbächer *et al.,* 1977; Uchimiya and Wildman, 1979). Steer (1975) showed that the Fraction I proteins from *Avena* species could be placed in two groups based on their electrophoretic mobility at pH 8.4 in 3–5% polyacrylamide gels and that the electrophoretic mobility of Fraction I protein from synthesized amphiploid hybrids was identical to the mobility of the protein from the maternal parent species. The Fraction I protein from the C genome diploids had a lower mobility than the Fraction I proteins from the A genome diploids and from the tetraploid and hexaploid species and it was concluded that the A genome diploids, but not the C genome diploids, were involved on the maternal side of the crosses giving rise to the tetraploid and hexaploid species (Steer, 1975).

Reichenbächer *et al.* (1977) have shown that immuno-electrophoresis, using antisera to Fraction I proteins from *Triticum aestivum* and *Hordeum vulgare,* can be used to distinguish species in several genera of Graminae. They showed that Fraction I protein from *T. monococcum* had a greater electrophoretic mobility than the proteins from *T. dicoccum* and *T. aestivum,* which had identical electrophoretic mobilities, and pointed out a correlation between the electrophoretic mobility of Fraction I protein and the isoelectric points of the large subunit polypeptides.

Electrophoresis of Fraction I protein offers a simple method for determining species relationships in the Gramineae and the electrophoretic mobility provides another maternally inherited phenotypic character, which is little influenced by the charge on the small subunits. The isoelectric focusing data demonstrated the similarity of the single subunit polypeptide in species of *Triticum* (Chen *et al.,* 1975a) and *Avena* (Steer and Kernoghan, 1977), and hence the small subunits would not be expected to contribute to the differences in electrophoretic mobility of complete Fraction I proteins.

In other genera where differences in the patterns of small subunit polypeptides between species have been shown by isoelectric focusing, it might be expected *a priori* that the interpretation of electrophoretic data would be more complex. Indeed this does appear to be the case in the genus *Nicotiana* where conflicting modes of inheritance of the electrophoretic mobility of Fraction I protein have been demonstrated (Hirai, 1977; Uchimiya and Wildman, 1979). Hirai (1977) used crossed immunoelectrophoresis to demonstrate that the electrophoretic mobility at pH 5.8 of Fraction I protein from the hybrid *N. bonariensis* × *N. langsdorffii* was intermediate between the electrophoretic mobilities of Fraction I proteins from the two parental species. However these species were specially chosen to minimize any effects of the large subunits; the two species and the hybrid were shown to have identical large subunit polypeptide patterns by isoelectric focusing (Hirai, 1977). However, in a similar examination by immunoelectrophoresis of the electrophoretic mobility of the Fraction I proteins from the hybrid *N. gossei* × *N. tabacum* and its parent species, Uchimiya and Wildman (1979) showed that the mobilities of the proteins from the hybrid and its maternal parent, *N. gossei,* were similar but different from the mobility of the protein of the paternal parent, *N. tabacum.* In this case, the two parent species had previously been shown to contain completely different large and small subunits by isoelectric focusing in 8 M urea (Sakano *et al.,*

1974a), and hence it appears that the charge on the large subunits predominates to determine differences in the electrophoretic mobility of complete Fraction I proteins.

From this limited number of analyses it appears that electrophoresis of Fraction I proteins is a promising technique for determining species relationships but more studies are required to determine its general applicability. One advantage of the technique is that pure Fraction I proteins are not required; Fraction I protein is the major protein in leaf extracts and can be easily detected after electrophoresis by a general protein stain or by immunochemical methods (Hirai, 1977; Reichenbächer *et al.*, 1977).

SEROLOGY

It has been known for over 20 years that the Fraction I proteins from diverse groups of plants are serologically related (Dorner *et al.*, 1958), but only recently has the immunochemical analysis of Fraction I proteins been applied to studies of plant phylogeny (Gray, 1977, 1978; Murphy, 1978). Murphy (1978) compared Fraction I proteins from over 50 species of angiosperms and gymnosperms by quantitative microcomplement fixation using antisera to Fraction I protein from tobacco and spinach, *Spinacia oleracea*. This showed there were close antigenic similarities between Fraction I protein from tobacco and Fraction I proteins from species in the Solanaceae, Nolanaceae, Cuscutaceae and Convolvulaceae. There were also close similarities between the protein from spinach and the proteins from other species in the Chenopodiaceae. Outside these families there were considerable differences between the tobacco or spinach proteins and Fraction I proteins from other species. However, this study failed to extract the maximum amount of information of phylogenetic relevance from the Fraction I proteins, for Gray and Kekwick (1974) had previously reported that the majority of the antigenic sites on the protein were located on the large subunits. The significance of this location of the antigenic sites was established when Gray (1978) showed by immunodiffusion that the serological reactions of Fraction I proteins from 11 interspecific hybrids in the genus *Nicotiana* resembled the reactions of the maternal species but not the paternal species. It was also shown that changes in ploidy level did not affect the serological reactions of Fraction I protein and hence immunodiffusion could be used to determine the direction of crosses giving rise to amphiploid species. Using this approach, further support was obtained for *N. sylvestris* as the maternal parent in the cross with *N. tomentosiformis* which gave rise to *N. tabacum*, and for the involvement of species in the section *Noctiflorae* on the maternal side of the crosses giving rise to the *Nicotiana* species found in Australia and certain Pacific Islands (Gray, 1977). It was also suggested that *N. rustica* originated following the cross *N. undulata* ♀ × *N. paniculata* ♂ and that *N. arentsii* originated following the cross *N. wigandioides* ♀ × *N. undulata* ♂ (Gray, 1977). The direction of these crosses could not be determined by isoelectric focusing of Fraction I proteins in 8 M urea because of the similarities in the large subunit polypeptide patterns of the parent species (Chen and Wildman, 1980). This clearly indicates that immunodiffusion and isoelectric focusing are revealing different features of the structure of Fraction I protein, as has been pointed out by Gatenby (1978).

As with electrophoresis, further studies are required to determine the usefulness of immunochemical analyses of Fraction I proteins in phylogenetic studies on other groups of plants. However, the attraction of the immunodiffusion technique lies in its extreme simplicity and its use of crude leaf extracts. Pure Fraction I protein is required only for the preparation of antibodies.

CONCLUSION

Fraction I protein is able to provide information of phylogenetic interest that is not available from any other single technique or approach. The analysis of Fraction I proteins by isoelectric focusing in 8 M urea has been used to determine the origins of a large number of amphiploid species and must now be regarded as a well established technique. Electrophoresis and immunodiffusion have considerable potential for the analysis of Fraction I proteins and should be investigated further. The results obtained with Fraction I protein should be checked by the analysis of other proteins which are the products of both nuclear and cytoplasmic genomes; the chloroplast coupling factor and the mitochondrial cytochrome oxidase would appear to be worthy of investigation.

ACKNOWLEDGEMENTS

I am especially grateful to Professor S. G. Wildman for introducing me to the phylogenetic potential of Fraction I protein. I am also grateful to Sam Wildman, Kevin Chen, Tony Gatenby and Martin Steer for providing manuscripts prior to publication.

REFERENCES

Baker, T. S., Eisenberg, D., Eiserling, F. A. and Weissman, L. (1975). The structure of Form I crystals of D-ribulose 1,5-diphosphate carboxylase. *J. Mol. Biol.* **91**, 391–399.

Baker, T. S., Suh, S. W. and Eisenberg, D. (1977). Structure of ribulose 1,5-bisphosphate carboxylase-oxygenase: Form III crystals. *Proc. Nat. Acad. Sci. U.S.* **74**, 1037–1041.

Bottomley, W. and Whitfeld, P. (1979). Cell-free transcription and translation of total spinach chloroplast DNA. *Eur. J. Biochem.* **93**, 31–39.

Chan, P. H. and Wildman, S. G. (1972). Chloroplast DNA codes for the primary structure of the large subunit of Fraction I protein. *Biochim. Biophys. Acta* **277**, 677–680.

Chan, P. H., Sakano, K., Singh, S. and Wildman, S. G. (1972). Crystalline Fraction I protein: preparation in large yield. *Science* **176**, 1145–1146.

Chen, K. and Meyer, V. G. (1979). Mutation in chloroplast DNA coding for the large subunit of Fraction I protein correlated with male sterility in cotton. *J. Hered.* **70**, 431–433.

Chen, K. and Sand, S. A. (1979). *Nicotiana* chromosome coding for a specific polypeptide of the small subunit of Fraction I protein. *Science* **204**, 179–180.

Chen, K. and Wildman, S. G. (1980). Relative change in Fraction I protein composition among species of *Nicotiana, Gossypium* and the Lemnaceae is consistent with previous ideas that these angiosperms had evolved before continental drift. *Pl. Syst. Evol.* In press.

Chen, K., Gray, J. C. and Wildman, S. G. (1975a). Fraction I protein and the origin of polyploid wheats. *Science* **190**, 1304–1306.

Chen, K., Kung, S. D., Gray, J. C. and Wildman, S. G. (1975b). Polypeptide composition of Fraction I protein from *Nicotiana glauca* and from cultivars of *Nicotiana tabacum,* including a male sterile line. *Biochem. Genet.* **13**, 771–778.

Chen, K., Kung, S. D., Gray, J. C. and Wildman, S. G. (1976a). Subunit polypeptide composition of Fraction I protein from various plant species. *Pl. Sci. Lett.* **7**, 429–434.

Chen, K., Johal, S. and Wildman, S. G. (1976b). Role of chloroplast and nuclear genes during evolution of Fraction I protein. *In* "Genetics and Biogenesis of Chloroplasts and Mitochondria" (T. Bücher, W. Neupert, W. Sebald and S. Werner, eds), pp. 3–11. Amsterdam, North Holland.

Chen, K., Johal, S. and Wildman, S. G. (1976c). Phenotypic markers for chloroplast DNA genes in higher plants and their use in biochemical studies. *In* "Nucleic Acids and Protein Synthesis in Plants" (L. Bogorad and J. H. Weil, eds), pp. 183–194. Plenum Press, New York.

Chen, K., Wildman, S. G. and Lu, R. (1979). Amino acid analyses and *N*-terminal sequence study of two segregating small subunit polypeptides of Fraction I protein. *Pl. Physiol.* **63**, Suppl. 154.

Clausen, R. and Goodspeed, T. H. (1925). Interspecific hybridisation in *Nicotiana*. II. A tetraploid *glutinosa-tabacum* hybrid, an experimental verification of Winge's hypothesis. *Genetics* **10**, 278–284.

Coen, D. M., Bedbrook, J. R., Bogorad, L. and Rich, A. (1977). Maize chloroplast DNA fragment encoding the large subunit of ribulose bisphosphate carboxylase. *Proc. Nat. Acad. Sci. U.S.* **74**, 5487–5491.

Dorner, R. W., Kahn, A. and Wildman, S. G. (1958). Proteins of green plants. VIII. The distribution of Fraction I protein in the plant kingdom as detected by precipitin and ultracentrifuge analyses. *Biochim. Biophys. Acta* **29**, 240–245.

Ellis, R. J., Highfield, P. E. and Silverthorne, J. (1977). The synthesis of chloroplast proteins by subcellular systems. Proc. 4th Int. Congr. Photosynthesis (D. O. Hall, J. Coombs and T. W. Goodwin, eds), pp. 497–506. Biochemical Society, London.

Gatenby, A. A. (1978). A comparison of the polypeptide isoelectric points and antigenic determinant sites of the large subunit of Fraction I protein from *Lycopersicon esculentum, Nicotiana tabacum* and *Petunia hybrida. Biochim. Biophys. Acta* **534**, 169–172.

Gatenby, A. A. and Cocking, E. C. (1977). Polypeptide composition of Fraction I protein subunits in the genus *Petunia. Pl. Sci. Lett.* **10**, 97–101.

Gatenby, A. A. and Cocking, E. C. (1978a). Fraction I protein and the origin of the European potato. *Pl. Sci. Lett.* **12**, 177–181.

Gatenby, A. A. and Cocking, E. C. (1978b). The evolution of Fraction I protein and the distribution of the small subunit polypeptide coding sequences in the genus *Brassica. Pl. Sci. Lett.* **12**, 299–303.

Gatenby, A. A. and Cocking, E. C. (1978c). The polypeptide composition of the subunits of Fraction I protein in the genus *Lycopersicon. Pl. Sci. Lett.* **13**, 171–176.

Gibbons, G. C., Strobaek, S., Haslett, B. and Boulter, D. (1975). The *N*-terminal amino acid sequence of the small subunit of ribulose 1,5-diphosphate carboxylase form *Nicotiana tabacum. Experientia* **31**, 1040–1041.

Gillham, N. (1978). "Organelle Heredity". Raven, New York.

Goldthwaite, J. J. and Bogorad, L. (1971). A one-step method for the isolation and determination of leaf ribulose 1,5-diphosphate carboxylase. *Anal. Biochem.* **41**, 57–66.

Goodspeed, T. H. (1954). The genus *Nicotiana. Chronica Bot.* **16**, 1–536.

Gray, J. C. (1977). Serological relationships of Fraction I proteins from species in the genus *Nicotiana. Pl. Syst. Evol.* **128**, 53–69.

Gray, J. C. (1978). Serological reactions of Fraction I proteins from interspecific hybrids in the genus *Nicotiana*. *Pl. Syst. Evol.* **129**, 177–183.

Gray, J. C. and Kekwick, R. G. O. (1974). An immunological investigation of the structure and function of ribulose 1,5-bisphosphate carboxylase. *Eur. J. Biochem.* **44**, 481–489.

Gray, J. C. and Wildman, S. G. (1976). A specific immunoabsorbent for the isolation of Fraction I protein. *Pl. Sci. Lett.* **6**, 91–96.

Gray, J. C., Kung, S. D., Wildman, S. G. and Sheen, S. J. (1974). Origin of *Nicotiana tabacum* L. detected by polypeptide composition of Fraction I protein. *Nature* **252**, 226–227.

Gray, J. C., Kung, S. D. and Wildman, S. G. (1978). Polypeptide chains of the large and small subunits of Fraction I protein from tobacco. *Archs. Biochem. Biophys.* **185**, 272–281.

Harlan, J. R. and deWet, J. M. J. (1975). On Ö Winge and a prayer: the origins of polyploidy. *Bot. Rev.* **41**, 361–390.

Hirai, A. (1977). Random assembly of different kinds of small subunit polypeptides during formation of Fraction I protein macromolecules. *Proc. Nat. Acad. Sci. U.S.* **74**, 3443–3445.

Holder, A. A. (1978). Peptide mapping of the ribulose bisphosphate carboxylase large subunit from the genus *Oenothera*. *Carlsberg Res. Commun.* **43**, 391–399.

Iwai, S., Tanabe, Y. and Kawashima, N. (1976). Origin of sequence heterogeneity of the small subunit of Fraction I protein from *Nicotiana tabacum*. *Biochem. Biophys. Res. Commun.* **73**, 993–996.

Johal, S. and Bourque, D. P. (1979). Crystalline ribulose 1,5-bisphosphate carboxylase-oxygenase from spinach. *Science* **204**, 75–77.

Johnson, B. L. (1976). Polyploid wheats and Fraction I protein. *Science* **192**, 1252.

Kawashima, N. and Wildman, S. G. (1970). Fraction I protein. *A. Rev. Pl. Physiol.* **21**, 325–358.

Kawashima, N. and Wildman, S. G. (1972). Studies on Fraction I protein IV. Mode of inheritance of primary structure in relation to whether chloroplast or nuclear DNA contains the code for a chloroplast protein. *Biochim. Biophys. Acta* **262**, 42–49.

Kawashima, N., Tanabe, Y. and Iwai, S. (1976). Origin of *Nicotiana tabacum* detected by primary structure of Fraction I protein. *Biochim. Biophys. Acta* **427**, 70–77.

Kung, S. D. (1976). Tobacco Fraction I protein: a unique genetic marker. *Science* **191**, 429–434.

Kung, S. D., Sakano, K. and Wildman, S. G. (1974). Multiple peptide composition of the large and small subunits of *Nicotiana tabacum* Fraction I protein ascertained by fingerprinting and electrofocusing. *Biochim. Biophys. Acta* **365**, 138–147.

Kung, S. D., Gray, J. C., Wildman, S. G. and Carlson, P. S. (1975a). Polypeptide composition of Fraction I protein from parasexual hybrid plants in the genus *Nicotiana*. *Science* **187**, 353–355.

Kung, S. D., Sakano, K., Gray, J. C. and Wildman, S. G. (1975b). The evolution of Fraction I protein during the origin of a new species of *Nicotiana*. *J. Mol. Evol.* **7**, 59–64.

Kung, S. D., Lee, C. I., Wood, D. D. and Moscarello, M. A. (1977). Evolutionary conservation of chloroplast genes coding for the large subunits of Fraction I protein. *Pl. Physiol.* **60**, 89–94.

Lowe, R. H. (1978). Crystallisation of Fraction I protein from tobacco by a simplified procedure. *FEBS Letters* **78**, 98–100.

Melchers, G., Sacristan, M. D. and Holder, A. A. (1978). Somatic hybrid plants of potato and tomato regenerated from fused protoplasts. *Carlsberg Res. Commun.* **43**, 203–218.

Murphy, T. M. (1978). Immunochemical comparisons of ribulose bisphosphate carboxylases using antisera to tobacco and spinach enzymes. *Phytochemistry* **17**, 439–443.

Nishimura, M. and Akazawa, T. (1973). Further proof for the catalytic role of the larger subunit in the spinach leaf ribulose 1,5-diphosphate carboxylase. *Biochem. Biophys. Res. Commun.* **54**, 842–848.

O'Connell, P. B. H. and Brady, C. J. (1979). Ribulose 1,5-bisphosphate carboxylase of higher plants: only one type of large subunit? *Proc. Aust. Biochem. Soc.* **12**, 25.

Paulsen, J. M. and Lane, M. D. (1966). Spinach ribulose diphosphate carboxylase I. Purification and properties of the enzyme. *Biochemistry* **5**, 2350–2357.

Poulsen, C., Martin, B. and Svendsen, I. (1979). Partial amino acid sequence of the large subunit of ribulosebisphosphate carboxylase from barley. *Carlsberg Res. Commun.* **44**, 191–199.

Raven, P. H. and Axelrod, D. I. (1974). Angiosperm biogeography and past continental movements. *Ann. Missouri Bot. Garden* **61**, 539–673.

Reichenbächer, D., Richter, J. and Spaar, D. (1977). Differences in the electrophoretic mobility of Fraction I protein and their possible utilisation in genetics and plant breeding. *Biochem. Physiol. Pflanzen* **171**, 299–306.

Roy, H., Costa, K. A. and Adari, H. (1978). Free subunits of ribulose 1,5-bisphosphate carboxylase in pea leaves. *Pl. Sci. Lett.* **11**, 159–168.

Sakano, K. (1975). Inheritance of Fraction I proteins in *Nicotiana*. *Kagaku to Seibutsu* **13**, 269–272.

Sakano, K., Kung, S. D. and Wildman, S. G. (1974a). Identification of several chloroplast DNA genes which code for the large subunit of *Nicotiana* Fraction I proteins. *Mol. Gen. Genet.* **130**, 91–97.

Sakano, K., Kung, S. D. and Wildman, S. G. (1974b). Change in the solubility of crystalline Fraction I proteins correlated with change in the composition of the small subunit. *Pl. Cell Physiol.* **15**, 611–617.

Sears, E. R. (1969). Wheat cytogenetics. *A. Rev. Genet.* **3**, 451–468.

Steer, M. W. (1975). Evolution in the genus *Avena*: inheritance of different forms of ribulose diphosphate carboxylase. *Can. J. Genet. Cytol.* **17**, 337–344.

Steer, M. W. and Kernoghan, D. (1977). Nuclear and cytoplasmic genome relationships in the genus *Avena*. Analyses by isoelectric focusing of ribulose biphosphate carboxylase subunits. *Biochem. Genet.* **15**, 273–286.

Steer, M. W. and Thomas, H. (1976). Evolution of *Avena sativa:* origin of the cytoplasmic genome. *Can. J. Genet. Cytol.* **18**, 796–771.

Strobaek, S., Gibbons, G. C., Haslett, B., Boulter, D. and Wildman, S. G. (1976). On the nature of the polymorphism of the small subunit of ribulose 1,5-diphosphate carboxylase in the amphidiploid *Nicotiana tabacum. Carlsberg Res. Commun.* **41**, 335–343.

Uchimiya, H. and Wildman, S. G. (1979). Non-translation of foreign genetic information for Fraction I protein under circumstances favourable for direct transfer of *Nicotiana gossei* isolated chloroplasts into *N. tabacum* protoplasts. *In vitro.* **15**, 463–468.

Uchimiya, H., Chen, K. and Wildman, S. G. (1977). Polypeptide composition of Fraction I protein as an aid in the study of plant evolution. *Stadler Symp.* **9**, 83–99.

Uchimiya, H. and Wildman, S. G. (1978). Evolution of Fraction I protein in relation to origin of amphidiploid *Brassica* species and other members of the Cruciferae. *J. Hered.* **69**, 299–303.

Uchimiya, H., Chen, K. and Wildman, S. G. (1979a). A microelectrofocusing method for determining the large and small subunit polypeptide composition of Fraction I proteins. *Pl. Sci. Lett.* **14**, 387–394.

Uchimiya, H., Chen, K. and Wildman, S. G. (1979b). Evolution of Fraction I protein in the genus *Lycopersicon. Biochem. Genet.* **17**, 333–341.

Uchimiya, H., Chen, K. and Wildman, S. G. (1979c). Genetic behaviour of information coding for the small subunit polypeptides of *Lycopersicon* Fraction I protein. *Pl. Sci. Lett.* **17**, 63–66.

Wildman, S. G., Chen, K., Gray, J. C., Kung, S. D., Kwanyuen, P. and Sakano, K. (1975). Evolution of ferredoxin and Fraction I protein in the genus *Nicotiana. In* "Genetics and Biogenesis of Chloroplasts and Mitochondria" (P. S. Perlman, C. W. Birky and T. J. Byer, eds), pp. 309–329. Ohio State University Press, Columbus.

10 | Haemoglobins and the Systematic Problems set by Gobioid Fishes

P. J. MILLER and M. Y. EL-TAWIL[1]

Department of Zoology, Bristol University, Bristol BS8 1UG, England

R. S. THORPE

Department of Zoology, The University, Aberdeen AB9 2TN, Scotland

C. J. WEBB[2]

Department of Zoology, Bristol University, Bristol BS8 1UG, England

Abstract: Phyletic classification of the gobioid fishes presents a range of problems at both tokogenetic and phylogenetic levels. A number of these are outlined, including conflicting phylogenetic schemes based on suggested synapomorphies in either skeletal or modified lateral-line systems. The choice and usefulness of haemoglobin electropherograms in teleost classification is discussed, and patterns illustrated for 28 species of gobioid fishes. Geographical

---

[1] Present address: Marine Research Centre, Bab El-Bahr, P.O. Box 315, Tripoli, Libya.

[2] Authors listed in alphabetical order of surname root.

Systematics Association Special Volume No. 16, "Chemosystematics: Principles and Practice", edited by F. A. Bisby, J. G. Vaughan and C. A. Wright, 1980, pp. 195–233, Academic Press, London and New York.

variation of *Pomatoschistus microps* and *Gobius niger* and genetic relationships within the sympatric *P. minutus* species complex are considered with reference to haemoglobin polymorphism. Dividing the species examined for haemoglobin into seven basic morphological groups, the haemoglobin features of cladistic groupings from the opposing phylogenetic schemes are compared in terms of average band number and mobility and also by numerical analysis, based on coincidence in band number and mobility and relative concentration of haemoglobin. No close correspondence between haemoglobin similarity and phylogenetic relationship by either system was apparent. Comparison with habitats yielded similar inconclusive results. Further use of haemoglobin characters is discussed.

INTRODUCTION

The gobioid fishes form a numerous and diverse suborder (Gobioidei) of higher teleosts, probably derived from percoid acanthopterygian stock (Miller, 1973b), although a paracanthopterygian origin has been suggested (Freihofer, 1970). A recent estimate (Hoese, 1971) puts the number of gobioid species at between 1300 and 1700, in 270–330 genera, comprising at least 5%, perhaps nearer 10%, of all recent teleosts (Nelson, 1976).

Being mostly small fishes, neglected by fishermen and collectors, the gobies have long been regarded as a particularly difficult group for the systematic ichthyologist (Smith, 1958). However, a usable phenetic classification of most gobies, providing a fairly unequivocal system for the definition and identification of morphological categories, can be achieved in the face of obstacles which may be laborious to overcome but not inherently insurmountable. Such difficulties include the familiar ones of inadequate material and the need for time, patience and skill in the detailed examination of minute characters, often requiring special staining techniques and dissection of the skeleton. An added complication is that many previous workers on gobioid fishes have omitted to describe features of the modified lateral-line system and of the skeleton whose value in the systematics of the group was clearly demonstrated at least fifty years ago (Regan, 1911; Sanzo, 1911). Progress in conventional gobioid systematics could be greatly speeded if new and existing taxa were supported by morphological data as comprehensive as those recorded by Hoese and Allen (1977), Prince Akihito and Meguro (1977) and Miller and Wongrat (1979). While biochemical features can be incorporated into such a phenetic classification, there is no reason to expect their use to make a more profitable contribution than that of conventional characters still awaiting investigation in many taxa by more convenient means.

However, a modern goal of phyletic classification, if envisaged for gobioid fishes, raises more perplexing difficulties for whose resolution biochemical data may be thought to hold special promise. According to the principles discussed by Hennig (1966), to devise a hologenetic system for the gobies, based on the totality of their genetic relationships, it becomes necessary to consider the ontogenetic succession of individual developmental stages, tokogenetic (reproductive) relationships between individuals in the formation of genetically demarcated populations, and the phylogenetic pattern of ancestry preceding the current existence of these populations as what are termed "biological species".

Ontogenetic problems in gobioid systematics chiefly concern the matching of postlarvae, easily recognizable as a developmental stage, with subsequent demersal juvenile or adult members of the same species (Russell, 1976). There is much work to be done on this topic, with great opportunity for deployment of biochemistry, but not as an area likely to advance general gobioid classification. In the phyletic classification of recent populations, there is a more widespread need to trace the extent of tokogenetic relationships for the purpose of delimiting biological species, reproductively isolated from one another and with evolutionary potential as separate phyletic lines in the future. For the gobies, in many cases, careful examination yields attributional characters which may be used to link populations into what might be biological, as well as evident morphological, species. However, a number of both allopatric and sympatric population complexes present essentially tokogenetic questions in which magnitude of genetic interchange cannot be satisfactorily judged by the usual morphological criteria. Some examples among Atlantic–Mediterranean gobiids (Miller, 1973a) may be briefly described as a background to the haemoglobin studies reported below.

The question of gene-flow between populations is raised in a number of species where a wide range in geographical distribution is accompanied by more or less marked allopatric variation in average morphology between these populations customarily regarded as belonging to a single species. Geographical trends in meristic characters, frequently seen in teleosts (Garside, 1970) and earlier noted in the Californian gobiid *Gillichthys mirabilis* Cooper (Barlow, 1963), have also been found in the eastern Atlantic common goby, *Pomatoschistus microps* (Krøyer) (El-Tawil, 1974). Regional differentiation, involving head squamation as well as meristic features, exists between eastern Atlantic and Mediterranean populations of the black goby, *Gobius niger* L., with those of the latter basin sometimes designated as a separate taxon, *G. jozo* L. (Fage, 1915; De Buen, 1928; Miller and El-Tawil, 1974). From morphological study alone, it is impossible to decide to what extent these geographical differences are the result of genetic isolation between the populations, or of direct environmental influences, varying locally, on individual ontogeny. The susceptibility of meristic characters to the latter has been reported for a number of teleosts (Garside, 1970), although Fonds (1970, 1971, 1973), who investigated the effect of environmental factors on vertebral number in the gobiids *Pomatoschistus minutus* (Pallas) and *P. lozanoi* (De Buen), concluded that this character was largely determined by heredity in these species.

Detailed study of various gobioid faunas has revealed the existence of sympatric populations whose morphological distinctiveness seems slight or is bridged by intermediate individuals, and whose phyletic status is thus uncertain. In the north-eastern Atlantic, this situation is well illustrated by the marine sand gobies of the *Pomatoschistus minutus* complex, occurring on soft deposits of the shallower continental shelf and larger estuaries (Fonds, 1973; Webb and Miller, 1975; Webb, 1980). Three nominal forms have been recognized: *P. minutus,* a more offshore *P. norvegicus* (Collett) and *P. lozanoi,* intermediate between the first two taxa in a number of external features. However, specimens have been obtained which are morphologically intermediate between *lozanoi* and *minutus*, and between *lozanoi* and *norvegicus* (Webb and Miller, 1975; Webb, 1980). An extreme view, that *P. lozanoi* represented a hybrid form between *P. minutus* and *P. norvegicus* (Fonds, 1973), has led to attempts at determining the extent of gene-flow between the three major taxa and their ranking in tokogenetic terms (see below).

At a higher level of taxonomic endeavour, the arrangement of gobiid species in a phylogenetic classification depends on closeness of common ancestry, involving the recognition of shared specializations (synapomorphies) judged to be derived from more generalized (plesiomorphic) character states in a common ancestor and distinguished from a condition of homoplasy, the attainment of similar features by convergence, parallelism, or homoiology (Hennig, 1966; Griffith, 1972). At present, the reconstruction of gobioid phylogeny depends on evidence from the comparative morphology of living forms, in the absence of a good fossil record. Here a fundamental difficulty arises when a scheme based on head patterns of free neuromast organs (sensory papillae) belonging to the modified lateral-line system, a source of characters elaborated within the group, contrasts in primary dichotomy with one dependant on osteological criteria, provided by minor skeletal deletions, with high likelihood of homoplasy but the only features of the body with the chance of fossilization (Miller, 1973b; Birdsong, 1975; Miller and Wongrat, 1979). Dividing the species whose haemoglobins have been examined into seven morphological groups (Table I), conflicting phylogenies may be prepared (Fig. 1) and their incongruence serves to epitomize the dilemma for gobioids as a whole.

The obvious need for further sources of tokogenetic and phylogenetic information turned our attention towards the study of gobioid proteins, if only by electrophoretic methods. Most protein systems have a simple genetic basis, with encodement at merely one to a few loci, and the frequent occurrence of polymorphism, involving codominant alleles, permits easy comparison of allele presence and frequency between different populations as a guide to the probability of genetic interchange, without recourse to the often impracticable alternative of breeding experiments. In addition, a now classical axiom of biochemical taxonomy, that proteins may afford a closer insight into genomic affinities (Sibley, 1962), suggested their possible value as indicating the course of gobioid phylogeny. Initially, a general survey of electrophoretic properties displayed by haemoglobins and muscle-myogens in this group was made by El-Tawil (1974). Later, examination of haemoglobins and selected enzymes played an essential part in a detailed study of the *Pomatoschistus minutus* complex (Webb, 1977, 1980). In continuation of this work, over the past several years, electrophoretic data for gobioid haemoglobins has been gradually accumulated as fishes have become available by collecting or from commercial aquarist sources, where gobies, unfortunately, are not a popular group

Table I. Morphological Groups for Genera for which Haemoglobin Electrophoretic Data have been obtained; Characters mostly from Takagi (1964), Wongrat (1977) and Miller (unpublished).

Characters	*Group A*	*Group B*	*Group C*	*Group D*	*Group E*	*Group F*	*Group G*
Skeleton:							
branchio-stegal rays (S1[c])	6	6	5[a]	5[a]	5[a]	5[a]	5[a]
endopterygoid (mesopterygoid) (S2)	present	present	absent[a]	absent[a]	absent[a]	absent[a]	absent[a]
post-cleithra (S3)	present	present	absent[a]	absent[a]	absent[a]	absent[a]	absent[a]
supratemprals (S4)	present	present	absent[a]	absent[a]	absent[a]	absent[a]	absent[a]
hypercoracoid[b] (scapula) (S5)	A	A	B[a]	D[a]	D[a]	B–D[a]	B–D[a]
epurals (S6)	2	2	2	2	1[a]	1[a]	1[a]
Lateral-line system:							
Suborbital papillae[d] (L)	longitudinal	transverse[a]	abbreviate[a]	longitudinal	longitudinal	abbreviate[a]	transverse[a]
Genera studied, and formal classification (Miller, 1973b)	Eleotrinae: *Batanga*, *Dormitator*	Eleotrinae: *Bostrychus*, *Eleotris*	Tridentigerinae: (= Rhinogobiinae of Takagi, 1964) *Tukugobius*	Gobionellinae, Apocrypteini: *Apocryptichthys;* Periophthalmini: *Periophthalmus*	Gobiinae, Gobiini: *Lesueurigobius* *Oligolepis*	Gobiinae, Gobinni: *Brachygobius*, *Buenia*, *Gobiusculus*, *Pomatoschistus*, *Stigmatogobius*, *Valenciennea*	Gobiinae, Gobiini: *Cryptocentrus*, *Gobius*, *Padogobius*, *Zosterisessor;* ptereleotrini: *Vireosa*

[a] Apomorphic (derived) character state.

[b] Hypercoracoid types A–D as illustrated by Prince Akihito (1969).

[c] Abbreviation for Fig. 1.

[d] Pattern types as defined by Aurich (1938), Harry (1948), and Wongrat (1977); "abbreviate", introduced here and exemplified by *Buenia jeffreysii* (see Miller, 1963, Fig. 21), may be derived in basic form by neoteny or progenesis in the ontogeny of either longitudinal or transverse patterns at an early stage in determination of papillae numbers.

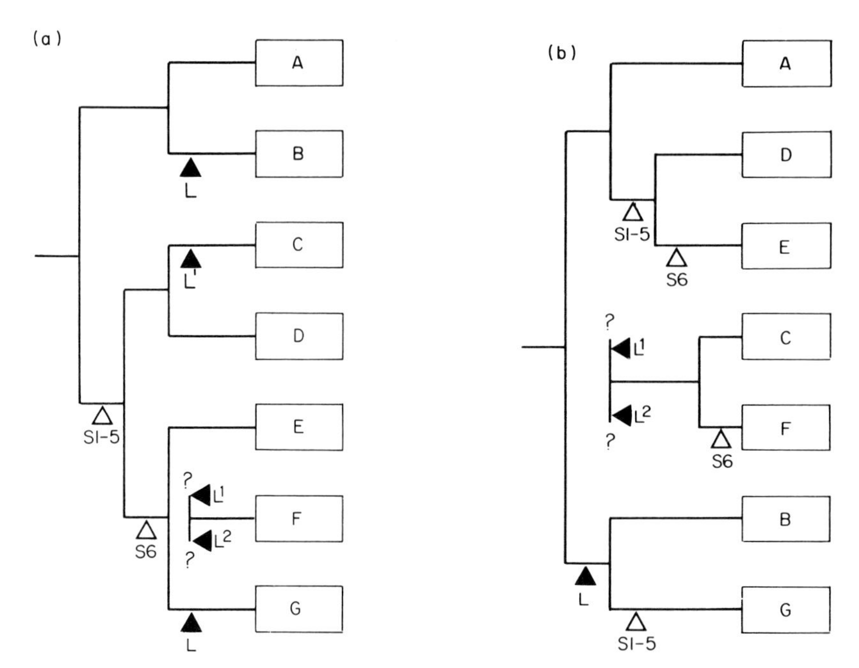

Fig. 1. Alternative schemes for phylogenetic relationships of gobioid morphological groups defined in Table I; (a) synapomorphy in skeletal characters, homoplasy in patterns of suborbital papillae; (b) synapomorphy in patterns of suborbital papillae, homoplasy in skeletal characters. △, ▲, apomorphic states for skeletal (S1-6) and suborbital papillae patterns (L) respectively; note alternative derivation of abbreviate suborbital papillae pattern from longitudinal (L^1) or transverse (L^2) patterns.

for domestic fish-keeping. Acquisition of exotic species has therefore been infrequent and limited in number. Haemoglobin was selected for this purpose because of reports, for other teleost groups (see below), of differences between species or of correspondence with higher categories, and the known occurrence of specific differences and polymorphism in gobiids investigated by Raunich *et al.* (1967), Cucchi and Callegarini (1969) and Fonds (1973). This protein system can be obtained by relatively easy and rapid sampling of blood by cardiac puncture without damage to morphological characters, especially sensory papillae, needed for further study. Because of its location in erythrocytes, haemoglobin can then be isolated without difficulty, while the characteristic red colour enables success of extraction and electrophoresis to be conveniently judged. Since gobies are easily maintained in aquaria, the effects of storage for varying periods on the electrophoretic patterns of haemoglobins may be excluded.

It is important to note that this work on gobioid haemoglobins has been performed as an ancillary and incidental long-term project related to zoological study of conventional systematic characters in the group by one of us (PJM). Consequently, no attempt has been made in further analysis of haemoglobin structure or chemical behaviour. However desirable, the possibility of such work would have been limited by shortage of material in many of the species obtained. All assumptions of homology between haemoglobins, as represented by electrophoretic bands, have been made entirely on the basis of mobility, with full recognition of all the disadvantages of this simple approach as discussed by Avise (1974). Some of the tokogenetic and phylogenetic questions raised earlier have been reviewed below in the light of haemoglobin data, but it must be emphasized that this area of gobioid systematics would be well worth much more extensive biochemical attention.

TELEOST HAEMOGLOBINS

1. Structure and Genetics

Most vertebrate haemoglobins have a tetrameric structure (Riggs, 1970; Wilkins, 1970). In mammals, these tetramers are generally composed of two distinct polypeptide chains (α and β) which are encoded at two different loci in the genome (Braunitzer *et al.*, 1964). Fish haemoglobins, on the other hand, can contain two, three, or even four different polypeptides and up to eight structural genes may be involved in the haemoglobin synthesis of some species (Tsuyuki and Ronald, 1970, 1971; Wilkins, 1970). Although there is evidence that all haemoglobin polypeptide chains are derived from a common ancestral form (e.g. Perutz *et al.*, 1968), the homology of fish polypeptides with those from higher vertebrates is unclear (Wilkins, 1970; Ronald and Tsuyuki, 1971). Several fish species do seem to possess two different globin families to which have been applied the traditional α–β notation (Manwell and Baker, 1967, 1970; Ronald and Tsuyuki, 1971). However, the original designations (Ingram, 1961) of α and β for the chains of mammalian haemoglobins have come to represent those polypeptides with

Val-Leu and Val-Leu-His N-terminal sequences and, since the subunits of fish haemoglobins for which sequence data are available all possess Ser-Leu or valine at the N-terminal (Hashimoto and Matsuura, 1962; Buhler, 1963; Hilse and Braunitzer, 1968), various authors (such as Wilkins, 1970; Perez and Maclean, 1975, 1976a) have recommended the use of an alternative nomenclature for fish globins.

Many fish species possess multiple haemoglobins (De Ligny, 1969; Manwell and Baker, 1970; Riggs, 1970). Multiple components in haemolysates usually represent tetrameric combinations of different globin molecules and reflect the occurrence of multiple loci or allelic variants (Wilkins, 1968, 1970; Perez and Maclean, 1975, 1976a), although apparent multiple haemoglobins can also arise as technical artefacts, caused, for example, by polymerization or the formation of methaemoglobin, particularly during storage (Yamanka *et al.*, 1965a, b, 1967; Sharp, 1969, Riggs, 1970; Perez and Maclean, 1975, 1976a).

Haemoglobin polymorphism has been reported in a number of fishes (De Ligny, 1969; Manwell and Baker, 1970) and is often determined by co-dominant alleles (e.g. Sick, 1961, 1965a, b; Wilkins, 1971; Fyhn and Sullivan, 1974). This situation can also result from a differential ontogenetic expression of loci encoding different subunits (Manwell, 1957, 1958a, b, 1963a, b; Vanstone *et al.*, 1964; Koch *et al.*, 1966; Wilkins and Iles, 1966; Wilkins, 1972; Perez and Maclean, 1974, 1976b).

2. *Systematic Value*

Work on the use of haemoglobin polymorphism to delimit panmictic teleost populations has been reviewed by De Ligny (1969). In an assessment of the contribution of protein electrophoresis to scorpaenid systematics, Tsuyuki *et al.* (1968) suggested that for rockfish, as well as fishes in general, haemoglobin electropherograms are most useful diagnostically at the species and intraspecific levels and have much less value as indicators of phylogeny. However, not all fish species have a characteristic electrophoretic array of haemoglobins and, in some groups such as the scombroids, haemoglobin electropherograms may show a fair correspondence with higher systematic categories which could be monophyletic (Sharp, 1969, 1973).

The validity of multiple haemoglobins as indicators of phyletic relationship depends to a large extent on their adaptive significance. Most previous studies on fish haemolysates, particularly those of migratory species, have indicated that different haemoglobin components have different functional properties (Hashimoto *et al.*, 1960; Yamaguchi *et al.*, 1962, 1963; Binotti *et al.*, 1971; Powers, 1972; Poluhowich, 1972). However, as Perez and Maclean (1976a) noted, most of these observations were made on haemolysates that separated or were grouped into only two components. Where more than two haemoglobin components have been studied the majority, or all, of the components have identical properties (Gillen and Riggs, 1972, 1973a, b; Weber and De Wilde, 1976). Perez and Maclean (1976a) emphasize that if all the components are functionally equivalent, then it must be concluded that multiple components result from the random fixation of neutral alleles. These authors cite the presence of three different haemoglobin components in young stages of the roach and rudd and the later loss of one of these haemoglobins in adults of the species (Perez and Maclean, 1974) as constituting an example where multiple haemoglobins are the result of mutations which have become fixed in a species not because they represent an advantage but simply by a non-directional mechanism such as genetic drift. Variation which occurs without prejudicing the function of a protein may, as Heslop-Harrison (1968) suggests, "preserve the record of accidental evolutionary change and provide some of the evidence for reconstructing part at least of the réseau of evolutionary change".

GOBIOID HAEMOGLOBINS

1. Material and Methods

Examples of gobioid species were obtained by collecting in marine and estuarine habitats around the British Isles, or purchased from aquarist dealers. Prior to use, fishes were held in laboratory aquaria by the usual techniques of maintenance.

The species involved are listed alphabetically as follows, with area of natural occurrence, habitat, source of material, and number of electropherograms prepared: (1) *Apocryptichthys cantoris* Day*;

*As used by Koumans (1953), but a junior synonym of *Oxuderces dentatus* Eydoux and Souleyet according to Springer (1978).

Indo-Pacific; brackish water (b); aquarist suppliers (a.s.); 4. (2) *Batanga lebretonis* (Steindachner); West Africa; fresh water (f.); a.s.; 8. (3) *Bostrychus africanus* (Steindachner); West Africa; f.; a.s.; 4. (4) *Brachygobius nunus* (Hamilton Buchanan); Indo-Pacific; f.; a.s.; 8. (5) *Buenia jeffreysii* (Günther); eastern Atlantic boreal; marine (m.); Oldbury, Bristol Channel; 1. (6) *Cryptocentrus pavoninoides* (Bleeker); Indo-Pacific; m., cryptobenthic (c.); a.s.; 1. (7) *Dormitator maculatus* (Bloch); central and northern South America; f.; a.s.; 8. (8) *Eleotris vittata* Dumèril; West Africa; f.; a.s.; 5. (9) *Gobius couchi* Miller and El-Tawil; western English Channel and southern Ireland; m., c., Helford, Cornwall; 20. (10) *G. niger* L.; eastern Atlantic boreal to Mediterranean and Black Sea; m.; Helford and Plymouth Sound; 35. (11) *G. paganellus* L.; as (10); m.; c.; 69. (12) *Gobiusculus flavescens* (Fabricius); eastern Atlantic boreal; m.; St Anthony, Cornwall and Plymouth Sound; 13. (13) *Lesueurigobius friesii* (Malm); eastern Atlantic boreal to Mediterranean; m.; c.; Loch Torridon, Scotland; 1. (14) *Oligolepis acutipennis* (Valenciennes); Indo-Pacific; f.; a.s.; 1. (15) *Padogobius martensi* (Günther); Adriatic; other data from Raunich *et al.* (1967). (16) *Periophthalmus gracilis* Eggert; Indo-Pacific; b., c.; a.s.; 1. (17) *Pomatoschistus lozanoi* (De Buen); eastern Atlantic boreal; m.; see text; 37. (18) *P. microps* (Krøyer); as (17); b.; 1177. (19) *P. minutus* (Pallas); as (17), also Mediterranean and Black Sea; 102. (20) *P. norvegicus* (Collett); as (19) but not Black Sea; 33. (21) *P. pictus* (Malm); as (17); St Anthony and Plymouth Sound; 9. (22) *Stigmatogobius sadanundio* (Hamilton-Buchanan); Indo-Pacific; f., c.; a.s.; 5. (23) *Stigmatogobius* sp.; f., c. (?); a.s.; 5. (24) *Thorogobius ephippiatus* (Lowe); eastern Atlantic boreal to Mediterranean; m., c.; Loch Torridon. (25) *Tukugobius* sp.; eastern Pacific. Philippines to Japan; f.; a.s.; 1. (26) *Valenciennea strigata* (Broussonet); Indo-Pacific; m., c.; 1. (27) *Vireosa zebra* (Fowler); as (26), 1. (28) *Zosterisessor ophiocephalus* (Pallas); Mediterranean and Black Sea; b.; other data from Raunich *et al.* (1967).

For preparation of haemoglobin extracts, fishes were terminally anaesthetized with MS 222 (Sandoz) and blood samples collected by puncture of the heart with a small (27G) hypodermic needle containing a few drops of 5% trisodium citrate solution as an anticoagulant. The blood was then transferred to a cooled tube and centrifuged at 3000 r min^{-1} (504 g) for 10 min (Tsuyuki and Gadd,

1963; Tsuyuki *et al.*, 1965). Supernatant fluid was removed and red cells washed twice with 2.5% sodium chloride solution. After the second wash, haemolysis was effected with one small drop of cold reservoir buffer, the sample being kept for 1 h under refrigeration until haemolysis was completed. The haemolysate, without further storage, was centrifuged for 2 min and the supernatant solution of haemoglobin applied to the gel.

Electrophoretic mobility of haemoglobins was studied by horizontal thin-layer starch gel electrophoresis (Smith, 1968), using 13% hydrolysed starch (BDH) and tris-EDTA-borate buffer solution (Smithies, 1959; Tsuyuki *et al.*, 1966) in a modified Shandon U77 tank described by Miller and El-Tawil (1974). Electrophoresis was performed at 25–35 V cm^{-1} for 2.0–3.5 h. To ensure cooling, the electrophoretic tank was situated in a refrigerator and water at 4°C circulated from external equipment through an aluminium platen beneath the gel. After electrophoresis, haemoglobins were revealed more precisely by staining the gel with 1% naphthalene black 12B (BDH) for 1 min and destaining with four washes of a 5:1:5 mixture of methyl alcohol, glacial acetic acid and water. Destained gels were photographed and stored under refrigeration for future reference.

For presentation of results (Fig. 2), mobility was measured from slit to centre of each haemoglobin band, and standardized with reference to the fastest band observed in *Pomatoschistus microps*, an easily obtainable species generally run as control. For multiple haemoglobins, relative concentration for the individual bands was arbitrarily estimated from band width, agreeing well with subjective visual assessment of staining intensity. For numerical analysis, a data matrix was constructed, with species and morphs as OTUs (numbered as in Fig. 2) and, as characters, occurrence of major haemoglobin bands, within 24 "mobility groups" of 0.05 units (*P. microps* fastest band), from 1.44 to −0.34. For multiple haemoglobins, relative concentration for individual bands was arbitrarily estimated from band width, agreeing well with subjective visual assessment of staining intensity, and ultimately expressed as percentage of total for major bands (i.e. those containing more than 9% of all haemoglobin as first estimated from the electropherogram). A complete table of haemoglobin mobilities and relative concentrations in the gobioid fishes listed above has been deposited in the Zoology Library of the British Museum (Natural History).

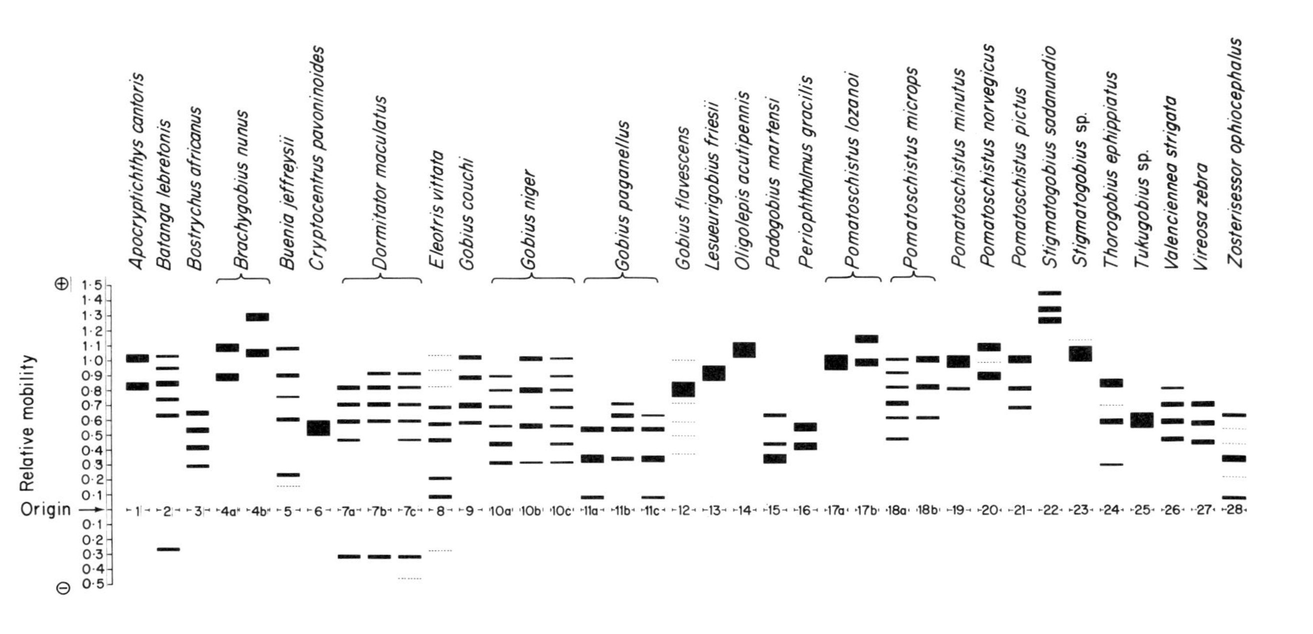

Fig. 2. Diagram of electrophoretic separation of haemoglobins in gobioid fishes listed in the text. Phenotypes numbered at origin; band width in proportion to relative concentration; broken lines indicate faint bands.

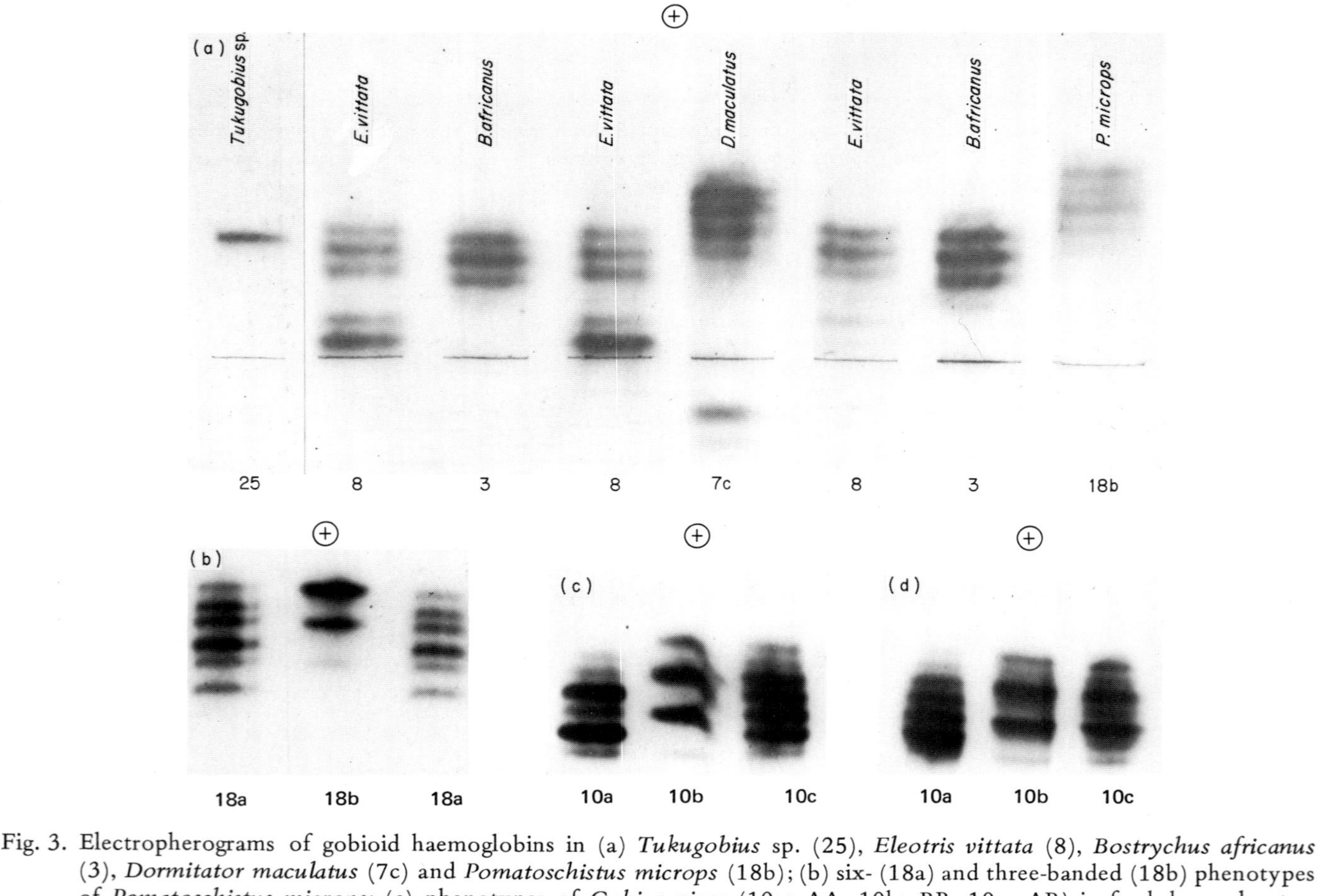

Fig. 3. Electropherograms of gobioid haemoglobins in (a) *Tukugobius* sp. (25), *Eleotris vittata* (8), *Bostrychus africanus* (3), *Dormitator maculatus* (7c) and *Pomatoschistus microps* (18b); (b) six- (18a) and three-banded (18b) phenotypes of *Pomatoschistus microps*; (c) phenotypes of *Gobius niger* (10a: AA; 10b: BB; 10c: AB) in fresh haemolysates, (d) phenotypes of *G. niger* in same samples after refrigerator storage of haemolysate for 24 h.

2. Tokogenetic studies

(a) Correspondence with morphological species. Inspection of Figs. 2 and 3 shows a wide variety in haemoglobin electropherograms, with each morphological species displaying a more or less distinctive pattern, or complex of patterns, in haemoglobin mobility. The number of bands ranges from one to seven. In one case, contrasting results for individuals of *Stigmatogobius*, from different aquarist sources but all keying-out to *S. sadanundio* (Hamilton-Buchanan) in Koumans (1953), led to closer morphological inspection of the fishes and the discovery of slight but consistent differences in coloration which might support recognition of an additional species. When abundant material was available, variation in haemoglobin patterns within a morphological species was explicable in terms of polymorphism, not "sibling species".

(b) Intraspecific polymorphism. This has been found previously in European gobiids by Raunich *et al.* (1967) for *Gobius niger* (as *G. jozo*), in *Padogobius martensi* (Günther) (as *Gobius fluviatilis* Nardo) by Cucchi and Callegarini (1969) and noted also for *G. paganellus* L. by Miller and El-Tawil (1974). In the present material, intraspecific polymorphism in haemoglobin patterns has been recorded in *G. paganellus, G. niger, Pomatoschistus microps* and *P. lozanoi*, with the possibility of its occurrence in *Brachygobius doriae* (Günther) and *Dormitator maculatus* (Bloch), although the number of individuals obtained in the last two species was very limited.

Intraspecific haemoglobin polymorphism has been extensively studied in the common goby *Pomatoschistus microps* by one of us (El-Tawil, 1974). Two haemoglobin phenotypes were recognized in the populations of *P. microps* investigated, one of these (Figs. 2 and 3, 18a) consisting of six major anodic zones and the other (Figs. 2 and 3, 18b) of three such zones. It is postulated that this polymorphism is determined by co-dominant alleles (A and A′) and that individuals with the three-banded phenotype were homozygous for one of these (A′) while those with six bands were heterozygotes. A third phenotype, representing the AA homozygote, was not encountered among the specimens studied and its absence could be due to lethality or strong selection acting against it at the

Table II. Distribution of Haemoglobin Genotypes among *Pomatoschistus microps* from south-west England and Wales

Locality	phenotype: genotype:	— AA	*18a* AA'	*18b* $A'A'$	N	pA'	X^2 (*1*)	P
Dale, Dyfed	O	—	38	99	137	0.8613	3.5255	$0.10 > P > 0.05$
	E	2.6	32.7	101.6				
Swansea, West	O	—	25	46	71	0.8239	3.2402	$0.10 > P > 0.05$
Glamorgan	E	2.2	20.6	48.2				
Clevedon, Avon	O	—	7	39	46	0.9239	0.3408	$0.70 > P > 0.50$
	E	0.3	6.5	39.3				
Burnham-on-Sea,	O	—	51	27	78	0.6731	18.3825	$P < 0.001$
Somerset	E	8.3	34.3	35.3				
St Anthony,	O	—	29	324	353	0.9598	0.6529	$0.50 > P > 0.30$
Cornwall	E	0.6	27.8	324.6				
Plymouth Sound	O	—	60	432	492	0.9390	2.0372	$0.20 > P > 0.10$
Devon	E	1.8	56.4	433.8				
Total	O	—	210	967	1177	0.9108	11.3390	$P < 0.001$
	E	9.4	191.2	976.4				

O, observed; E, expected.

localities sampled. In most populations, distribution of phenotypes was in accordance with the Hardy-Weinberg Law (Table II). Haemoglobin polymorphism in the congeneric *P. lozanoi* is described below in connection with reproductive isolation between sympatric species.

A more complex type of haemoglobin polymorphism was described in another gobiid, *Gobius niger*, by Raunich *et al.* (1967). These authors observed six variant haemoglobin phenotypes and suggested that the polymorphism is determined by three co-dominant alleles at the same locus (or three closely-linked groups of loci). However, as noted by Miller and El-Tawil (1974), only three haemoglobin phenotypes (Fig. 2, 10a–c), explicable with a hypothesis of control by only two co-dominant alleles (A and B), were detected when fresh blood from specimens of the western English Channel was investigated, but phenotypes rather similar to the fundamental phenotypes described by Raunich and his co-workers could be generated if haemolysates were stored before electrophoresis (El-Tawil, 1974), phenotypes 10a (AA), 10b (BB) and 10c (AB) becoming comparable to those termed A, B and C by Raunich *et al.* (1967). Genetic similarity between the Atlantic *G. niger* and the nominal *G. jozo* of the Mediterranean may thus be greater than has been supposed.

(c) Allopatric populations within a morphological species. The small size and restricted vagility of gobiid fishes may be thought to promote a situation of genetic isolation and local differentiation between geographically separated populations of a single morphological species (Rosenblatt, 1963). Within the relatively small area of south-western British Isles, haemoglobin polymorphism has been employed for genetic comparison between populations of *Pomatoschistus microps*, which show some local differentiation in the meristic character of mean number of pectoral fin-rays (El-Tawil, 1974).

The distribution of haemoglobin genotypes among *P. microps* from these different localities are compared in Table II with the distributions calculated according to the Hardy-Weinberg Law. At St Anthony, Plymouth and Clevedon there is good agreement between the observed and expected distribution of genotypes. At Dale Fort and Swansea, although there is a difference between the observed and expected distributions, this is not very significant,

Table III. Annual Distribution of the Haemoglobin Genotypes in Year Classes of *Pomatoschistus microps* from St Anthony, Cornwall and Plymouth Sound, Devon

Year class	*Date of collection*	*phenotype:* *genotype:*	— AA	*18a* AA'	*18b* $A'A'$	N	pA'	X^2 (*1*)	P
St Anthony									
1970	12.4.71	O	—	13	170	183	0.9645	0.2202	$0.70 > P > 0.50$
		E	0.2	12.5	170.2				
1971	24.4.72	O	—	10	78	88	0.9432	0.3394	$0.70 > P > 0.50$
		E	0.3	9.4	78.3				
1972	3.4.73	O	—	6	76	82	0.9634	0.1070	$0.80 > P > 0.70$
		E	0.1	5.8	76.1				
Total		O	—	29	324	353	0.9589	0.6529	$0.50 > P$ 0.30
		E	0.6	27.8	324.6				
Plymouth Sound									
1970	22.4.71	O	—	5	27	32	0.9219	0.2363	$0.70 > P > 0.50$
		E	0.2	4.6	27.2				
1971	8.2.72	O	—	50	336	386	0.9352	1.8264	$0.20 > P > 0.10$
		E	1.6	46.8	337.6				
1972	13.5.73	O	—	5	69	74	0.9662	0.1084	$0.80 > P > 0.70$
		E	0.1	4.8	69.1				
Total		O	—	60	432	492	0.9390	2.1457	$0.20 > P > 0.10$
		E	1.8	56.4	433.8				

O, observed; E, expected.

while at Burnham-on-Sea there is a highly significant difference due to a large excess of heterozygotes. The frequency of the most common allele (A′) varied between localities, being lowest at Burnham-on-Sea (0.6731) and highest at St Anthony (0.9589), although no clear geographical trend was discernible. Comparison with meristic variation (El-Tawil, 1974) also revealed the Burnham-on-Sea population to be more distinct than the remainder. One problem to be faced in such work on *P. microps* is the possibility of seasonal migrations (Miller, 1975) selectively affecting gene frequency.

Variation in time was also investigated by analysing samples of *P. microps* in three successive year-classes from Plymouth and St Anthony (Table III). No significant variation was detected between the observed and expected genotypic distributions in both localities. The frequency of the most common allele (A′) was more or less similar in the three year classes in samples obtained from St Anthony, while the frequency of this allele increased from 0.9219 in 1970 to 0.9662 in 1972 in year-class samples from Plymouth Sound.

(d) Sympatric populations. Small gobiids of the European genus *Pomatoschistus* have posed a number of problems in species definition, reflected in the lengthy synonymies listed by Miller (1973a). The *Pomatoschistus minutus* complex has already been mentioned, and the electrophoretic behaviour of haemoglobins has been included among morphological, karyological, and biochemical features examined by one of us in an attempt to establish the biological status of the three forms concerned (Webb, 1980). Haemoglobin was obtained from fishes collected in the Severn estuary, western English Channel and the North Sea (totalling 102 *P. minutus,* 37 *P. lozanoi* and 33 *P. norvegicus*).

Haemolysates from all members of the *P. minutus* complex separated on electrophoresis into one or more major anodic zones. Irrespective of their sex, size or place of capture, haemolysates from specimens of *minutus* separated into two zones migrating at 81% and 98% of the rate of the *P. microps* standard described above, and the faster of these contained more haemoglobin. Haemolysates from all specimens of *norvegicus* also separated into two major zones, although these contained roughly equal amounts of haemoglobin and migrated at 89% and 108% of the standard rate.

Thus, the faster zone in *norvegicus* migrated more rapidly than either of the *minutus* zones, while the slower zone of *norvegicus* had a mobility intermediate between those of the *minutus* zones. Haemolysates from *lozanoi* either separated into a single zone with the same mobility as the faster *minutus* zone or into two zones of roughly the same concentration, the slower of which again migrated at the same rate as the faster *minutus* zone, while the faster zone migrated at 114% of the standard rate, i.e. faster than any of the zones in *minutus* and *norvegicus*. The occurrence of these types was not sex-linked and although the mean standard length of fishes with a single zone (42.2 ± 0.94 mm) and those with two zones

Table IV. Distribution of Haemoglobin Genotypes among *Pomatoschistus lozanoi* from the Dutch North Sea and Burnham-on-Crouch, Essex

	pheno-type:	*17a*	*17b*	–				
	geno-type:	B^1B^1	B^1B^2	B^2B^2	N	pB^1	$X^2\,(1)$	P
Dutch North Sea	O	15	10	–	25	0.80	1.5625	$0.25 > P > 0.10$
	E	16	8	1				
Burnham-on-Crouch, Essex	O	5	3	–	8	0.81	0.4233	$0.75 > P > 0.50$
	E	5.3	2.4	0.3				
Total	O	20	13	–	33	0.80	2.0293	$0.25 > P > 0.10$
	E	21.3	10.4	1.3				

O, observed; E, expected.

(44.9 ± 0.71 mm) was significantly different ($p \simeq 0.05$–0.02), there was no obvious difference in their degree of sexual maturity. Both types of haemoglobin were present in samples from the Dutch North Sea and Burnham-on-Crouch, but a small sample of *lozanoi* from Oldbury-on-Severn all had the double-banded type.

If it is assumed that the major zones observed from haemolysates of fishes of the *minutus* complex represent tetrameric combinations of at least two distinct polypeptide chains, then it follows that the haemoglobin of *minutus* and *norvegicus*, which invariably separates into two main components, is determined by three monomorphic loci (termed HbA, HbB, and HbC), while the behaviour of *lozanoi* haemoglobin, which occurs as a single or double-banded form, is consistent with a model of control by two loci (HbA and HbB), at one of which (HbB) two co-dominant alleles (B^1 and B^2) segregate. The electrophoretic behaviour of haemoglobin from the *minutus* complex suggests that *minutus* and *lozanoi* have the same allele at HbA and that *minutus* is monomorphic at HbB for one of the alleles (B^1) which segregates at this locus in *lozanoi*. On the other hand, it seems that *norvegicus* and *lozanoi* are allelically distinct at either the HbA or HbB locus while *minutus* and *norvegicus* must have different alleles at a minimum of two out of their three haemoglobin loci. Although the mean length of *lozanoi* with the single and double-banded form of haemoglobin was different, there was no obvious relationship between haemoglobin phenotype and maturity (i.e. ontogenetic correlation), and the hypothesis that the polymorphism observed in *lozanoi* is determined by two allelic genes is supported (see Koehn, 1972; Williams *et al.*, 1973; Avise and Smith, 1974; Johnson, 1975) by the distribution of supposed genotypes in samples from Burnham-on-Crouch and the Dutch North Sea which is consistent with the distribution calculated according to the Hardy-Weinberg Law (Table IV).

From all evidence obtained, Webb (1977, 1980) concluded that, although *lozanoi* was intermediate between *minutus* and *norvegicus* in a number of its features and may interbreed in the wild with both of these taxa, this form is genetically distinct and, along with them, deserves full status as a biological species. The results from haemoglobin studies support this view on both phenetic and genetic grounds. First, haemoglobin electropherograms are different for the three taxa, in keeping with differences observed between more clearly defined morphological species. Secondly, haemoglobin synthesis appears to be controlled by a different number of loci and/or different alleles at some of these loci in the different taxa of the *minutus* complex.

3. Phylogenetic Studies

As noted earlier, it is possible to divide the present species, whose haemoglobin has been examined, into seven morphological groups (Table I) and to construct different phylogenetic arrangements for these (Fig. 1), according to whether osteological (Fig. 1a) or

Table V. Number of Haemoglobin Bands in Gobioid Species of Morphological Groups (Table I) and Cladistic Groupings of Fig. 1.

Morphological groups (Table I):	*A*	*B*	*C*	*D*	*E*	*F*	*G*
No. of species	2	2	1	2	2	12	7
$\overline{x}$	6.0	4.5	1.0	2.0	1.0	2.83	3.57
Range	—	4, 5	—	—	—	1–6	1–7
S. D.	—	0.71	—	—	—	1.53	1.81
Cladistic groups							
Fig. 1a:	*AB*	*CDEG*	*EG*	*CD*			
Fig. 1b:					*ADE*	*BG*	*DE*
No. of species	4	12	9	3	6	9	4
$\overline{x}$	5.25	2.67	3.00	1.67	3.00	3.78	1.50
Range	4–6	1–7	1–7	1, 2	1–6	1–7	1, 2
S. D.	0.96	1.78	1.94	0.58	2.37	1.64	0.58

sensory papillae (Fig. 1b) characters are thought more likely to be synapomorphic than homoplastic. By various means, ranging from determination of average number of bands or mobility to sophisticated analysis of patterns in band position and haemoglobin distribution, comparisons have been made between the haemoglobin features of species within these groups, in the hope that one or the other phylogenetic hypothesis may be supported by agreement with trends in haemoglobin similarities. It should be noted that the species studied form only a minute component of this large suborder, and that intraspecific polymorphism, however valuable a phenomenon for tokogenetic studies, must obscure phylogenetic indications from merely electrophoretic mobilities. In this preliminary exercise, morphs have been kept separate.

(a) Number of bands. Maximum number of haemoglobin bands found for each species is used in calculating mean number of bands for species grouped by successive dichotomies in the two phylogenetic schemes, as well as for the seven basic morphological groups. Results (Table V) indicate significant differences between fishes with primitive and advanced osteology respectively, at the first dichotomy in scheme (a) (into AB and CDEG) and, in scheme (b), at the osteological dichotomy in species with longitudinal papillae (A and DE). However, numbers of species in groups A and B are very limited, and the number of bands displayed by these fall within the range seen in groups F and G.

(b) Average mobility of haemoglobins. Average mobility of electrophoretically distinct haemoglobin bands occurring within the species grouped by successive dichotomies and within each basic morphological group are shown in Table VI. Average mobility differs significantly between species segregated by the first dichotomy of the phylogeny based on sensory papillae patterns (Fig. 1b), but not between groupings AB, CDEG, CD and EG of that depending on osteology (Fig. 1a). Among gobiids with derived skeletal characters, average mobility of haemoglobins with abbreviate papillae patterns (F) is significantly different from that in Group G, with transverse papillae, but not from that of longitudinal group E represented, however, by only two mobilities. Again, no cladistic grouping is completely separated in range of mobility values.

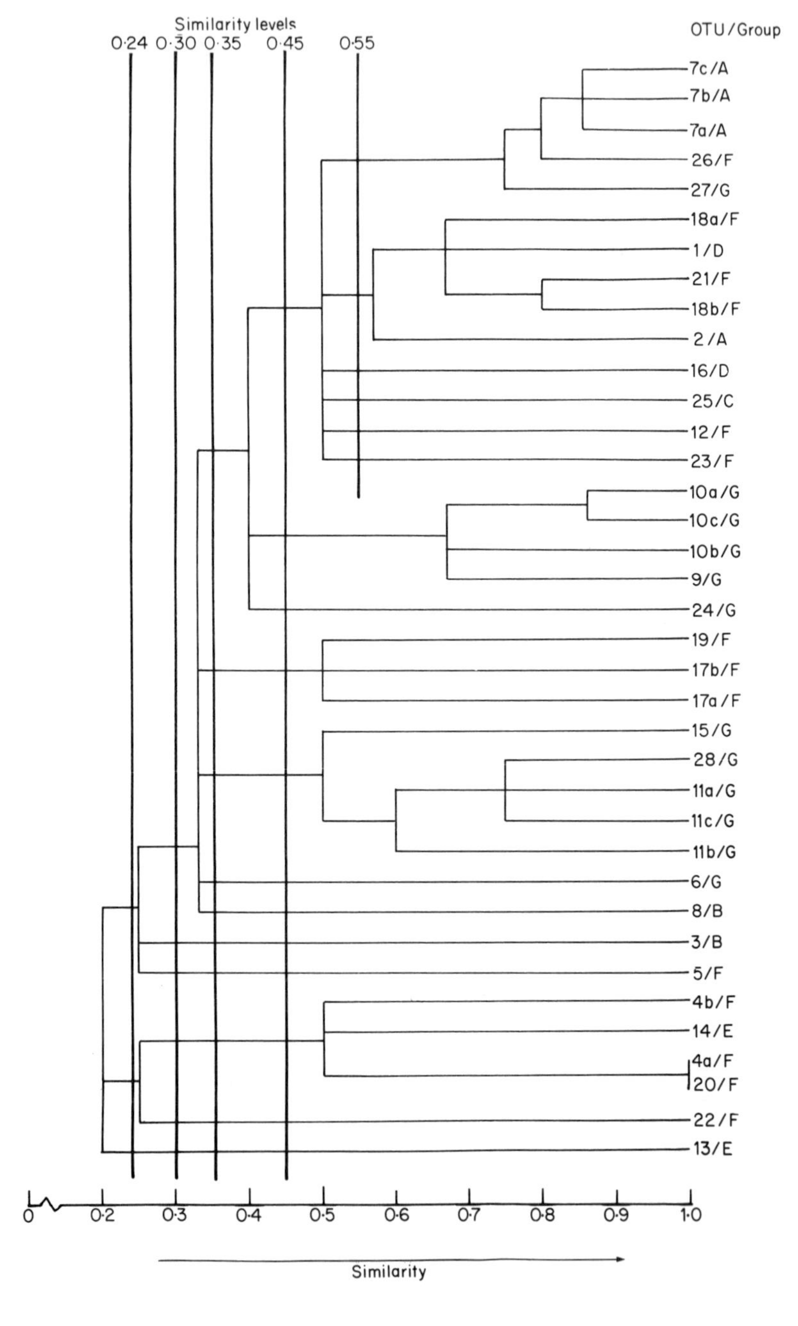

Fig. 4. Single-linkage cluster analysis of similarity values, using OTUs. Morphological group (A–G) indicated for OTUs (see Table I).

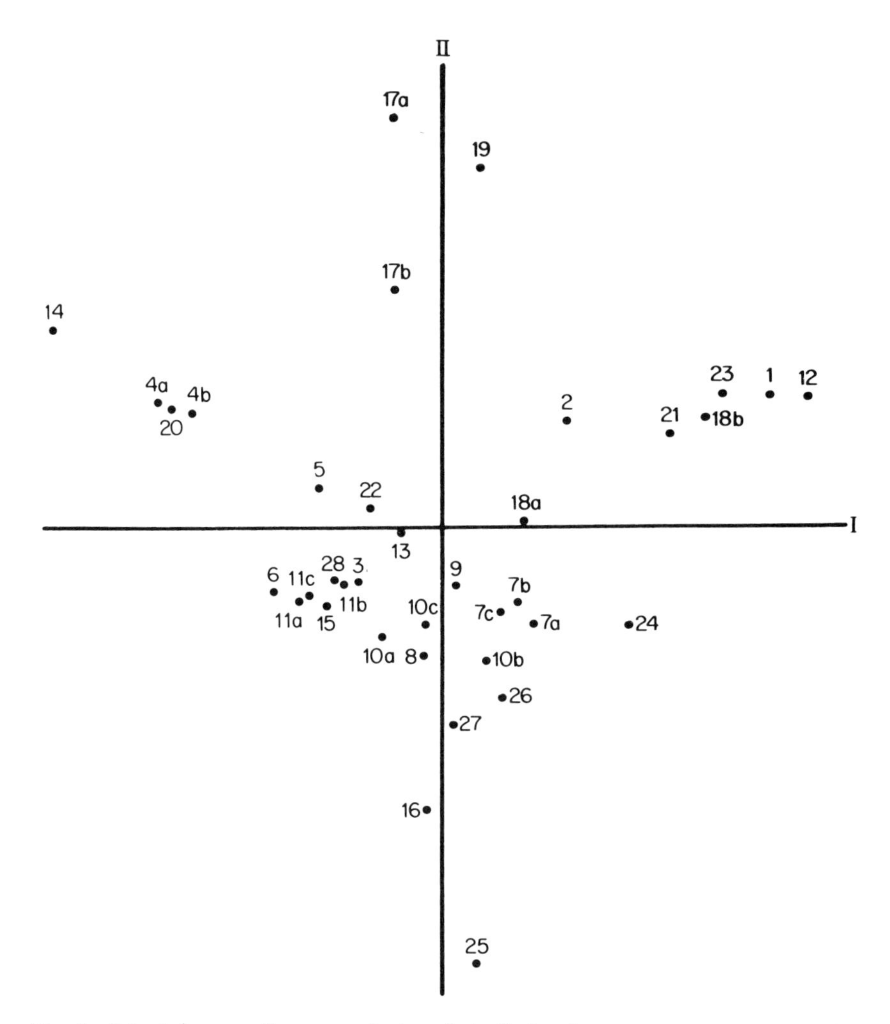

Fig. 5. Principle coordinate analysis of similarity between OTUs. Scatter diagram of OTUs against first two coordinates (expressing one third of variation).

(c) Polarity. Only the two species forming group A possess a haemoglobin component whose direction of migration is cathodic under the experimental conditions used (Fig. 2, 2 and 7; Fig. 3a, 7c) Cathodic haemoglobin bands are also found in many lower teleosts, percoids and derivatives (Tsuyuki *et al.*, 1965), and their occurrence among gobies may be viewed as retention of a primitive feature,

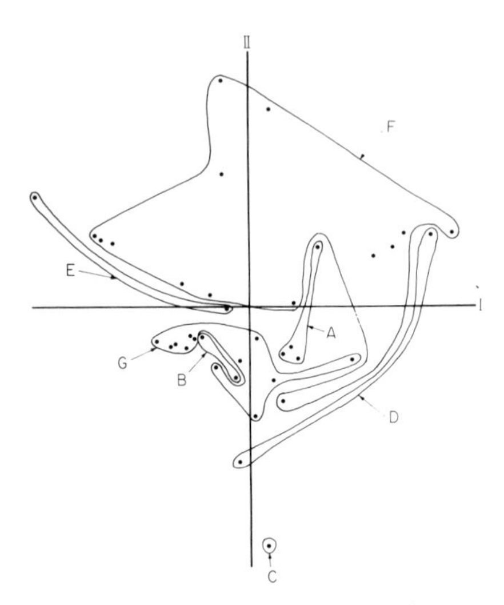

Fig. 6. Clustering of OTUs belonging to morphological groups (A–G) defined in Table II, in scatter diagram from principal coordinate analysis.

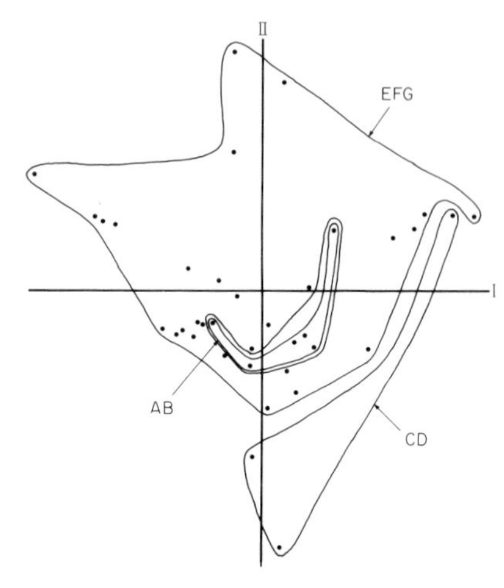

Fig. 7. Clustering of OTUs belonging to cladistic groupings of Fig. 1(a) in scatter from principle coordinate analysis.

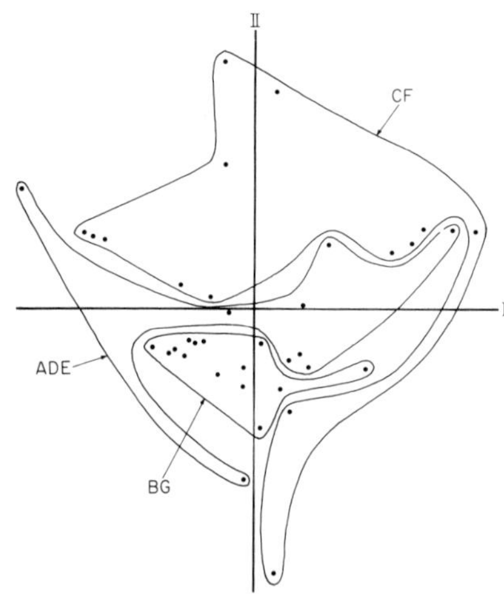

Fig. 8. Clustering of OTUs belonging to cladistic groupings of Fig. 1(b) in scatter diagram from principle coordinate analysis.

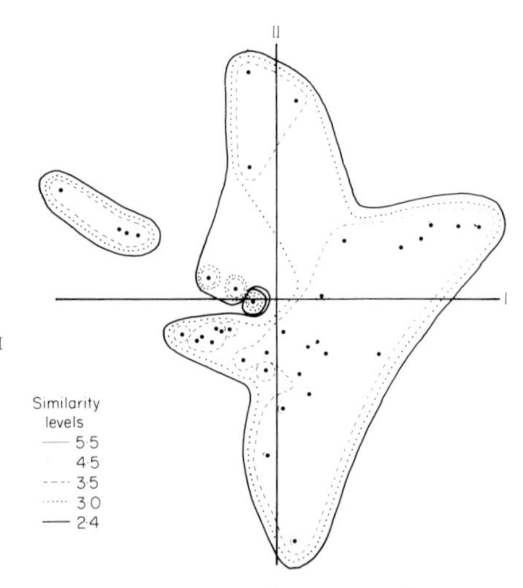

Fig. 9. Nested clustering of OTUs, according to similarity levels of Fig. 3, in scatter from principle coordinate analysis. Similarity contours as indicated on diagram.

in keeping with the osteology of the two eleotrines concerned (Wongrat, 1977). Loss of cathodic bands could therefore be regarded as a specialization, but one which would need to have occurred on more than one phyletic line in both the suggested phylogenies.

(*d*) *Numerical analysis of mobility patterns.* The patterns of haemoglobin mobility were analysed by two different phenetic methods for indicating similarity between the species and morphs of Fig. 2, considered as separate OTUs (operational taxonomic units).

(i) Single-linkage cluster analysis. Similarity in the mobility pattern of a pair of OTUs was assessed as: numbers of bands with identical mobility *divided* by total number of band positions for that pair of species. This similarity value ranges from zero (totally dissimilar – no band position in common) to one (totally similar – all band positions in common). The measurement of similarity does not take into account (i) differences in haemoglobin concentration within the bands, or (ii) the larger number of negative similarities when neither of a pair of species has a band in one of the 24 positions. The resultant similarity matrix between OTUs can be summarized by the very simple method of single-linkage cluster analysis (Sneath and Sokal, 1973) which results in a dendrogram (Fig. 4).

(ii) Principle coordinate analysis. This second analysis is entirely unrelated to the first since a different measure of similarity (or rather dissimilarity) between OTUs is employed, as well as a different logical basis for summarizing the matrix. The "distance" between species is assessed by taking into account the relative concentration of the bands and the negative similarities (i.e. no band at a position in either species). The concentration of a band (estimated as described earlier) is expressed as a percentage of the total amount of haemoglobin present in all the bands (scale 0–100%), and the average taxonomic distance between the OTUs is computed in the 24-dimensional space (from 24 band mobilities providing the characters).

This measured distance or dissimilarity gives zero between totally similar OTUs and progressively greater distances between progressively more dissimilar OTUs. The OTUs can be envisaged as points in space with similar species being close together in the cloud of points and dissimilar species far apart. The first axis or principle coordinate (Gower, 1966) through the cloud of points can be found so that it expresses as much of the variation between species as possible. Subsequent axes, at right angles to the first (and one another), express progressively less variation between OTUs. Thus, the principle coordinate analysis can be used to summarize in fewer dimensions the variation between OTUs in the 24-dimensional space. The OTUs can be plotted against the principle coordinates in a scatter diagram (Fig. 5). In this ordination, similar OTUs will tend to be closer together and dissimilar OTUs farther apart. However, it should be borne in mind that the first two coordinates express only one third of the variation between OTUs in this analysis.

Both phenetic analyses agree in showing that the data are "weakly structured". The groupings are diffuse and poorly defined rather than there being compact clusters widely separated from one another. The seven morphological groups defined in Table I are not clearly separated (Fig. 6). Groupings according to dichotomies in the two phylogenetic schemes are similarly not supported by the scatter of points (Figs 7, 8).

When the data are weakly structured, different phenetic analyses may give very different results but those from the two analyses here are somewhat comparable. This can be demonstrated by taking the clusters at the various similarity levels shown in the dendrogram (Fig. 4) and plotting these as contours, enclosing the groupings from cluster analysis, on the ordination diagram from the principle coordinate analysis (Fig. 9). This diagram shows that the cluster analysis groupings conform to the similarities indicated by the ordination diagram insofar as the contours do not overlap one another at any of the five levels chosen. This degree of conformity between the two entirely different methods indicates that the pattern of similarities revealed are not just meaningless by-products of the numerical methods used.

(iii) Numerical cladistics. As well as the cluster and ordination methods for summarizing a matrix, it is possible to construct the most parsimonious phylogenetic tree and rootless network from the average taxonomic distance matrix, as used in the principle coordinate analysis described above. Both a Fitch and Margoliash

Table VI. Relative Mobilities of Haemoglobin Bands occurring in Gobioid Morphological Groups (Table I) and Cladistic Groupings of Fig. 1.

Morphological groups (Table I):	*A*	*B*	*C*	*D*	*E*	*F*	*G*
No. of band mobilities	10	9	1	4	2	33	29
$\bar{x}$	0.77	0.44	0.60	0.71	0.99	0.89	0.57
Range	0.47–1.03	0.09–0.69	–	0.42–1.02	0.91, 1.07	0.24–1.44	0.08–1.02
S. D.	0.18	0.20	–	0.27	0.11	0.26	0.24
Cladistic groups							
Fig. 1a:	*AB*	*CDEG*	*EG*	*CD*			
Fig. 1b:					*ADE*	*BG*	*DE*
No. of band mobilities	19	36	31	5	16	38	6
$\bar{x}$	0.61	0.61	0.60	0.69	0.78	0.54	0.80
Range	0.09–1.03	0.08–1.07	0.08–1.07	0.42–1.02	0.42–1.07	0.08–1.02	0.42–1.07
S. D.	0.25	0.25	0.26	0.24	0.20	0.24	0.26

(1967) network (Fig. 10) and a Wagner network (Farris, 1972) (Fig. 11) were computed, their "goodness of fit" being indicated by their percentage standard deviation (Fitch and Margoliash, 1967). Although these networks are not identical, many resemblances can be seen. Several alternative Fitch and Margoliash networks were also generated, with only slightly higher values for percentage standard deviation, but these were so similar as to make no difference to the present study.

Again, as inspection of Figs. 10 and 11 will show, these networks display little correspondence with either of the suggested phylogenetic schemes. Wagner networks were also computed using Manhattan, rather than Euclidean, distances, but once again they had little correspondence with either phylogenetic hypothesis.

CONCLUSIONS

Electrophoretic behaviour of haemoglobins would appear to be a feature worth examination in gobioid systematics concerned either with tokogenetic problems at the species level or those involving genetic comparison between allopatric populations within a postulated species. An account of electrophoretic patterns for haemoglobins, as well as for other proteins, would seem a worthwhile part of any modern description of a gobioid species, if the material available permits extraction in a suitable condition.

For answering questions of gobioid phylogeny, examination of haemoglobin electropherograms has not contributed much conclusive information, despite the employment of a number of different techniques of comparison. Lack of clear and consistent agreement in patterns of haemoglobin mobility with either of the proposed phylogenetic hypotheses might suggest that mobility patterns reflect adaptiveness in haemoglobins (discussed above) for particular environmental conditions, with the consequent chance of parallel evolution of similar haemoglobin properties in different phyletic lines. The natural habitats of the species studied are noted under Material and Methods. Groupings into marine, brackish and freshwater species, and into a cryptobenthic category (Miller, 1979), have therefore been indicated on the scatter of OTUs obtained by principle coordinate analysis (Fig. 12), but no obvious separation is revealed. It should be noted that this result may be due to

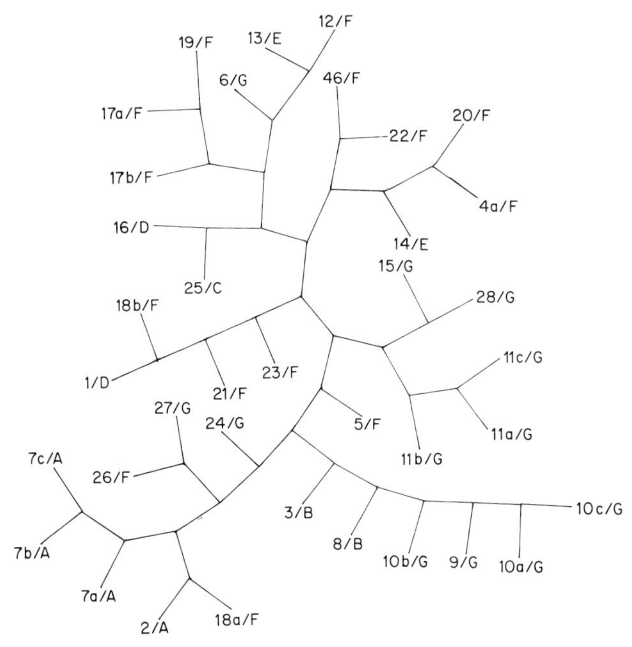

Fig. 10. Fitch-Margoliash rootless network derived from average taxonomic distance matrix used in principle coordinate analysis. Morphological groups as Table I. Percentage standard deviation 20.6.

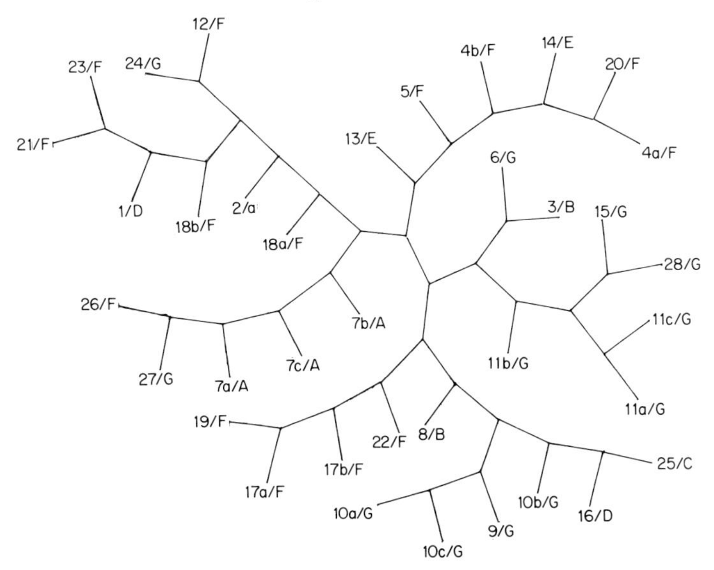

Fig. 11. Wagner rootless network derived and lettered as Fig. 10. Percentage standard deviation 29.5.

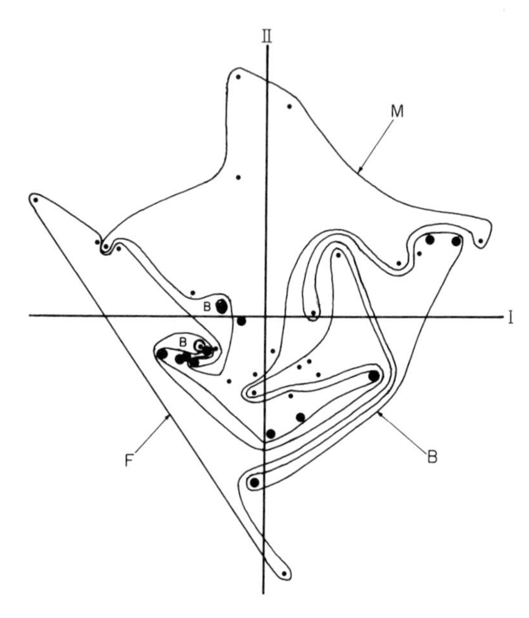

Fig. 12. Clustering of OTUs according to habitat in scatter diagram from principle coordinate analysis. M, marine; F, freshwater; B, brackish; cryptobenthic OTUs indicated by larger solid circle.

lack of precise data about microhabitat comparability, before an element of randomness or neutralism is suggested for these electrophoretic features of gobioid haemoglobins. Compared with the size of the suborder, the number of species which have been available for this work is very low and, with study of merely electrophoretic mobility of haemoglobins in a much larger proportion of the group, it is possible that more obvious correlation with phylogeny or habitat will be indicated. This does not require a high level of biochemical technology, and would seem to hold much promise as an introductory research programme to modern systematics in tropical areas where gobioid fishes are most diverse.

Electrophoretic mobility is now accepted as only a very partial measure of similarity between proteins (Avise, 1974). If biochemical criteria are to play an important role in the phylogenetic systematics of gobies, especially if most species remain inaccessible for study of these, then it may appear that a much more detailed investigation of structure in available haemoglobins, and other proteins, will be the only profitable course of research. For a more limited measure of genetic affinity, electrophoretic variation at a larger number of loci, encoding many enzymes as well as haemoglobins, might also be attempted in a similar way to that already exploited in other groups of teleosts (see, for example, Avise *et al.*, 1977). Whatever more elaborate project is undertaken for the gobioid fishes, this must take the form of an interdisciplinary exercise between the zoologist, in defining phyletic problems, the biochemist, in obtaining the necessary data base, and the statistician, in complex evaluation of the results. The emphasis here is on collaboration. In the context of the present symposium, systematics may be expected to be revitalized by biochemistry, but not revolutionized to the extent of excluding the traditional order.

ACKNOWLEDGEMENTS

For research studentships thanks are due to the Libyan Government (MYE-T) and to the Science Research Council (CJW). Further detailed acknowledgements have been made by El-Tawil (1974) and Webb (1977). We are grateful to Mrs Guise and Drs Friday and Peacock for computing some of the phyletic networks.

REFERENCES

Akihito, Prince (1969). A systematic examination of the gobiid fishes based on the mesopterygoid, postcleithra, branchiostegals, pelvic fins, scapula and suborbital. *Jap. J. Ichthyol.* **16**, 93–114.

Akihito, Prince and Meguro, K. (1977). Five species of the genus *Callogobius* found in Japan and their relationships. *Jap. J. Ichthyol.* **24**, 113–127.

Aurich, H. J. (1938). Die Gobiiden. *Int. Rev. Hydrobiol.* **38**, 125–183.

Avise, J. C. (1974). Systematic value of electrophoretic data. *Syst. Zool.* **23**, 465–481.

Avise, J. C. and Smith, M. H. (1974). Biochemical genetics of sunfish. I. Geographic variation and subspecific intergradation in the bluegill, *Lepomis macrochirus*. *Evolution* **28**, 42–56.

Avise, J. C., Straney, D. O. and Smith, M. H. (1977). Biochemical genetics of sunfish IV. Relationships of centrarchid genera. *Copeia* **1977**, 250–258.

Barlow, G. W. (1963). Species structure of the gobiid fish *Gillichthys mirabilis* from coastal sloughs of the eastern Pacific. *Pacific Sci.* **17**, 47–72.

Binotti, I., Giovenco, S., Giardina, B., Antonini, E., Brunori, M. and Wyman, J. (1971). Studies on the functional properties of fish haemoglobins. II. The oxygen equilibrium of the isolated haemoglobin components from trout blood. *Archs Biochem. Biophys.* **142**, 274–280.

Birdsong, R. S. (1975). The osteology of *Microgobius signatus* Poey (Pisces:

Gobiidae), with comments on other gobiid fishes. *Bull. Florida State Mus., Biol. Sci.,* **19,** 135–187.
Braunitzer, G., Hilse, K., Rudloff, V. and Hilschmann, H. (1964). The hemoglobins. *Adv. Protein Chem.* **19,** 1–71.
Buhler, D. R. (1963). Studies on fish hemoglobins. Chinook salmon and rainbow trout. *J. biol. Chem.* **238,** 1665–1674.
Cucchi, C. and Callegarini, C. Analisi elettroforetica delle emoglobine e delle globine di due distinte popolazioni di *Gobius fluviatilis* (Teleostea, Gobiidae). *Rc. Ist. lomb. Sci. Lett.,* Ser. B **103,** 276–280.
De Buen, F. (1928). El *Gobius niger* L. en aguas atlánticas y mediterráneas de Europa. *Not. Res. Inst. Esp. Oceanogr. Ser.* 2, No. 27, 32 pp.
De Ligny, W. (1969). Serological and biochemical studies on fish populations. *Oceanogr. Mar. Biol. Ann. Rev.* **7,** 411–513.
El-Tawil, M. Y. (1974). "The Application of Protein Electrophoresis to Problems of Phylogeny and Intraspecific Variation in Gobioid Fishes". Ph. D. Thesis, University of Bristol.
Fage, L. (1915). Sur quelques *Gobius* méditerranéens (*G. Kneri* Stndr., *G. elongatus* Canestr., *G. niger* L.). *Bull. Soc. zool. Fr.* **40,** 164–175.
Farris, J. S. (1972). Estimating phylogenetic trees from distance matrices. *Am. Nat.* **106,** 645–668.
Fitch, W. M. and Margoliash, E. (1967). Construction of phylogenetic trees. *Science* **155,** 279–284.
Fonds, M. (1970). Remarks on the rearing of gobies (*Pomatoschistus minutus* and *P. lozanoi*) for experimental purposes. *Helgoländer wiss. Meeresunters.* **20,** 620–628.
Fonds, M. (1971). The seasonal abundance and vertebral variation of *Pomatoschistus minutus minutus* and *lozanoi* (Gobiidae) in the Dutch Wadden Sea. *Vie Milieu,* Suppl. 22, 393–408.
Fonds, M. (1973). Sand gobies in the Dutch Wadden Sea (*Pomatoschistus,* Gobiidae, Pisces). *Neth. J. Sea Res.* **6,** 417–478.
Freihofer, W. C. (1970). Some nerve patterns and their systematic significance in paracanthopterygian, salmoniform, gobioid, and apogonid fishes. *Proc. Cal. Acad. Sci.* Ser. 4, **38,** 215–263.
Fyhn, U. E. H. and Sullivan, B. (1974). Haemoglobin polymorphism in fishes. I. Complex phenotypic patterns in the toadfish, *Opsanus tau. Biochem. Gen.* **11,** 373–385.
Garside, E. T. (1970). Structural responses. In "Marine Ecology" (O. Kinne, ed.), Vol. 1. Environmental Factors, Pt 1, Ch. 3, pp. 561–573.
Gillen, R. G. and Riggs, A. (1972). Structure and function of the haemoglobins of the carp, *Cyprinus carpio. J. biol. Chem.* **247,** 6039–6046.
Gillen, R. G. and Riggs, A. (1973a). The haemoglobins of a freshwater teleost, *Cichlasoma cyanoguttatum* (Baird and Giscard). II. Subunit structure and oxygen equilibria of the isolated components. *Archs Biochem. Biophys.* **154,** 348–359.
Gillen, R. G. and Riggs, A. (1973b). Structure and function of the isolated

hemoglobins of the American eel, *Anguilla rostrata. J. biol. Chem.* **248**, 1961–1969.

Gower, J. C. (1966). Some distance properties of latent root and vector methods used in multivariate analysis. *Biometrika* **53**, 325–338.

Griffiths, G. C. D. (1972). The phylogenetic classification of Diptera: Cyclorrapha. *Ser. Entomologica* 8, 1–340.

Harry, R. R. (1948). The gobies of the Indo-Malayan eleotrid genus *Bunaka. Proc. Calif. Zool. Club* **1**, 13–18.

Hashimoto, K. and Matsuura, F. (1962). Comparative studies on two hemoglobins of salmon – IV. N-terminal amino acid. *Bull. Jap. Soc. scient. Fish.* **28**, 914–919.

Hashimoto, K., Yamaguchi, Y. and Matsuura, F. (1960). Comparative studies on two hemoglobins of salmon – IV. Oxygen dissociation curve. *Bull. Jap. Soc. scient. Fish.* **26**, 827–834.

Hennig, W. (1966). "Phylogenetic Systematics". University of Illinois Press, Urbana.

Heslop-Harrison, J. (1968). Chairman's summing-up. *In* "Chemotaxonomy and Serotaxonomy" (J. G. Hawkes, ed.), Systematics Association Special Volume No. 2, pp. 279–284. Academic Press, London and New York.

Hilse, K. and Braunitzer, G. (1968). Die Aminosäresequenz der α-Ketten der beiden Hauptkomponenten des Karpfenhämoglobins. *Hoppe-Seyler's Z. Physiol. Chem.* **349**, 433–450.

Hoese, D. F. (1971). "A Revision of the Eastern Pacific Species of the Gobiid Fish Genus *Gobiosoma*, with a Discussion of Relationships of the Genus". Ph. D. Thesis, University of California, San Diego.

Hoese, D. F. and Allen, G. R. (1977). *Signigobius biocellatus,* a new genus and species of sand-dwelling coral reef gobiid fish from the western tropical Pacific. *Jap. J. Ichthyol.* **23**, 199–207.

Ingram, V. M. (1961). Gene evolution and the haemoglobins. *Nature* **189**, 704–708.

Johnson, M. S. (1975). Biochemical systematics of the atherinid genus *Menidia. Copeia* **1975**, 662–691.

Koch, H. J. A., Bergstrom, E. and Evans, J. C. (1966). A size-correlated shift in the proportion of the haemoglobin components of the Atlantic salmon (*Salmo salar* L.) and the sea trout (*Salmo trutta* L.) *Meded. K. vlaam. Acad.* **28**, 1–20.

Koehn, R. K. (1972). Genetic variation in the eel: a critique. *Mar. Biol.* **14**, 179–181.

Koumans, F. P. (1953). Gobioidea. *The Fishes of the Indo-Australian Archipelago.* Vol. 10.

Manwell, C. (1957). Alkaline denaturation of haemoglobin of postlarval and adult *Scorpaenichthys marmoratus. Science* **126**, 1175–1176.

Manwell, C. (1958a). A "fetal maternal shift" in the ovoviviparous spiny dogfish *Squalus suckleyi* (Girard). *Physiol. Zool.* **31**, 93–100.

Manwell, C. (1958b). Ontogeny of haemoglobin in the skate, *Raja binoculata. Science* **128**, 419–420.

Manwell, C. (1963a). The blood proteins of cyclostomes. A study in phylogenetic and ontogenetic biochemistry. *In* "The Biology of *Myxine*" (A. Brodal and R. Fange, eds.), pp. 372–455. Universitets forlaget, Oslo.

Manwell, C. (1963b). Fetal and adult haemoglobins of the spiny dogfish *Squalus suckleyi*. *Archs Biochem. Biophys.* **101**, 504–511.

Manwell, C. and Baker, C. M. A. (1967). Polymorphism of turbot hemoglobin: a "hybrid" hemoglobin molecule with three kinds of polypeptide chains. *Am. Zool.* **7**, 214.

Manwell, C. and Baker, C. M. A. (1970). "Molecular Biology and the Origin of Species". Sidgwick and Jackson, London.

Miller, P. J. (1973a). Gobiidae. *In* "Check-List of the Fishes of the North-Eastern Atlantic and of the Mediterranean". (J. C. Hureau and Th. Monod, eds.), **1**, 483–515; **2**, 320–321. UNESCO.

Miller, P. J. (1973b). The osteology and adaptive features of *Rhyacichthys aspro* (Teleostei: Gobioidei) and the classification of gobioid fishes. *J. Zool. Lond.* **171**, 397–434.

Miller, P. J. (1975). Age-structure and life-span in the common goby, *Pomatoschistus microps*. *J. Zool. Lond.* **177**, 425–448.

Miller, P. J. (1979). Adaptiveness and implications of small size in teleosts. *Symp. Zool. Soc. Lond.* **44**, 263–306.

Miller, P. J. and El-Tawil, M. Y. (1974). A multidisciplinary approach to a new species of *Gobius* (Teleostei: Gobiidae) from southern Cornwall. *J. Zool. Lond.* **174**, 539–574.

Miller, P. J. and Wongrat, P. (1979). A new goby (Teleostei: Gobiidae) from the South China Sea and its significance for gobioid classification. *Zool. J. Linn. Soc. Lond.* **67**, 239–257.

Nelson, J. S. (1976). "Fishes of the World". Wiley, New York.

Perez, J. E. and Maclean, N. (1974). Ontogenetic changes in haemoglobins in roach, *Rutilus rutilus* (L.) and rudd, *Scardinius erythrophthalmus* (L.) *J. Fish Biol.* **6**, 479–482.

Perez, J. E. and Maclean, N. (1975). Multiple globins and haemoglobins in four species of grey mullets (Mugilidae, Teleosta). *Comp. Biochem. Physiol.* **53B**, 465–468.

Perez, J. E. and Maclean, N. (1976a). Multiple globins and haemoglobins in the bass *Dicentrarchus labrax* (L.) (Serranidae: Teleostei). *J. Fish Biol.* **8**, 413–417.

Perez, J. E. and Maclean, N. (1976b). The haemoglobins of the fish *Sarotherodon mossambicus* (Peters): functional significance and ontogenetic changes. *J. Fish Biol.* **9**, 447–455.

Perutz, M., Muirhead, H., Cox, J. and Goaman, L. (1968). Three-dimensional Fourier synthesis of horse oxyhaemoglobin at 2.8 Å resolution: the atomic model. *Nature* **219**, 131–139.

Poluhowich, J. J. (1972). Adaptive significance of eel multiple haemoglobins. *Physiol. Zool.* **45**, 215–222.

Powers, D. A. (1972). Hemoglobin adaptation for fast and slow water habitats in sympatric catostomid fishes. *Science, N. Y.* **177**, 360–362.

Raunich, L., Battaglia, B., Callegarini, C. and Mozzi, C. (1967). Il polimorfismo emoglobinico del genere *Gobius* della Laguna di Venezia. *Atti Ist. Veneto Sci.* **125**, 87–105.

Regan, C. T. (1911). The osteology and classification of the gobioid fishes. *Ann. Mag. nat. Hist.*, Ser. 8, 8, 729–733.

Riggs, A. (1970). Properties of fish haemoglobins. *In* "Fish Physiology" (W. S. Hoar and D. J. Randall, eds), Vol. 4, pp. 209–252. Academic Press, London and New York.

Ronald, A. P. and Tsuyuki, H. (1971). The subunit structures and the molecular basis of the multiple hemoglobins of two species of trout, *Salmo gairdneri* and *S. clarki clarki*. *Comp. Biochem. Physiol.* **39B**, 195–202.

Rosenblatt, R. H. (1963). Some aspects of speciation in marine shore fishes. *Systematics Association Publication* No. **5**, 171–180.

Russell, F. S. (1976). "The Eggs and Planktonic Stages of British Marine Fishes". Academic Press, London and New York.

Sanzo, L. (1911). Distribuzione della papille cutanee (organi ciatiforme) e suo valore sistematico nei Gobi. *Mitt. Zool. Sta. Neapel.* **20**, 249–328.

Sharp, G. D. (1969). Electrophoretic studies of tuna haemoglobins. *Comp. Biochem. Physiol.* **31**, 749–755.

Sharp, G. D. (1973). An electrophoretic study of hemoglobins of some scombroid fishes and related forms. *Comp. Biochem. Physiol.* **44B**, 381–388.

Sibley, C. G. (1962). The comparative morphology of protein molecules as data for classification. *Syst. Zool.* **11**, 108–118.

Sick, K. (1961). Haemoglobin polymorphism in fishes. *Nature, Lond.* **192**, 894–896.

Sick, K. (1965a). Haemoglobin polymorphism of cod in the Baltic and the Danish Belt Sea. *Hereditas* **54**, 19–48.

Sick, K. (1965b). Haemoglobin polymorphism of cod in the North Sea and the North Atlantic Ocean. *Hereditas* **54**, 49–73.

Smith, I. (ed.) (1968). "Chromatographic and Electrophoretic Techniques". 2nd Edn. Heinemann, London.

Smith, J. L. B. (1958). The fishes of the family Eleotridae in the western Indian Ocean. *Ichthyol. Bull. Rhodes Univ.* **11**, 137–163.

Smithies, O. (1959). An improved procedure for starch-gel electrophoresis. Further variations in the serum proteins of normal individuals. *Biochem. J.* **71**, 585–587.

Sneath, P. H. A. and Sokal, R. R. (1973). "Numerical Taxonomy". Freeman, San Francisco.

Springer, V. G. (1978). Synonymization of the family Oxudercidae, with comments on the identity of *Apocryptes cantoris* Day (Pisces: Gobiidae). *Smiths. Contr. Zool.* **270**, 1–14.

Tsuyuki, H. and Gadd, R. E. A. (1963). The multiple hemoglobins of some members of the Salmonidae family. *Biochim. biophys. Acta* **71**, 219–221.

Tsuyuki, H. and Ronald, A. P. (1970). Existence in salmonid hemoglobins of molecular species with three and four different polypeptides. *J. Fish. Res. Bd Can.* **27**, 1325–1328.

Tsuyuki, H., Roberts, E. and Vanstone, W. E. (1965). Comparative zone electrophoresis of muscle myogens and blood haemoglobins of marine and freshwater vertebrates and their application to biochemical systematics. *J. Fish. Res. Bd Can.* **22**, 203–213.

Tsuyuki, H., Roberts, E., Kerr, R. H. and Ronald, A. P. (1966). Micro starch gel electrophoresis. *J. Fish. Res. Bd Can.* **23**, 929–933.

Tsuyuki, H., Roberts, E., Lowes, R. H., Hadaway, W. and Westrheim, S. J. (1968). Contribution of protein electrophoresis to rockfish (Scorpaenidae) systematics. *J. Fish. Res. Bd Can.* **25**, 2477–2501.

Vanstone, W. E., Roberts, E. and Tsuyuki, H. (1964). Changes in the multiple haemoglobin patterns of some Pacific salmon, genus *Oncorhynchus* during the parr-smolt transformation. *Can. J. Physiol. Pharmac.* **42**, 697–703.

Webb, C. J. (1977). "Systematics of the *Pomatoschistus minutus* complex (Teleostei: Gobioidei)". Ph. D. Thesis, University of Bristol.

Webb, C. J. (1980). Systematics of the *Pomatoschistus minutus* complex (Teleosteii Gobioidei). *Phil. Trans. Roy. Soc.* (B). In press:

Webb, C. J. and Miller, P. J. (1975). A redescription of *Pomatoschistus norvegicus* (Collett, 1903) (Teleostei: Gobioidei) based on syntype material. *J. Fish Biol.* **7**, 735–747.

Weber, R. E. and De Wilde, J. A. M. (1976). Multiple haemoglobins in plaice and flounder and their functional properties. *Comp. Biochem. Physiol.* **54B**, 433–437.

Wilkins, N. P. (1968). Multiple haemoglobins of the Atlantic salmon (*Salmo salar*). *J. Fish. Res. Bd Can.* **25**, 2651–2663.

Wilkins, N. P. (1970). The subunit composition of the haemoglobins of the Atlantic salmon (*Salmo salar* L.). *Biochim. biophys. Acta.* **214**, 52–63.

Wilkins, N. P. (1971). Haemoglobin polymorphism in cod, whiting and pollack in Scottish waters. *In* "Special Meeting on the Biochemical and Serological Identification of Fish Stocks (W. De Ligny, ed.). *Cons. Internat. Explor. Mer: Rapports et Proces-Verbaux* **161**, 60–63.

Wilkins, N. P. (1972). Biochemical genetics of the Atlantic salmon (*Salmo salar* L.) I. A review of recent studies. *J. Fish Biol.* **4**, 487–504.

Wilkins, N. P. and Iles, T. D. (1966). Haemoglobin polymorphism and its ontogeny in herring (*Clupea harengus*) and sprat (*Sprattus sprattus*). *Comp. Biochem. Physiol.* **17**, 1114–1158.

Williams, G. C., Koehn, R. K. and Mitton, J. B. (1973). Genetic differentiation without isolation in the American eel, *Anguilla rostrata. Evolution* **27**, 192–204.

Wongrat, P. (1977). "Systematics, Comparative Anatomy, and Phylogeny of Eleotrine Gobies (Teleostei: Gobioidei)". Ph. D. Thesis, University of Bristol.

Yamaguchi, K., Kochiyama, Y., Hashimoto, K. and Matsuura, F. (1962). Studies on multiple hemoglobins of eel. III. Oxygen dissociation curve and relative amounts of components F and S. *Bull. Jap. Soc. scient. Fish.* **28**, 192–200.

Yamaguchi, K., Kochiyama, Y., Hashimoto, K. and Matsuura, F. (1963). Studies on two hemoglobins of loach. II. Oxygen dissociation curve. *Bull. Jap. Soc. scient. Fish.* **29**, 180–188.

Yamanka, H., Yamaguchi, K. and Matsuura, F. (1965a). Starch gel electrophoresis of fish hemoglobins. I. Usefulness of cyanmethemoglobin for electrophoresis. *Bull. Jap. Soc. scient. Fish.* **31**, 827–832.

Yamanka, H., Yamaguchi, K. and Matsuura, F. (1965b). Starch gel electrophoresis of fish hemoglobins. II. Electrophoretic patterns of hemoglobin of various fishes. *Bull. Jap. Soc. scient. Fish.* **31**, 833–839.

Yamanka, H., Yamaguchi, K., Hashimoto, K. and Matsuura, F. (1967). Starch gel electrophoresis of fish haemoglobins. III. Salmonid fishes. *Bull. Jap. Soc. scient. Fish.* **33**, 195–203.

11 The Evaluation of Present Results and Future Possibilities of the Use of Amino Acid Sequence Data in Phylogenetic Studies with Specific Reference to Plant Proteins

D. BOULTER

Durham University, South Road, Durham DH1 3LE, England

Abstract: Amino acid sequence data sets for cytochrome *c*, plastocyanin and ferredoxin of flowering plants have now been accumulated. However, these are too incomplete and "noisy" to allow firm phylogenetic inferences to be made, even though some interesting similarities occur. Two courses of action should now be followed, firstly the development of more powerful data handling methods which overcome the limitations of existing ones and secondly sequence data from several different proteins from the same organisms should be determined; a method is suggested whereby suitable proteins could be quickly evaluated. Several examples are given in which molecular data might suggest taxonomic re-investigations.

INTRODUCTION

Initial results obtained by using amino acid sequences of vertebrate cytochrome *c* led to an outline of the phylogeny of the vertebrates which was similar to that derived from fossil evidence (Dayhoff, 1972). This very encouraging start was soon to change to a less satisfactory one as the results from other proteins were assembled. Amino acid sequence data sets of different proteins did not always

Systematics Association Special Volume No. 16, "Chemosystematics: Principles and Practice", edited by F. A. Bisby, J. G. Vaughan and C. A. Wright, 1980, pp. 235–240, Academic Press, London and New York.

lend themselves to the same phylogenetic interpretation or agree with the accepted phylogeny obtained mainly from fossil or morphological characters (de Jong *et al.*, 1977).

Whilst it is clear that facets of protein structure should provide excellent taxonomic characters, it now appears that they are subject to the same limitations of interpretation, when used to reconstruct phylogenetic relationships, as are other characters of present day organisms. The main reason for this in the case of amino acid sequence data sets is that they contain many parallel substitutions and in addition exhibit other features which may give rise to distortions of the correct phylogeny (see Boulter *et al.*, 1979, for discussion).

With very "robust" data, e.g. the vertebrate cytochrome *c* amino acid sequences, the present data handling methods, whether these be matrix or parsimony methods, give an accurate phylogenetic result even though only one protein from each member of the taxonomic group is used; for many other groups including the flowering plants, this is not so. Thus parsimony and compatibility methods are not statistically consistent if parallel substitutions are common in a data set, but are satisfactory when the latter are rare. Alternatively, maximum likelihood procedures, whilst giving satistically consistent results, suffer from the need to have an agreed probabilistic model of character evolution and this is usually not forthcoming. The limitations and future developments in data handling methodology have been discussed recently by Felsenstein (1978).

POSSIBLE FUTURE DEVELOPMENTS

Due to the availability of automatic sequence methods (see for example Haslett *et al.*, 1978), one cheerful aspect of the problem is that acquisition of amino acid sequence data has been speeded up considerably. In the near future we can expect that gene cloning and DNA sequencing will also be used in phylogenetic studies; these methods are even faster and have the additional advantage that they apply directly to the genetic information although the occurrence of "split" genes in eukaryotes is a complication.

There are now two possible courses of action open to the molecular biologist interested in flowering plant phylogeny, both of which should be followed. First, more sequence data should be accumulated using several different proteins from the same organisms and secondly more powerful data handling methods should be devised which give rise to fewer of the distortions mentioned previously.

In following the first course, it would be very advantageous to be able to decide, with a relatively small amino acid sequence data set, the potential usefulness of a particular protein. To this end, D. Peacock and co-workers have made use of the compatibility concept (LeQuesne, 1969) to determine the average percentage parallelism within a protein data set; this will vary for a particular protein with the species group involved. With eight sequences this method (Peacock *et al.*, unpublished) can give a good approximation as to the relative usefulness of different proteins in different groups.

Cytochrome *c* and plastocyanin of flowering plants have about 40% parallel substitutions and consequently cannot be used to show phylogenetic relationships of plants from widely separated families (Table I). Whether a suitable protein exists for widespread comparisons of flowering plants remains to be demonstrated, otherwise comparisons must be restricted initially within familial or tribal limits and only subsequently can interfamilial relationships be explored.

Table I. Proportion of parallelism in Molecular Data Sets

Protein	*Species Group*	*Parallelism 8 species sample (%)*
Cytochrome *c*	flowering plants (26)	38 ± 7
Plastocyanin[b]	flowering plants (77)	32 ± 4
Plastocyanin[b]	composites (21)	—
Myoglobin	mammals (15)	31 ± 5
Alpha - haemoglobin	vertebrates (16)	28 ± 13
Cytochrome *c*	vertebrates (40)	24 ± 8
Plastocyanin[b]	legumes (10)	14[a]

[a]All species used as one sample.
[b]First 40 residues used.
Numbers in brackets refer to number of sequences. Table data by courtesy of Dr D. Peacock.

However, a protein may vary in its usefulness from group to group; thus the average percentage parallelism is about 40% in the cytochrome *c* data set of flowering plants as compared to 30% in the vertebrates when the data are normalized to a 15 species sample.

Again, the plastocyanin sequences within the flowering plants generally have an average percentage parallelism of the same order as those within the Compositae, whereas in the legumes the average percentage parallelism is less than half that figure.

PRESENT POSITION WITH REGARD TO THE FLOWERING PLANTS

At least three separate intracellular genetic systems occur in higher plants, namely the nucleus, chloroplast and mitochondrion. A protein may be coded for by the genetic information of any of these and in some cases, for example, ribulosebiphosphate carboxylase, by two, i.e. the large sub-unit by the chloroplast and the small sub-unit by the nucleus. The precise evolutionary origins of these different genetic systems is still uncertain, for example it has been suggested that the organelles may have originated by endosymbiosis; furthermore it may be that the evolutionary pressures on organelle DNA differ from those of nuclear DNA and this could lead to different rates of evolution of two proteins from the same group of organisms according to their intracellular coding site. Since, however, these organelles have been established in the plant kingdom long before the origin of the flowering plants and since the evidence points to the fact that the chloroplasts from all flowering plants most likely originated via a common chloroplast ancestor, the use of chloroplast or mitochondrial proteins is perfectly justified in the present context.

With respect to the presently accepted flowering plant schema, for example those of Cronquist (1968), Takhtajan (1969) and Thorne (1968), it could be argued that the evidence for them has not been sufficiently well documented to allow a critical assessment as to their accuracy; of crucial importance is the evaluation of the more recently acquired angiosperm fossil data most of which is of pollen origin (Doyle, 1978). Whilst the present plastocyanin and cytochrome *c* data sets are too incomplete and "noisy" to allow mutual comparisons to be made, more recently acquired ferredoxin data (Fig. 1) show some important similarities to the cytochrome *c* data with respect to the position of *Spinacia* and the intermediate position of the monocotyledons relative to the dicotyledons.

Even so, at present it would be premature to make phylogenetic suggestions about the flowering plants with the available data and methodology; these must await improvements in data handling

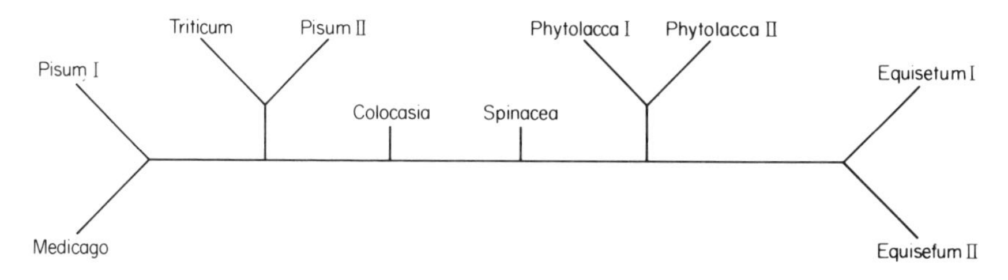

Fig. 1. Affinity tree of partial amino acid sequences of ferredoxin. I and II refer to two different ferredoxins – first 40 residues used. (Boulter *et al.*, unpublished data)

methods and the acquisition of additional useful sequence data. The latter must not be collected randomly, but from appropriate proteins as described earlier. For example, acquisition of more plastocyanin amino acid sequence data from taxonomically widely distributed flowering plants would simply introduce more "noise" into the data handling procedures. However, the molecular data can be used as an intellectual stimulus and in some cases as a basis for considering possible taxonomic re-investigations (Bremer and Wanntorp, 1978) as, for example, the position of the Centrospermae relative to the other flowering plants in view of the cytochrome *c* data set (Boulter, 1976, and also higher plant ferredoxins, Boulter, *et al.*, unpublished and Fig. 1), the relationship of the Compositae and Solanaceae (see the plastocyanin data set in Boulter *et al.*, 1979), the relatively isolated position of the genus *Phaseolus* relative to other legumes investigated (plastocyanin data set, Boulter *et al.*, 1979) and the very close relationship between the tribes of Chicoreae and the Cynareae in the Compositae (plastocyanin data set, Boulter *et al.*, 1978).

REFERENCES

Boulter, D. (1976). The evolution of plant proteins with special reference to higher plant cytochromes *c*. *In* "Commentaries in Plant Science" (H. Smith, ed.), pp. 77–91. Pergamon, Oxford.

Boulter, D., Gleaves, J. T., Haslett, B. G. and Peacock, D. (1978). The relationships of 8 tribes of the compositae. *Phytochemistry* 17, 1585–1589.

Boulter, D., Peacock, D., Guise, A., Gleaves, J. T. and Estabrook, G. (1979). Relationships between the partial amino acid sequences of plastocyanin from members of ten families of flowering plants. *Phytochemistry* **18**, 603–608.

Bremer, K., and Wanntorp H. -E. (1978). Phylogenetic systematics in botany. *Taxon* **27**, 317–329.

Cronquist, A. (1968). "The evolution and classification of flowering plants". Thomas Nelson, London.

Dayhoff, M. O. (1972). "Atlas of Protein Sequence and Structure". Vol. 5. National Biomedical Research Foundation, Silver Spring, Maryland, U.S.A.

Doyle, J. A. (1978). Origin of angiosperms. *Ann. Rev. Ecol. Syst.* **9**, 365–392.

Felsenstein, J. (1978). Cases in which parsimony or compatibility methods will be positively misleading. *Syst. Zool.* **27**, 401–410.

Haslett, B. G., Evans, I. M. and Boulter, D. (1978). Amino acid sequence of plastocyanin from *Solanum crispum* using automatic methods. *Phytochemistry* **17**, 735–739.

de Jong, W. W., Gleaves, J. T. and Boulter, D. (1977). Evolutionary changes of α-crystallin and the phylogeny of mammalian orders. *J. mol. Evol.* **10**, 123–135.

LeQuesne, W. J. (1969). A method of selection of characters in numerical taxonomy. *Syst. Zool.* **18**, 201–205.

Takhtajan, A. (1969). "Flowering Plants – Origin and Dispersal". Oliver and Boyd, Edinburgh.

Thorne, R. F. (1968). Synopsis of a putatively phylogenetic classification of flowering plants. *Aliso* **6**, 57–66.

12 Problems in the use of DNA for the Study of Species Relationships and the Evolutionary Significance of Genomic Differences

G. A. DOVER

Department of Genetics, Cambridge University, Cambridge CB2 3EH, England

Abstract: Until recently, it has been thought that the non-coding sequences of higher organisms were not subject to biological constraints and hence were free to drift. This assumption was based on measurements of sequence divergence between pairs of species that indicated that the degree of divergence reflected the time elapsed between the species: a parameter that went into the construction of phylogenies.

It is now known that single copy and repetitive families of DNA that are shared between groups of related species show considerable modulation in size and degree of heterogeneity. In addition there is often a fine intercalation of sequences that gives each genome a unique and characteristic arrangement.

The relative conservation of families of sequences and their unusual wide distribution does not facilitate the arrangement of simple bifurcating lineages in phylogenies that are based on ancestral and diverged characters. However, species relationships can be described in terms of common "libraries" of sequences that are free to undergo arbitrary amplifiation and interspersion within each genome.

The data are presented from "libraries" of similar sequences of middle-repetitive and high-repetitive (satellite) DNA that differ in abundance and organization in several species groups. Reference is made to models of quirks of DNA replication as suggestions for the concomitant amplification and interspersion of sequences.

An argument is developed that the fine differences in genome organization,

Systematics Association Special Volume No. 16, "Chemosystematics: Principles and Practice", edited by F. A. Bisby, J. G. Vaughan and C. A. Wright, 1980, pp. 241–268, Academic Press, London and New York.

although born of accidental causes, may generate sufficient biological novelty that will allow for greater flexibility during species differentiation.

INTRODUCTION: GENOMES AND SPECIES

The intrinsic interest in the science of systematics must reside in the light it throws on the phylogenetic relationships between species; and by this I am referring to the extrapolation of hypotheses from the phenetic data as to the possible course of evolution in a group of species. Naturally, we cannot reconstruct the myriad of events that have gone into shaping the contemporary pattern of relationships of a species group; nor can we realistically test, because of the magnitude of the evolutionary timescales, our hypotheses as to the prime determinative forces that have been involved. Nevertheless, the reconstruction of phylogenies is a flourishing business and the temporal ordering of characters, if not intuitively guessed at, is usually based either on the fossil record or on timing mechanisms intrinsic to some part of the organism, such as the rates of change of particular macromolecules.

My purpose in this chapter is first to assess the feasibility of using changes in the sequences of DNA in the reconstruction of phylogenies and secondly, to examine the significance of the large and often abrupt changes that are observed in the genomes of closely related species.

With regards to the first point it is clear that for the majority of groups there is little or no fossil evidence from which qualities of "primitive" or "advanced" may be extracted and imposed on the characters that have gone into the make-up of the phenetic relationships. In the absence of any independent measure of time we need to consider whether there is any constancy in the rate of change of DNA that would imply the working of a biological clock. In addition, we need to consider whether we can distinguish between primitive and derived character states (ancestral and diverged sequences) to satisfy the criteria for the construction of monophyletic lineages *sensu* Hennig (1966). The concept of biological clocks arose from the supposed constancies in the rates of substitutions of amino acids in protein sequences. This is not the place to examine in detail the validity of the clock concept with respect to proteins, for the issue is intricately enmeshed with the debate on the "selective" or "neutral" outcome of the substitutions. It is relevant, however, to the remarks I shall make concerning the changes occurring in DNA, to point out that although there might be some evidence that rates of protein change are constant, these rates are known to be very different for different proteins. These differences in rate have been explained as possibly due to the differences in the proportion of dispensible residues in the proteins and to differences in the degree of expendibility of the protein to the organism (Wilson *et al.*, 1977). Whatever might be the nature of the functional constraints, it clearly leaves open the question of whether the constancy of change is due only to stochastic processes, or whether irregular selective forces have also played their part. In assessing the rates of divergence of the majority of sequences of non-coding DNA, a similar confusion arises, not only because of the possibility of confounding drift with selection, but also because there is, probably, a third factor operating that uniquely contributes to the observed levels of DNA divergence. This other component of sequence change, that has the capacity to periodically generate sequence homogeneity (see later), confuses the whole issue of primitive and derived states and needs very careful consideration before cladistic lineages can be drawn from DNA sequence similarities. Sneath and Sokal (1973) rightly emphasized the circularity often involved in the establishment of phylogenies in that judgements concerning phenetic similarities of characters have first to be made, for example, on organisms taken from different geological strata before a temporal order can be imposed on the phenetic data. Although at first sight it might appear that DNA sequences escape from this difficulty, in that it is often assumed that the non-coding sequences are drifting and diverging at random and hence have a built-in time measuring device, I shall develop the argument that the irregularities and unusual quirks of change of DNA are such that no simple answers can be derived from these sequences to the evolutionary problems we set ourselves.

Indeed these same peculiarities in the types of evolutionary change that have occurred in DNA relate to the second major point I have raised above in connection with the large and abrupt changes in the genomes of higher organisms. It is a sobering thought, but one that nevertheless tends to be forgotten and merits some reiteration, that there is no genetic theory of speciation. There is, however, a consensus of opinion that the differentiation of species is the continuation of the gradual accumulation of genetic differences that make up the adaptive responses of groups of individuals to different environments. This has been the majority view not only because Darwin, unnecessarily, equated gradualism in speciation with the concept of natural selection but also because there has been an overriding preoccupation this century with the minutae of phenotypic polymorphisms. Naturally this preoccupation has coloured our thinking concerning evolutionary changes, whether genetical or systematical. It might be, however, that there are biological differences between species that were the cause of species differentiation that are unrelated to the allelic differences between individuals and between populations. Such an idea of abrupt or spasmodic changes that have lead to species formation is not new and originated during Darwin's time because of the obvious and unaccountable discontinuities in the fossil record. It has enjoyed continual advocacy, in various forms, through the ideas of Mayr (1963), White (1973), Gould (1977) and Wilson *et al.* (1977). Particularly, it is an idea that holds no surprises for botanists who are well aware of the novel possibilities of species formation associated with polyploidy and other forms of accidental chromosome changes. The question we need to ask is whether there is a general phenomenon of abrupt and biologically significant changes in the genomes of plants and animals: changes that are still retained within the genomes and amenable to analysis.

A cursory examination of the genomes and chromosomes of higher organisms reveals that changes are numerous and often large. I do not intend to review all the data that are available from studies in different genera but instead to use selected examples of interspecific comparisons to indicate the proposed mechanisms of occurrence of differences in sequence organization: and to relate these to the points I have raised concerning the evolutionary ties between species.

GENERAL DATA AND DISCUSSION: DIVERGENCE AND CONSERVATION

The greater proportion of the DNA of higher organisms consists of non-coding sequences that are often discernible as families of similar sequences of defined length. Recent reviews of the spectrum of sequences of non-coding DNA of higher organisms are available: Davidson and Britten (1973); Skinner (1977); Jones (1978); Appels and Peacock (1978); John and Miklos (1979); plus a variety of papers on interspecific DNA differences in Smith (1972) and Cold Spring Harbor Symposia on Quantitative Biology (1973, 1977). There is a fine gradation of family size ranging from single member families (single-copy DNA) to the most abundant families of up to several million members, (highly-repetitive DNA). It is only recently, with the use of carefully controlled reannealing experiments of cloned sequences, that accurate sizes can be determined of the many families that go to make up the middle-repetitive DNAs. Previously, the experimental techniques that were available to separate and distinguish families have not been of high resolution, except for some of the highly repetitive DNAs that are amenable to isolation by caesium density centrifugation (satellite DNAs), and some proportion of the single-copy DNA. Because of these difficulties, the earlier literature on interspecific comparisons of family sequence and organization is hard to interpret.

Many of the earlier investigations into the differences in sequences of the non-coding single-copy DNA relied on the ability of single strands of near homologous sequences to reanneal (under a standard stringency of reaction) and to form a heteroduplex. The extent of heteroduplex formation is a measure of the relative abundancies in different genomes of homologous sequences. In addition, the thermal stability of the heteroduplex is a reflection of the degree of base-pair mismatching that occurs during its formation, which is a measure of the extent of sequence divergence between the reannealed sequences. Two papers that describe the usefulness of such techniques and also critically evaluate the inherent difficulties in interpretation are available (Laird *et al.*, 1969; Walker, 1969).

In general it became apparent, at this level of analysis, that the greater the phylogenetic distance between two species the greater is the extent of sequence divergence (mismatching) between ever decreasing amounts of heteroduplex. Such an apparent constant rate of change over long periods of time has been invoked recently as further evidence for the workings of a biological DNA clock (Wilson *et al.*, 1977). However, one or two seemingly paradoxical results disturbed the time-dependent process of divergence. In describing the linear relationship between sequence similarity and evolutionary divergence in a variety of comparisons, Hoyer *et al.* (1965) pointed out that at first glance there would appear to be "a random 'decay' process, the correspondance between genes (read DNA) decaying exponentially with a half-life of one hundred million years". However, there appeared in their data a component of DNA that was common to several distantly related orders of mammals, fish and birds which had persisted relatively unchanged over long periods of time. This persistance is unexpected on the assumption that changes are occurring at random throughout the genome. Hoyer *et al.* (1965) suggested that either there is a variable "distribution of mutability" in the genome or that new sequences are being added as a replenishment for lost "decayed" genes. Since that time several more examples of persistent genomic components that are common to distantly related species and that are relatively conserved in sequence have been described for both single-copy DNA and repetitive DNA (Graham and Skinner, 1973; Gall and Atherton, 1974; Angerer *et al.,* 1976; Mizuno *et al.*, 1976; Harpold and Craig, 1977, 1978; Peacock *et al.,* 1977; Gosden *et al.,* 1977; Fry and Salser, 1977; Flavell *et al.,* 1977; Moore *et al.,* 1978; Barnes *et al.,* 1978; Manuelidis and Wu, 1978; for review see Dover, 1977). The significance of the existence of such widely distributed homologous sequences, not all of which can be coding sequences, and their relevance to the problems I raised in the introduction to this chapter are discussed in the next section dealing with particular data.

Another interesting finding that I shall introduce now, for it has bearing on the problem of the rate of genome evolution and speciation, concerns the lack of correlation between the degree of morphological dissimilarities between species and the extent of DNA sequence divergence. For example, there is as great a degree of sequence divergence in species of frogs that are morphologically so similar that they are included in the one order (Anura) as there is between the morphologically very different, and more recently evolved orders, of placental mammals (King and Wilson, 1975). Furthermore, there is a greater degree of sequence divergence between morphologically indistinct sibling species of *Drosophila* than there is between man and chimpanzee. Although it is not always possible, in the absence of a measure of the elapsed time between different species, to calculate the rate of divergence in different groups, it is clear that careful consideration needs to be given to the variety of mechanisms (whether accidental or selective) that affect the levels of sequence similarity, before too much biological significance is extrapolated from the data.

PARTICULAR DATA AND DISCUSSION

I shall describe in detail selected examples of recent data on interspecific comparisons of satellite and middle-repetitive DNA, in order to illustrate the types of sequence changes that can occur during the evolution of the higher genome.

1. Species Distributions of Satellite DNAs and the Concept of a "Library"

The easiest components of DNA to isolate and to examine are the highly repetitive families of the satellite DNAs. Genomes can be fractionated into several satellite and main-band components by a variety of antibiotic caesium salt density gradients. The antibiotics have the property of differentially altering the buoyant densities of families that differ in sequence (%GC) and complexity; complexity is a parameter related to the kinetics of reassociation and can be understood as the minimum repeating length of reassociation of a family. Figure 1 illustrates the theoretical resolution of components a′, a″, b, c, d and main-band using neutral CsCl, actinomycin D, distamycin A and Hoechst 33258. Actinomycin D and Distamycin A can also discriminate between different main-band components (Thiery *et al.*, 1976; Barnes *et al.*, 1978; Dover, 1980). However, the sequence composition of these fractions is not understood and they are not included in Fig. 1. The ease of use of Hoechst 33258 is shown in Fig. 2 that depicts three UV fluorescing light-density bands of AT-rich satellites of *D. melanogaster.* Each fraction has been sedimented to equilibrium in neutral CsCl analytical gradients, after removal of the dye, and densities are given relative to a marker on the right of the figure. An example of a full analysis of the satellite and main-band components of *Drosophila simulans* in Actinomycin D and Distamycin A is given in Fig. 3 (all experimental methods can be obtained in Manuelidis, 1977 and Barnes *et al.*, 1978).

Using these analytical methods the genomes of seven of the sibling species of the *melanogaster* species subgroup have been fractionated and characterized with respect to the numbers, types and amounts of satellite DNAs (Barnes *et al.*, 1978; Dover, 1980). The seven genomes have been resolved into over 15 different satellites that are either unique to a species or common to two or more species. In addition, the karyotypes of the seven species have been compared

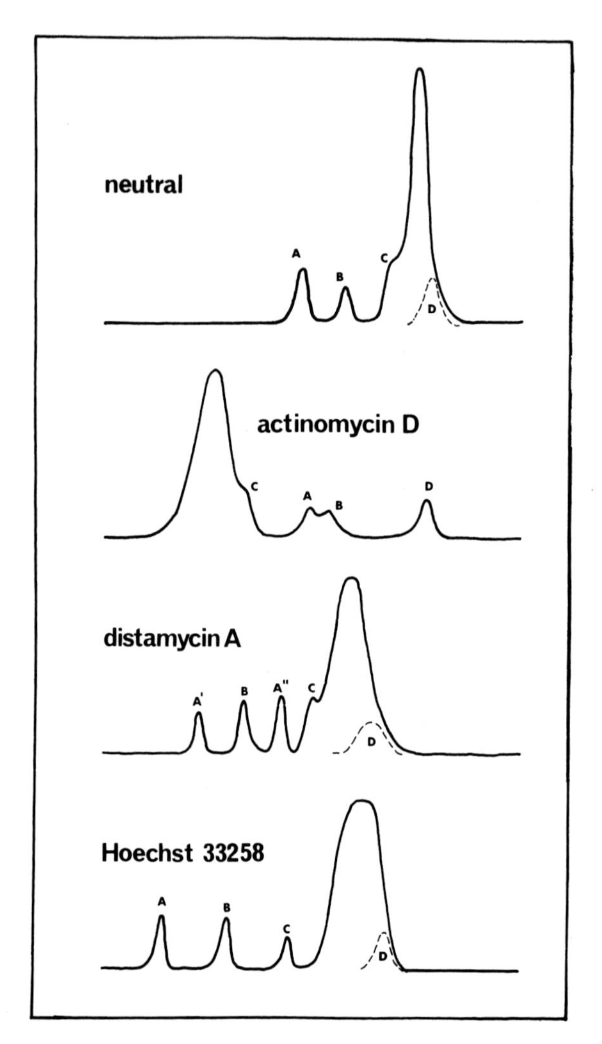

Fig. 1. A representation of differential separation of satellite DNA components a′, a″, b, c and d from main-band DNA using neutral, actinomycin D, distamycin A and Hoechst 33258 caesium chloride density gradients (see text). Peaks represent optical densities, under UV illumination of DNA molecules at equilibrium in gradients increasing in density from left to right. (For experimental methods see Barnes *et al.*, 1978 and Dover, 1980).

on the basis of polytene inversions and G-, C- and Q-staining reactions of regions of the chromosomes (Lemeunier and Ashburner, 1976; Lemeunier *et al.*, 1978); a pattern of relationships within the group

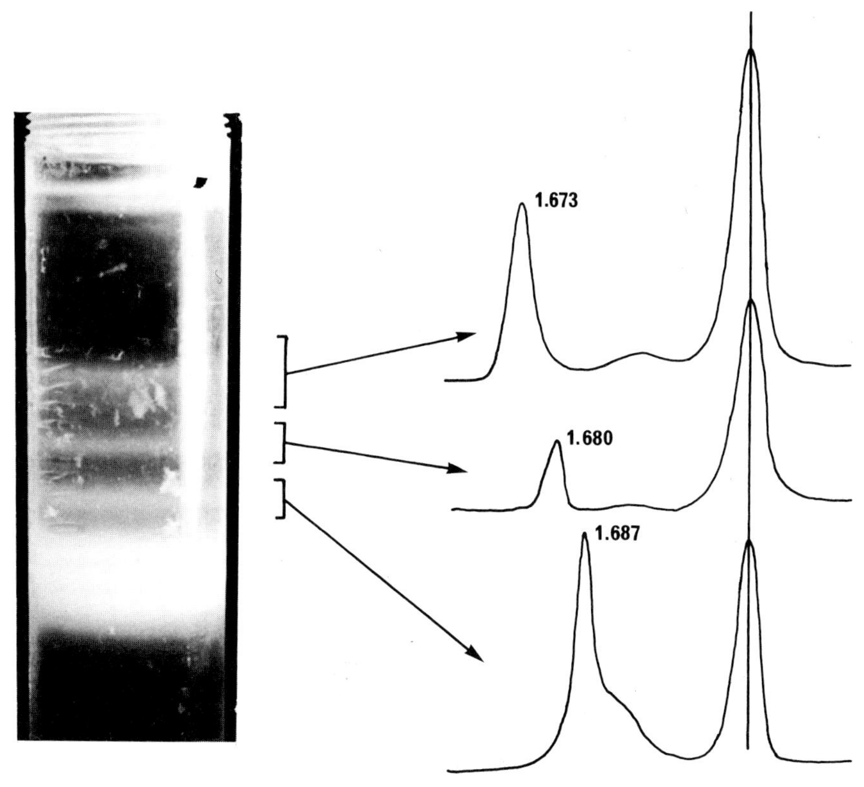

Fig. 2. Hoechst 33258 CsCl density gradient of total DNA of *D. melanogaster.* Three satellite DNA bands and a broad main-band DNA are fluorescing under UV illumination (left). Bouyant densities of individual bands (right) are obtained relative to the marker DNA *Pseudomonas aeruginosa* (vertical line) after removal of the Hoechst and sedimentation to equilibrium in an analytical centrifuge.

has been established. Given the data on the distribution of shared satellites in the genomes, we need to ask if it is feasible to construct a branching pattern of cladistic relationships between the species, using each satellite as a character with simple alternative states of presence/absence and basing the pattern of lineages on Hennig's method of ancestral and derived states.

In order to illustrate the difficulties inherent in such an exercise I have modified (Fig. 4) a theoretical cladogram given in Sneath and

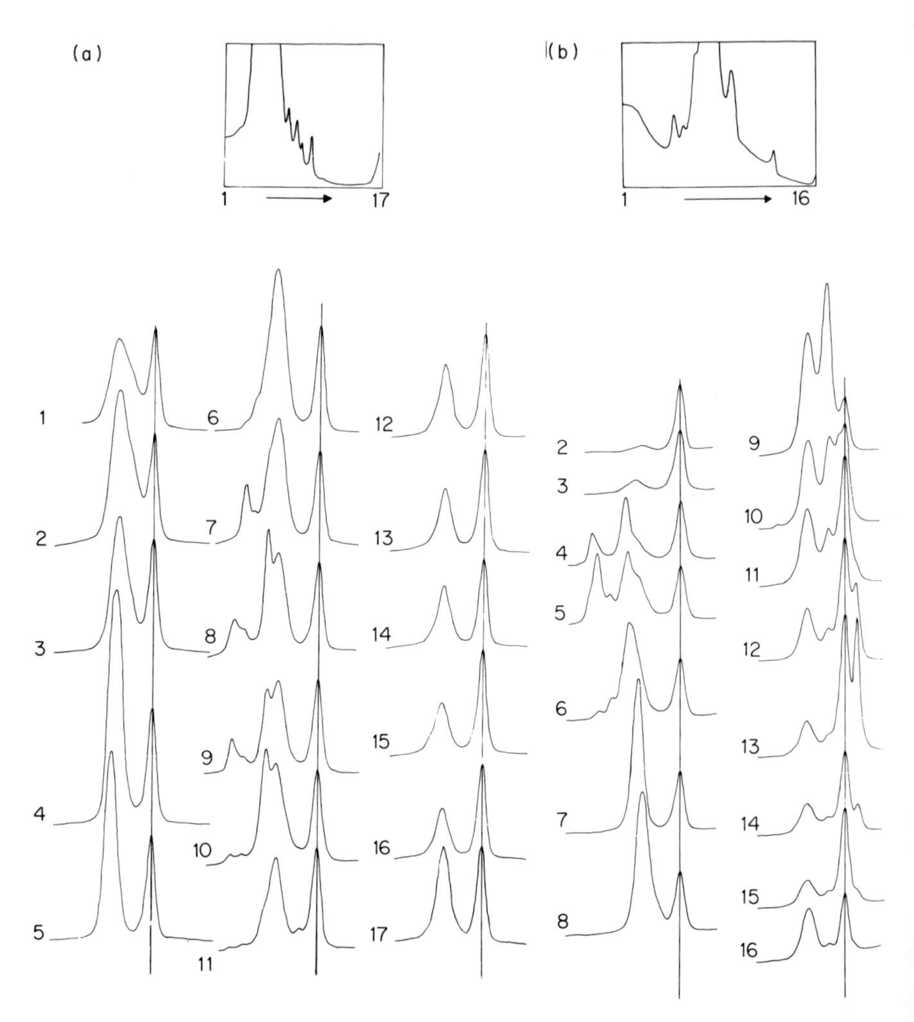

Fig. 3. An example of the combined use of actinomycin D- and distamycin A-CsCl density centrifugation for an analysis of highly repetitive components (satellites) of a genome (*Drosophila simulans*). Preparative gradients (cf. Fig. 1) are at the top: (a) actinomycin D; (b) distamycin A. Traces at the bottom are the sedimentation profiles in neutral CsCl "analytical" gradients (cf. Fig. 2) after removal of the antibiotic. Vertical line represents the marker DNA (Barnes *et al.*, 1978).

Sokal (1973) that is more in keeping with the observed distribution of satellites. With reference to Fig. 4 it is possible to delimit a satellite (character 1) that is common to all species and that could be

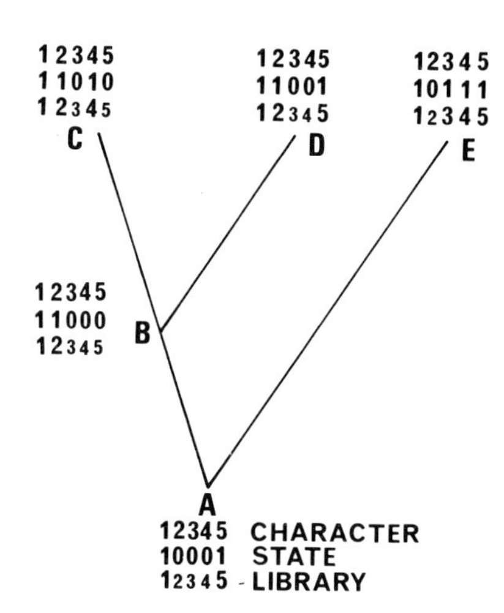

Fig. 4. Theoretical cladogram of species A, B, C, D and E adapted from Sneath and Sokal (1973). Characters 1–5 represent satellite DNA or middle-repetitive DNA families. State 1 (presence); state 0 (absence). Library sequences that are abundant (amplified) have large numerals. Library sequences in single or few copies have small numerals. Note characters 4 and 5 do not "fit" the cladogram based on the distributions of characters 1, 2 and 3 (see text).

described as ancestral. Satellite character 2, in being present in species B, C and D and not in A and E, can be assumed to have arisen in the lineage leading to B, C and D. Similarly, satellite 3 would have arisen in the lineage leading to E after divergence from B, C and D. However, the distribution of satellite characters 4 and 5 are unusual and novel with respect to the types of changes, depicted in Sneath and Sokal's original scheme. Satellite 4, for example, appears in species C and E and not D, B and A. If the five species are arranged as given, *according to the distribution of characters 1, 2 and 3,* then 4 could be described as a case of convergent evolution in the two more distantly related species of C and E. This is not the same as parallel evolution in which C and D might have converged on the same character state for satellite 4. The distribution of satellite 5 is even more unusual (*relative to the assumed cladogram of figure 4*) in that it appears in A (the ancestral species) and in both lineages leading to C, B, D and E, although it is missing in B and C. Clearly as the number of such supposed anomalies increases the construction of a phylogeny becomes increasingly arbitrary. However, one alternative way of qualifying the spatial relationships between the species is to return to the phenetic similarities contained in the data and to arrive at what could be described as a half-way-house between a phenetic and cladistic arrangement (Fig. 5). Each circle in the figure represents a satellite, the radius of which is proportional to the number of

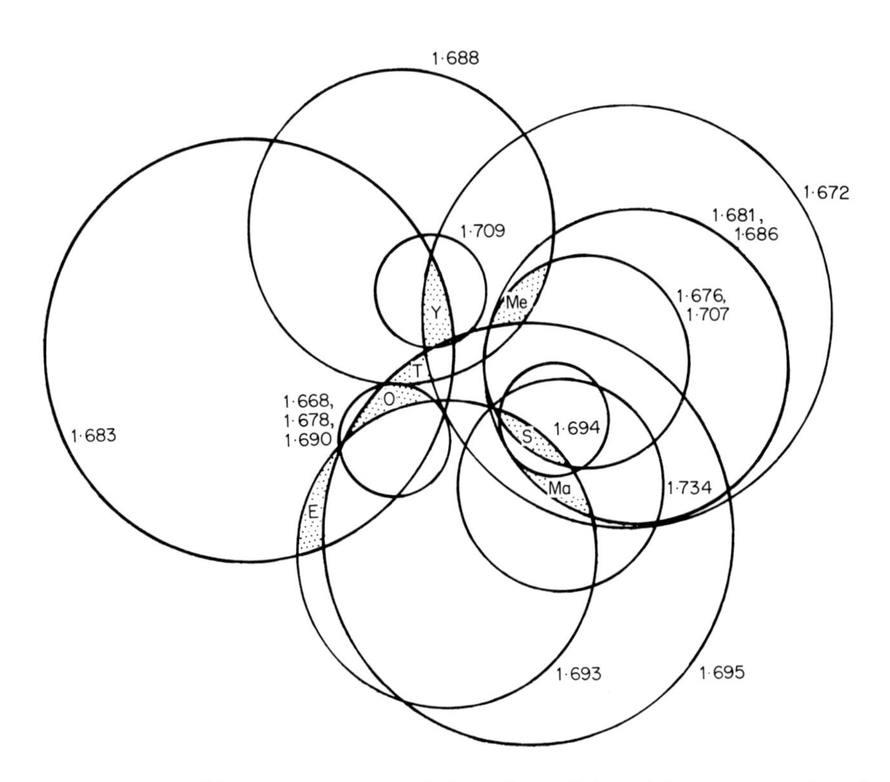

Fig. 5. One possible representation of the relationships of the seven species of the *melanogaster* species subgroup based on the numbers of satellite DNAs held in common. Circles represent satellites, the size of which are determined by the numbers of species sharing a satellite. Although the absolute positions of circles is not fixed, they are restricted to relative positions that in turn fix the relative spatial relationships between the species. (See text and also Barnes *et al.*, 1978, for a full discussion.) Numbers are buoyant densities. Me = *D. melanogaster;* Ma = *D. mauritiana;* S = *D. simulans;* Y = *D. yakuba;* T = *D. teissieri;* O = *D. orena;* E = *D. erecta.*

species sharing the satellite. By arranging a series of overlapping circles each species can be confined to the areas of intersection of those satellite circles which are represented in a species. Although the positions of overlap, the areas of overlap and the distances between the species are variable, the spatial relationships between the species are fixed. There is, of course, a circularity involved in the logic of this scheme because it is necessary to intuitively arrange the pattern of overlaps from prior knowledge of the combination of satellites in each species. It would have been better to have arrived, statistically, at this particular subset of overlaps from the total number of possible combinations of the given satellites in a number of theoretical species. The number of actual species involved, however, is too small for this procedure to be followed. I do not want to digress too far into what exactly it is that is being described in Fig. 5 for as I shall point out later only minimal significance can be attached to the sharing of a genomic component between two species, that is defined by its buoyant density alone. However, it is worth indicating that we have probably stumbled on a method of defining species relationships using a criterion of similarity of shared satellites that are able, without the use of cluster analysis, to reveal two subclusters. One group comprises the species *D. melanogaster, D. simulans* and *D. mauritiana* that on all other criteria of morphology, interfertility and cytology are very close together; and similarly the species relationships in the other group of *D. erecta, D. yakuba, D. tiessieri* and *D. orena* do not seriously disturb the relationships previously based on morphology and ecology. *D. mauritiana* would come close to being the "hypothetical median organism" and *D. orena* and *D. erecta* are distantly removed from the cluster centre, as they are on other criteria. Second and third glances, however, reveal alternative groupings which underlines the point, at this level of pictorial analysis, that there is a limit to the amount of dissection that can be performed. It should also be stressed that no dimension or direction of time can be imposed on the arrangement; although there is an interesting close parallel between the pattern based on satellites and that based on polytene inversions, to which I shall return later.

The inability to construct a branching monophyletic relationship on the lines of Fig. 4 has led us to examine the existence of "libraries" of sequences that are common to the genomes of all seven of the sibling species. The term "library" was first introduced by Fry and Salser (1977) to describe the sharing of satellites that are similar in sequence in distantly related rodents. The library is a useful analogy when precisely defined. Sequences that are held in common by a group of species, in single or a few copies, would be classified as library sequences in that they would represent the same books held in the different genomic libraries. The amplification of a library sequence to a repetitive family would be equivalent to the reprinting of a book (or new edition, if some sequence divergence had occurred) in the library of a particular genome.

Using this concept of a library, it is now possible to offer an explanation for the anomalies described in Fig. 4. For example, some of the satellite distributions (satellites 1, 2 and 3) would fix the overall shape of the cladogram in that they would be examples of derived states from previously amplified sequences in ancestral species. On the other hand the unusual distributions of satellite 4 and 5 could reflect a series of possibly arbitrary amplifications of library sequences in individual species. One could perhaps argue from the parallel pattern of species relationships derived from satellites, polytene inversions and morphology in the *melanogaster* subgroup that there is some order of time dependance to the amplification process and that satellites do not arise by chance. We would need, however, to increase the numbers of satellites and species in the analysis of the group before meaningful answers can be given to these sorts of suggestions.

Before we hazard a guess as to the effect libraries might have on the construction of phylogenetic relationships we need to have answers to some important questions. Are library sequences a general phenomenon? Where do individual libraries begin and end with respect to taxonomic groupings? How are the starting sequences, prior to amplification, maintained homogeneous over long periods of time?

2. Tests of Homology

In the *melanogaster* subgroup there is evidence for the existence of sequences in several genomes that are capable of entering a stable heteroduplex with a probe taken from a satellite component of one of the species (Peacock *et al.,* 1978; Dover, 1980). The homologous library sequences are not necessarily the "common" satellites that share the same buoyant density as the probe, for we have evidence that the homologous sequences in a particular species are differently composed both with respect to the organization of the probe sequences and the organization of the common satellite of that species (Dover, Strachan and Brown, 1980).

Similarities in sequence organization of satellite libraries can be revealed with the use of restriction endonucleases that cut double-stranded DNA at specific sites of dyad symmetries. If a restriction site occurs in a repeated family then a tandemly arranged array of repeats will generate a band of fragments of similar size in an electrophoretic gel. Individual satellite DNAs can have patterns of bands that are a reflection of the distribution of restriction sites within the arrays. From the distribution of sites and other sequence parameters of the repeats, an assessment can be made as to the possible course of evolution followed by a particular family. In this way, for example, Southern (1975a) and Horz and Zachau (1977) were able to calculate the numbers of cycles of amplification and intervening levels of divergence that might have given rise to the contemporary distribution of sites in the satellite DNA of *Mus musculus*. In Fig. 6, the results are presented of the very similar organization of library sequences in the closely related species of *Mus spretus* that are homologus to a probe taken from the satellite DNA of *M. musculus*. The levels of hybridization indicate that the "*musculus*" sequences in *M. spretus* are in relatively much reduced amounts; and in addition a variety of antibiotic caesium density gradients have failed to reveal the presence of an equivalent large satellite DNA in the *M. spretus* genome (Brown and Dover, 1980). The procedures adopted in these experiments were pioneered by E. Southern (1975b) and are popularly referred to as the "Southern" transfer or hybridization method. Total DNAs of *M. musculus* and *M. spretus* are cut independently by a variety of restriction enzymes that are known to produce specific patterns of fragments from the repeated family of the satellite probe. After digestion, the DNAs are electrophoresed and transferred to nitrocellulose paper where they are exposed, under hybridization conditions, to the radioactive probe. Exposure of the autoradiogram reveals bands at points of hybridization, and the sizes of the homologous sequences in *M. spretus* can be compared to those produced by the *M. musculus* probe alone when digested by the same enzymes. *M. musculus* satellite is cut almost

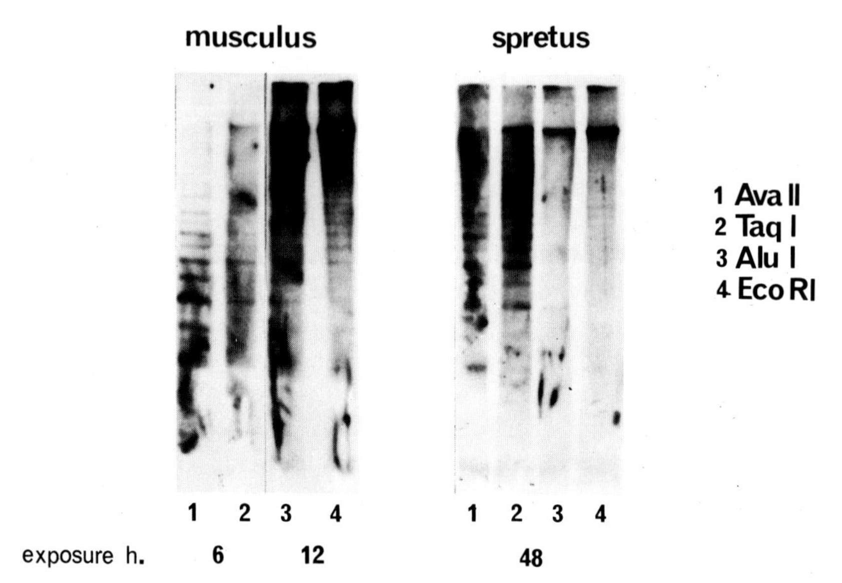

Fig. 6. Autoradiographs of "Southern" hybridization (see text and Southern, 1975b) of P^{32} nick-translated satellite DNA of *Mus musculus* against total DNA of *M. musculus* (left) or *M. spretus* (right) separated on a 1.8% agarose gel after cutting with restriction endonucleases Ava II (lane 1); Taq 1 (lane 2); Alu I (lane 3); ECoRI (lane 4). Times of exposure of the autoradiographs are indicated. Arrows indicate an ascending series of multimers based on a monomer of 240 base pairs. Lanes 1 are type A pattern and lanes 2, 3 and 4 (left) and lanes 2 and 4 (right) are type B pattern (see text) (Brown and Dover, 1980).

completely by Ava II to produce a series of bands referred to as a type A pattern (Southern, 1975a), (lane 1, Fig. 6). Restriction enzymes Taq 1, Alu 1 and ECoR 1 produce type B patterns (lanes 2, 3 and 4), in that they cut only smaller proportions of the satellite that are defined as segments which are known to be non-overlapping (Horz and Zachau, 1977). The *M. musculus* satellite probe is clearly able to detect the type A and type B patterns in the homologous reactions (lanes 1–4 left-hand side). Similarly, sequences in much smaller amounts, having type A and two of the type B patterns are detected in the DNA of the heterologous reaction with *M. spretus* (lanes 1–4 right-hand side). We offer this data as an example of libraries of sequences in different relative abundances in the genomes of two closely related species.

Libraries of sequences, detected by Southern hybridizations or by direct sequence comparisons, also exist in species of *Apodemus* (Brown and Dover, 1979); in species of primates (Manuelidis and Wu, 1978; Singer and Donehower, 1979); in the *D. virilis* species group (Gall and Atherton, 1974); in kangaroo rat and other rodents (Fry and Salser, 1977); in species of *Glossina* (tsetse fly) (Amos and Dover, 1980). In this latter group we find that the distribution of satellites is unexpected in that the genomes of two subspecies (lanes b and c, Fig. 7) do not contain homologous

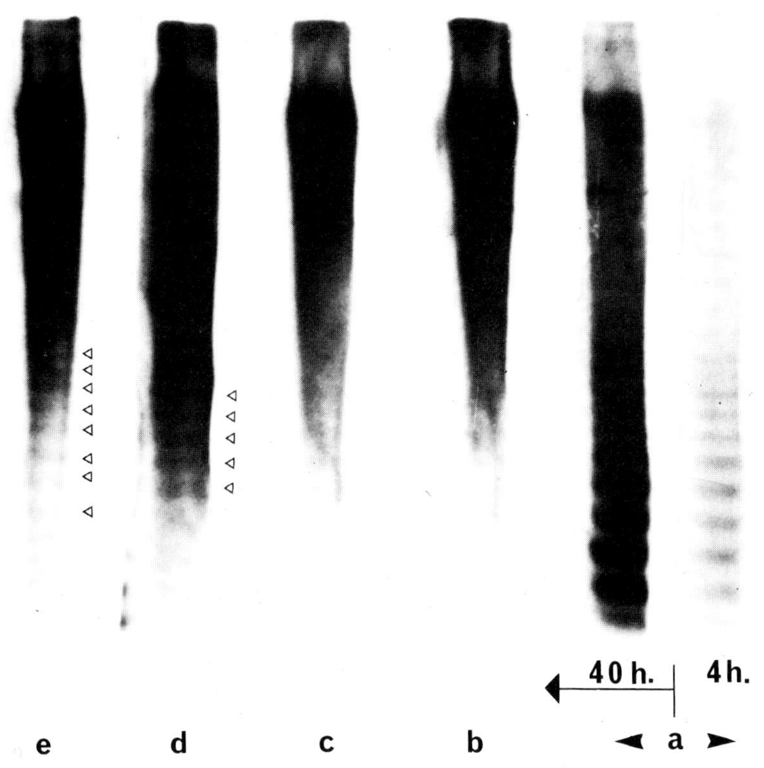

Fig. 7. Autoradiographs of "Southern" hybridizations (see text and Southern 1975b) of P^{32} nick-translated satellite DNA 1.685 of *Glossina morsitans morsitans* against total DNA of (a) *G. m. morsitans;* (b) *G. m. centralis;* (c) *G. m. submorsitans;* (d) *G. austeni;* (e) *G. pallidipes* separated on a 1.6% agarose gel after cutting with Taq I. Times of exposure are indicated in hours. Ascending series of fragment lengths similar to the homologous reaction (lane a) are visible in lanes d and e, but not b and c, (Amos and Dover, 1980) and unpublished results.)

sequences that are similarly organized to a probe taken from a third subspecies (lane a), at a time when more distantly related species are similarly organized as the probe (lanes d and e). The hybridization data reveal, however, that all the genomes contain some homologous sequences to the probe that remain at the origins of the lanes and must therefore be differently organized with respect to Taq 1 sites. The frequencies and organizations of these sequences are not known. The three subspecies can be distinguished morphologically only with difficulty and interbreed freely with various degrees of hybrid sterility (Curtis, 1972). It is possible that the distribution of these particular sequences in *Glossina* is an example of arbitrary sequence amplification on the lines of satellite characters 4 or 5 in Fig. 4. This data again underlines the inherent difficulties in assessing the significance of genomic similarities from the point of view of phylogenetic lineages. Until the evidence is at hand as to the precise taxonomic limits to a library of sequences, whether satellite or middle-repetitive (see below), the evolutionary significance of this phenomenon will remain obscure.

A more widely distributed library component has been detected using Southern transfers, in the genera *Mus, Rattus* and *Apodemus*. In addition, this conserved sequence is interesting in that the repeats are similarly organized with regard to restriction sites although they are probably differently dispersed amongst unrelated sequences in the separate genomes (Brown and Dover, in prep.). The interspersion of members of different families is an important feature of genome organization that is receiving intense study in two groups of species with particular reference to the middle repetitive DNAs (see below). The significance of the fine differences in genomic location of members of conserved families is not known although there is an interesting aspect of recent speculation on the reorganization of genomes and speciation to which these differences in interspersion may relate.

3. Modulation of Family Size and Interspersion of Sequences

Precise measurements, made with the use of cloned sequences as probes, of the size and divergence of common repetitive families in species of sea urchins, have revealed that considerable interspecific modulation in family size is occurring with varying degrees of sequence divergence within the families (Moore *et al.*, 1978; Klein *et al.*, 1978). Some families, that have changed little in sequence but vary in size, can be considered as library sequences on the lines described above. It is interesting to note that the overall sequence change in the repetitive families is less than that measured from the stabilities of heteroduplexes of the single-copy DNA: an interpretation of which is given later. Furthermore, Davidson and Britten (1973) and their colleagues (Moore *et al.*, 1978; Klein *et al.*, 1978) stress the striking observation that "the dominant repetitive sequence families in one genome typically have a higher frequency than in related species genomes" and yet the families are still relatively homologous despite the long periods of time since the species diverged and despite their subsequent amplification. The newly amplified families make up the majority of sequences in each genome and are known to be finely interspersed with single-copy DNA, in at least one of the genomes (Graham *et al.*, 1974).

On the basis of these observations it is possible that amplification and interspersion are related mechanisms of genomic change; a point previously stressed (Dover, 1977) in relation to the comparisons of interspecific sequence organization in the cereals (Flavell *et al.*, 1977), frogs (Galau *et al.*, 1976), salamanders (Mizuno and Macgregor, 1974) and in the satellite DNAs of several groups. Recent computer simulations, based on a process of unequal chromatid exchange first proposed by Smith (1976) have shown that amplification and concomitant interspersion are a theoretical possibility (Dover, Brown and Smith, in preparation). Species genomes may generate evolutionary novelty by a mixing of new with old sequences: a phenomenon clearly observed in the genomes of Gramineae (Flavell *et al.*, 1977; Rimpau *et al.*, 1978).

Flavell and his colleagues have been able to characterize the genomes of related Gramineae with respect to components that are held in common to differing extents amongst the species. The distribution of the components (library sequences) has permitted the establishment of a monophyletic lineage in which the component that is common to all the genomes is considered ancestral whilst the species distributions of the other components permits their derivation along simple bifurcating patterns. There is greater sequence homogeneity of a family within a species than there is for the same family between species: suggesting as with sea urchins that recurrent amplifications are acting as homogenizing mechanisms since species

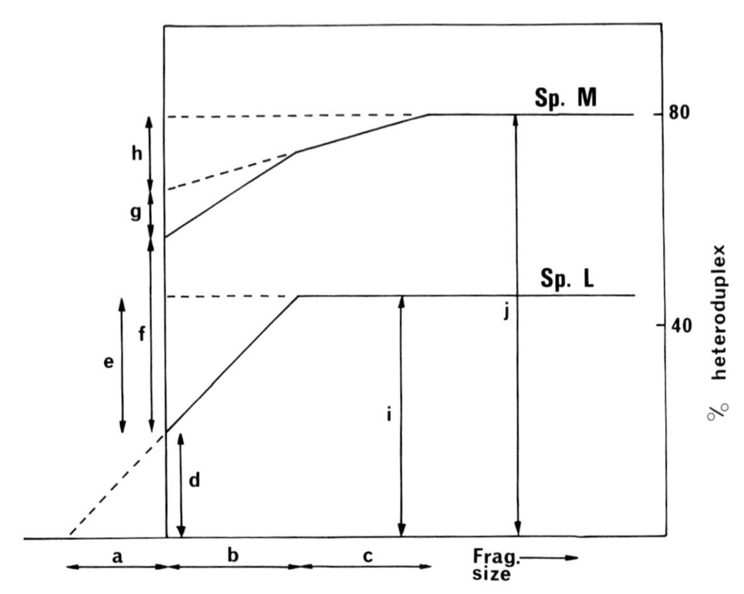

Fig. 8. Model hybridization curves of DNA of one species against DNA of two other species L and M with increasing fragment lengths (adapted from Rimpau *et al.*, 1978). % heteroduplex is the % of the species 1 DNA entering a duplex with L or M.

(d) % species 1 DNA hybridizing to DNA L
(d + f) % species 1 DNA hybridizing to DNA M
(i) % species 1 consisting of (d) sequences plus other sequences (e) within which they are interspersed.
(j) % species 1 DNA consisting of (d + f) sequences plus other sequences (g + h) within which they are interspersed.
(e) % species 1 DNA consisting of sequences (b) nucleotide pairs long separating sequences (d) that hybridize to L.
(g) % of species 1 DNA consisting of sequences (b) nucleotide pairs long separating some of sequences (d + f) that hybridize to M.
(h) % of species 1 DNA consisting of sequences (b + c) nucleotide pairs long separating some of sequences (d + f) that hybridize to M.
(a) mean length of sequences hybridizing to DNA L.

divergence. At the same time, it has been shown by the extensive use of hybridization of probes of individual library families to increasing lengths of fragments of total DNA from each genome, that sequence members of all the families are interspersed with each other and also with single-copy DNA, in arrangements that are different for each genome (Rimpau *et al.*, 1978). The rationale underlying these experiments is given in the legend to Figure 8, adapted from Rimpau *et al.* (1978).

Flavell has suggested that amplification of compound units containing repeated and non-repeated DNA might have given rise to the complex pattern of interspersion of ancestral and newly amplified sequences and I have illustrated this proposal in Fig. 9. In

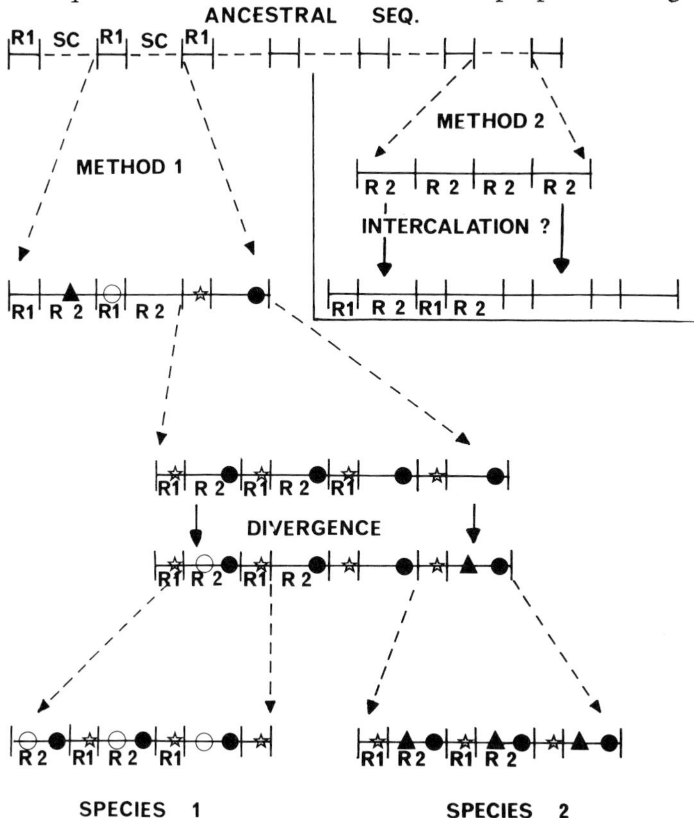

Fig. 9. Two possible routes by which an ancestral sequence of a repeated sequence family (R1) plus single-copy DNA (SC) evolves into families (R1) and (R2) in species 1 and 2, in the Gramineae. By the first method a compound unit of (R1 + SC) is amplified to generate unrelated repeated families (R1 and R2). Further rounds of divergence and amplification of differently diverged compound units can generate higher levels of homogeneity for R1 or R2 within the species than between the species. In the second method amplification of a repeated sequence is followed closely by its intercalation amongst other similarly amplified families (Dover *et al.*, in preparation).

addition to this suggestion, it is feasible that interspersion of a sequence is attendant on its amplification in that the mechanism of amplification would, whatever else happened, intersperse the new members with pre-existing genomic components (Dover, Brown and Smith, in preparation (See Fig. 9). This is not the place to enter into details of such models for they are intimately bound up with other data on the lengths of interspersed sequences and the varying levels of intraspecific and interspecific divergence of library families. I have adapted an informative representation of family divergence taken from Klein *et al.*, (1978), and included in the scheme one amplification event for sequence a_5 (Fig. 10). This illustrates the rise

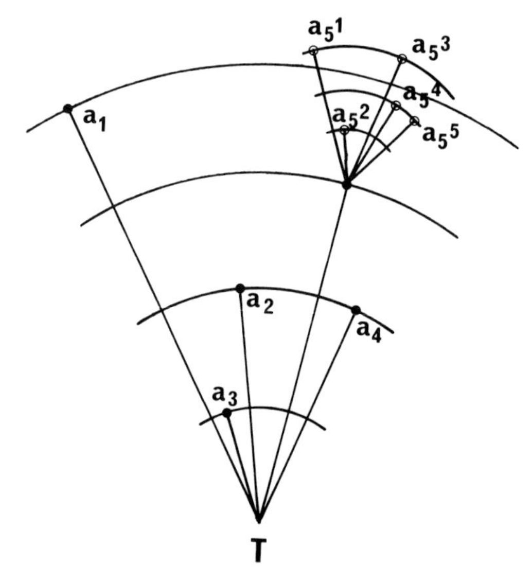

Fig. 10. Schematic representation of divergence and amplification of members of repeated sequence families (adapted from Klein *et al.*, 1978). Starting library sequence T is amplified into a repeated sequence family $a_1 - a_5$ that are free to diverge. Radial distances from T represent different degrees of divergence within each member. Member a_5 has undergone a further round of amplification to give rise to a new family $a_5{}^{1-5}$. This family is still related in sequence to family $a_1 - a_4$. Other members $a_1 - a_4$ are free to amplify in other genomes giving rise to differences in abundance of near related families in separate species. If $a_5{}^{1-5}$ also intercalate with other unrelated families (see Fig. 9) as part of the amplification process, patterns of sequence organization as observed in sea urchins and cereals can ensue (Dover, 1977; Dover *et al.*, 1980).

of a new family with a high degree of sequence homogeneity, that nevertheless retains sequence similarities with the other a_1-a_4 sequences, any one of which might have undergone amplification in other genomes. If the new amplified family became interspersed with members of unrelated families then patterns of sequence organization, as present in the genomes of Gramineae and sea urchins, might arise.

There are no shortages of suggestions for quirks of DNA replication that could act as mechanisms of amplification: including replicative loops, (Musich *et al.*, 1977); unequal chromatid exchange (Smith, 1976) and rolling circles (these latter are a proven mechanism of rDNA amplification during amphibian oogenesis, Rochaix *et al.*, 1974).

Whatever the mechanism, a recurrent process of amplification coupled with the disperal of the new sequences could be a general phenomenon of genomic growth in plants and animals. Such a process might explain the differences in homogeneity of sequences of a family within and between species and differences in abundance of families of satellite or middle-repetitive DNA. The important implication of this statement, and one that I would like to stress, is that the observed levels of close homology between species for certain components is a reflection of recent amplification events of very similar sequences held by the different genomes (library sequences), according to the scheme depicted in Fig. 10, rather than the evolutionary conservation of sequences by selection of unknown functions.

Another relevant aspect of the growth of genomes concerns the observation that the increase in size of chromosomes is proportional throughout their length within a species group. This has been documented fully in salamanders (Mizuno and Macgregor, 1974) and sweet-peas (Rees, 1972). The argument I have developed above, that amplification and interspersion are part and parcel of the same quirk of change could explain these proportionalities. This would imply that the sites of amplification are numerous and dispersed throughout the chromosomes with perhaps hot-spots of amplification at the satellite DNA locations of centromeres and telomeres. If the numbers of sites were fixed or if there was an upper limit to the growth of the genome in a particular species group then there might be, in this suggestion, an additional explanation for the fact that the amplification of sequences is not entirely haphazard. For example, in the *melanogaster* species subgroup the rate of occurrence of chromosome inversions is similar to the rate of appearance of new satellites, in that the species relationships established from these two phenomena are the same (Barnes *et al.*, 1978; Dover, 1980). There seems to be some order or limits imposed on the genomes with respect to their habits.

CONCLUDING REMARKS: GENOME REORGANIZATION AND BIOLOGICAL NOVELTY

What then are we to make of the variety of changes experienced by the genomes of higher organisms? From the point of view of phylogenetic systematics, I think the message is clear in what I have been describing above, that the sudden amplification of new DNA including single copy from similar sequences in related genomes, would impose severe constraints, if not entirely jeopardize, the construction of monophyletic scaffolds. Ancestry and derivation are not qualities that can always be extracted logically from observations of distribution and divergence, for these two parameters are too much at the whim of the mechanisms that go into their making. In this respect, DNA sequences are similar to proteins, in that although there might be, when analysing the broad levels of divergence of whole genomes, a rough fortuitous constancy in the rates of divergence, there are clearly other factors that cause severe fluctuations in rate. In the case of proteins these are related to the functional constraints of dispensibility and expendibility (Wilson *et al.*, 1977) and in the case of DNA sequences these are related to the propensities of the molecules to undergo recurrent amplification.

If some comfort were to be derived from differences in DNA towards an understanding of evolutionary relationships between species, I think that it would be of value only after a thorough investigation has been made on fluctuations in amount and organization, on the lines I have described above, and after the complete spectrum and taxonomic range of homologous sequences have been revealed.

I raised the issue in the introduction to this chapter of the biological significance of genomic changes and reorganization, not only from the point of view of species designations, but also from the fundamentally more important concern of speciation. In the absence of a genetic mechanism of speciation, there are in existence several modes of speculation concerning speciation, that have been usefully re-evaluated by White (1977). One line of thought that historically threads its way through the minds of several investigators concerns the role of genomic reorganization in the differentiation of species; what Mayr termed the "genetic revolution" (Mayr, 1963). Rapid genetic changes involved with the appearance of new species, the consequences of genome reorganization, has recently been re-emphasised as a strong possibility by Wilson in considering the "organismal" differences amongst primates and amongst other species groups (Wilson, 1976).

If what I have maintained above, concerning the accidental changes in genome composition proves to be a widespread evolutionary phenomenon, there is the interesting possibility that such accidental changes inadvertently introduce genomic novelties that are affecting the ontogenies of organisms. The persistence or success of a new genome (species) would then presumably be the outcome of the spread of the accidental changes and the selective forces operating at the time of occurrence, (Dover *et al.*, 1980).

There are very few studies that could be cited to support this suggestion of the relevance of genomic changes and biological differences between species. There is, however, in the studies of Britten and Davidson's group on the transcriptional activity of sea urchin genomes, the finding that the transciption and post-transcriptional processing of repeated sequence families is under stage specific control (Scheller *et al.*, 1978; Wold *et al.*, 1978). Given this observation there is a clear hint, emphasized by the authors, that although the large modulations in family size might not be significant, in that the overall morphology, behaviour and development of the species under study are very similar, there might be more subtle changes in the organization of families, in particular their interspersion, that affect the genes themselves and through that the developmental flexibility necessary for evolutionary change.

Also, it is reasonable to speculate that the particular genomic changes affected by the repetitive DNAs might affect the behaviour of chromosomes in germ-line processes, that would in turn affect the levels of genome incompatibilities in wide crosses. Some data and speculation surrounding this idea are presented in Dover *et al.* (1980); and a quantitative treatment of the rapid fixation of saltatory changes, due to sequence mobility between chromosomes, and the accidental inception of speciation, is in preparation.

REFERENCES

Amos, A., and Dover, G. A. (1980). The distribution of satellite DNAs between regular and supernumerary chromosomes in species of *Glossina* (tsetse): a two-step process in the origin of supernumeraries. *Chromosoma* (in press).

Angerer, R. C., Davidson, E. H. and Britten, R. J. (1976). Single-copy DNA and structural gene sequence relationships amongst four sea urchin species. *Chromosoma* **56**, 213–226.

Appels, R. and Peacock, W. J. (1978). The arrangement and evolution of highly repeated (satellite) DNA sequences with special reference to *Drosophila. Int. Rev. Cytol.* Suppl. **8**, 69–122.

Barnes, S. R., Webb, D. A. and Dover, G. A. (1978). The distribution of satellite and main-band DNA components in the *melanogaster* species subgroup of *Drosophila. Chromosoma* **67**, 341–363.

Brown, S. D. M. and Dover, G. A. (1979). Conservation of sequences in related genomes of *Apodemus:* constraints on the maintenance of satellite DNA sequences. *Nucl. Acid. Res.* **6**, 2423–2434.

Brown, S. D. M., and G. A. Dover, (1980) Conservation of segmental variants of satellite DNA of *Mus musculus* in a related species: *Mus spretus.* Nature 285: 47–49.

Curtis, C. F. (1972). Sterility from crosses between sub-species of the tsetse fly *Glossina morsitans. Acta Trop.* **29**, 250–268.

Davidson, E. H. and Britten, R. J. (1973). Organisation, Transcription and regulation in the animal genome. *Quart. Rev. Biol.* **48**, 565–613.

Dover, G. A. (1977). Variation in genome organisation in related species: an annotation. *In* "Chromosomes Today" (A. de la Chapelle and M. Sorsa, eds), Vol. 6. Elsevier, Amsterdam.

Dover, G. A. (1980). The evolution of 'common' sequences in closely-related insect genomes, *in* R. L. Blackman, G. M. Hewitt, and M. Ashburner, eds., Insect Cytogenetics. Roy. Ent. Soc. Lond. Symp. 10, Blackwell Sci. Publ. Oxford.

Dover, G. A., Strachan, T and Brown, S. D. M. (1980) The evolution of genomes in closely-related species. *in* Proc. 2nd Int. Congr. Systm. and Evol. Biol. Vancouver

Flavell, R. B., Rimpau, J. and Smith, D. B. (1977). Repeated sequence DNA relationships in four cereal genomes. *Chromosoma* **63**, 205–222.

Fry, K. and Salser, W. (1977). Nucleotide sequences in HS-α satellite DNA from kangeroo rat *Dipodomys ordii* and characterisation of similar sequences in other rodents. *Cell* **12**, 1069–1084.

Galau, G. A., Chamberlain, M. E., Hough, B. R., Britten, R. J. and Davidson, E. H. (1976). Evolution of repetitive and non-repetitive DNA. *In* "Molecular Evolution" (F. J. Ayala, ed.), pp. 200–224. Sinaner Press, Sunderland, Massachusetts.

Gall, J. G. and Atherton, D. D. (1974). Satellite DNA sequence in *Drosophila virilis. J. Mol. Biol.* **85**, 633–664.

Gosden, J. R., Mitchell, A. R., Seuanez, H. N., and Gosden, C. B. (1977). The distribution of sequences complementary to human satellite DNAs I, II, and IV in the chromosomes of chimp, gorilla and orangutan. *Chromosoma* **63**, 253–271.

Gould, S. J. (1977). "Ontogeny and Phylogeny". Harvard University Press, Massachusetts.

Graham, D. E. and Skinner, D. M. (1973). Homologies of repetitive DNA sequences among Crustacea. *Chromosoma* **40**, 135–152.

Graham, D. E., Neufield, B. R., Davidson, E. H. and Britten, R. J. (1974). Interspersion of repetitive and non-repetitive DNA sequence in the sea urchin genome. *Cell* **1**, 127–137.

Harpold, M. M. and Craig, S. P. (1977). The evolution of repetitive DNA sequences in sea urchins. *Nucl. Acid Res.* **4**, 4425–4438.

Harpold, M. M. and Craig, S. P. (1978). The evolution of non-repetitive DNA in sea urchins. *Different.* **10**, 7–11.

Hennig, W. (1966). "Phylogenetic Systematics". Urban University Illionis Press.

Horz, W. and Zachau, H. G. (1977). Characterisations of distinct segments in mouse satellite DNA by restriction nucleases. *Eur. J. Biochem.* **73**, 383–392.

Hoyer, B. H., Bolton, E. T., McCarthy, B. J. and Roberts, R. B. (1965). The evolution of polynucleotides. *In* "Evolving Genes and Proteins" (V. Bryson and H. J. Vogel, eds). Academic Press, New York and London.

John, B. and Miklos, G. L. G. (1979). Functional aspects of satellite DNA and heterochromatin. *Int. Rev. Cytol.* **58**, 1–114.

Jones, K. W. (1978). Speculations on the functions of satellite DNA in evolution. *Zeit. Morph. Anthrop.* **62**, 143–171.

King, M-C. and Wilson, A. C. (1975). Evolution at two levels in humans and chimpanzees. *Science* **188**, 107–188.

Klein, W. H., Thomas, T. L., Lai, C., Scheller, R. H., Britten, R. and Davidson, E. H. (1978). Characteristics of individual repetitive sequence families in the sea urchin genome studied with cloned repeats. *Cell* **14**, 889–900.

Laird, C. D., McConaughy, B. L. and McCarthy, B. J. (1969). Rate of fixation of nucleotide substitutions in evolution. *Nature* **224**, 149–154.

Lemeunier, F. and Ashburner, M. (1976). Relationships within the *melanogaster* species subgroup of the genus *Drosophila (Sophophora)*. II. Phylogenetic relationships between six species based upon polytene chromosome banding sequences. *Proc. R. Soc. B.* **193**, 275–284.

Lemeunier, F., Dutrillaux, B. and Ashburner, M. (1978). Relationship within the *melanogaster* species subgroup of the genus *Drosophila (Sophophora)*. III. The mitotic chromosomes and quinacrine fluorescent patterns of the polytene chromosomes. *Chromosoma* **69**, 349–361.

Manuelidis, L. (1977). A simplified method for the preparation of mouse satellite DNA. *Analyt. Biochem.* **78**, 561–568.

Manuelidis, L. and Wu, J. C. (1978). Homology between human and simian repeated DNA. *Nature* **276**, 92–94.

Mayr, E. (1963). "Animal Species and Evolution". Harvard University Press, Cambridge, Massachusetts.

Mizuno, S. and Macgregor, H. C. (1974). Chromosomes, DNA sequences and evolution in salamanders of the genus *Plethodon. Chromosoma* **48**, 239–296.

Mizuno, S., Andrews, C. and Macgregor, H. C. (1976). Interspecific "common" repetitive DNA sequence in salamanders of the genus *Plethodon. Chromosoma* **58**, 1–31.

Moore, G. P., Scheller, R. H., Davidson, E. H. and Britten, R. J. (1978). Evolutionary change in the repetition frequency of sea urchin DNA sequences. *Cell* **15**, 649–660.

Musich, P. R., Brown, F. L. and Maio, J. J. (1977). Mammalian repetitive DNA and the subunit structure of chromatin. *Cold Spring Harb. Symp. quant. Biol.* **42**, 1147–1160.

Peacock, W. J., Lohe, A. R., Gerlach, W. L., Dunsmuir, P., Dennis, E. S. and Appels, R. (1977). Fine structure and evolution of DNA in heterchromatin. *Cold Spring Harb. Symp. Quant. Biol.* **42**, 1121–1135.

Rees, H. (1972). DNA in higher plants. *Brookhaven Symp. Biol.* **23**, 394–418.

Rimpau, J., Smith, D. and Flavell, R. (1978). Sequence organisation analysis of the wheat and rye genomes by interspecies DNA/DNA hybridisation. *J. Mol. Biol.* **123**, 327–359.

Rochaix, J. D., Bird, A. and Bakken, A. (1974). Ribosomal RNA gene amplification by rolling circles. *J. Mol. Biol.* **87**, 473–487.

Scheller, R. H., Costantini, F. D., Kozlowski, M. R., Britten, R. J. and Davidson, E. H. (1978). Specific representation of cloned repetitive DNA sequences in sea urchin RNAs. *Cell* **15**, 189–203.

Singer, D. and Donehower, L. (1979). Highly repeated DNA of the baboon: organisation and sequences homologous to the highly repeated DNA of the African Green Monkey. *Science* In Press.

Skinner, D. M. (1977) Satellite DNA's. *Bil. Sci.* **27**, 790–795.

Smith, G. P. (1976). Evolution of repeated sequences by unequal crossing-over. *Science* **191**, 528–535.

Smith, H. H. (ed.) (1972). "Evolution of Genetic Systems". *Brookhaven Symp. Biol.* **23**. Gordon and Breach, New York.

Sneath, P. H. A. and Sokal, R. R. (1973). "Numerical Taxonomy." Freeman, San Francisco.

Southern, E. M. (1975a). Long range periodicities in mouse satellite NDA. *J. Mol. Biol.* **94**, 51–69.

Southern, E. M. (1975b). Detection of specific sequences among DNA fragments seperated by gel electrophoresis. *J. Mol. Biol.* **98**, 503–517.

Thiery, J. P., Macaya, G. and Bernardi, G. (1976). An analysis of eukaryotic genomes by density gradient centrifugation. *J. Mol. Biol.* **108**, 219–235.

Walker, P. M. B. (1969). The specificity of molecular hybridisation in relation to studies of higher organisms. *Progr. Nucl. Acid Res; Mol. Biol.* **9**, 301–326.

White, M. J. D. (1973). "Animal Cytology and Evolution". Third Edition. Cambridge University Press, Cambridge.

White, M. J. D. (1977). "Modes of speciation". Freeman, San Francisco.

Wilson, A. C. (1976). Gene regulation in evolution. *In* "Molecular Evolution" (F. J. Ayala, ed.), pp. 255–234. Sinaner Press, Sunderland, Massachusetts.

Wilson, A. C., Carlson, S. S. and White, T. J. (1977). Biochemical Evolution. *Ann. Rev. Gen.* **46**, 573–639.

Wold, B. J. Klein, W. H., Hough-Evans, B. R., Britten, R. J. and Davidson, E. H. (1978). Sea urchin embryo mRNA sequences expressed in the nuclear RNA of adult tissues. *Cell* **14**, 941–950.

13 | Interpretation and Analysis of Serological Data

G. CRISTOFOLINI

Istituto Botanico, Università di Trieste, via A. Valerio 30, 34100 Trieste, Italy

Abstract: Serological systematics is based upon measures of intensity of cross-reactions and self-reactions of antisera. The cross-reaction is measured in conventional absolute units, and then standardized. Usually, it is standardized by referring it to the intensity of self (reference) reaction. It is proposed that direct ratios (cross reaction)/(self reaction) and the arithmetic mean of them be adopted as the relative measure of serological similarity between species. The matrix of similarity coefficients is comparable to a matrix of similarities or euclidean distances from any other taxonomic data. Any type of ordination or hierarchic classification can be obtained from it. A special problem arises whenever there is no possibility of producing antisera to all taxa under study, a frequent occurrence in serobotanical work. In this case a distance between any pair of species can be computed. However, this interspecific distance is affected by an uncertainty, which is inversely proportional to the similarity of the species under study and the reference species. The uncertainty is due to ignorance about the identity of the reacting systems. Pre-absorption is often adopted to overcome this problem; technical limitations of pre-absorption are shortly discussed. The author stresses the possibility of joining quantitative measures and qualitative observations to identify reacting systems and to allow similarity measurements for species whose antiserum is not available.

INTRODUCTION

Systematic research consists generally of comparing taxa with regard to presence or absence of characters and to quantitative

---

Systematics Association Special Volume No. 16, "Chemosystematics: Principles and Practice", edited by F. A. Bisby, J. G. Vaughan and C. A. Wright, 1980, pp. 269–288, Academic Press, London and New York.

similarity of corresponding features. Taxa are then organized into a system which is defined as phenetic or phylogenetic, depending largely on the degree of scepticism or optimism in the systematist himself. Similarly, sero-systematics consists of organizing taxa into a system, which is based upon presence or absence of serological characters and the degree of similarity of corresponding immunological systems, with the exception that serological systematics (and in general, biochemical systematics) always claims to be phylogenetic. It must be said that such a firm belief is not always fully justified.

There has been a time, more than half a century ago, when somebody hoped that serological reactivity would be a sort of philosophers' stone of evolutionary systematics. The most representative result of such a tendency was the famous "Koenigsberger serodiagnostischer Stammbaum des Pflanzenreiches", that is the "serological phylogenetic tree of the plant kingdom" (Mez, 1925), the most elaborate and baroque offspring of the school of Koenigsberg. The whole diversity of plants, seaweeds, fungi and autotrophic micro-organisms was displayed and explained in terms of serological cross-reactivity. Unfortunately, the molecular bases of serology were scarcely understood at that time, the technique was rather approximate and data processing very arbitrary. The results were, therefore, hardly justified by the actual experimental evidence. Hence, a severe criticism was raised, and that phylogenetic tree was for a long time the gravestone of plant sero-systematics.

Indeed, if serological reactivity has to be employed as a contribution to biological systematics, it must be kept in mind that the way the data are processed is of basic importance in determining the completeness and the correctness with which such a great amount of information is exploited. Unfortunately, the development of serological data processing, at least insofar as serobotany is concerned, is far behind the development of serological techniques. Most serobotanical works are basically descriptive, and therefore renounce the possibility of completely exploiting the data themselves, or employ the serological evidence only to confirm or reject pre-existing systematical schemes. The potentiality of serological methods is very limited by this attitude. It should be noted, as stated by Sneath and Sokal (1973), that "one cannot convert a serological cross-reaction into a single character to be incorporated, with other characters, into a numerical analysis. The serological results are already a matrix of similarity coefficients", therefore "comparative serology must remain at present a separate technique for assessing phenetic relations". If this is true, a proper method of analysis must be developed, that takes in account the particular features of serological data and treats them in a fully autonomous way.

The following features of serological characters are especially relevant in respect of data processing: (1) The serological character is carried by active sites or determinants; every protein molecule carries several determinants, their number ranging from five or six up to a thousand and shows a rough direct proportionality to the molecular weight of the protein (Kabat and Mayer, 1961). (2) Every protein molecule usually presents one determinant with any given specificity, although protein molecules resulting from the aggregation of sub-units may well carry several equivalent determinants. (3) The antibody system is formed in its largest part by G-immunoglobulins having bivalent antibody activity. The presence of M-immunoglobulins and other types of immunoglobulins can be ignored in the usual sero-systematical praxis. (4) The immunological cross-reactivity has been shown to be proportional to the similarity of the amino acid sequences in lysozyme (Prager and Wilson, 1971a, 1971b), plastocyanin (Wallace and Boulter, 1967) and leghaemoglobins (Hurrell *et al.,* 1976, 1977). There are no reports contradicting these ones, at least as far as plant proteins are concerned, therefore, at present it is legitimate to assume as a working hypothesis that there is a positive correlation between serological cross-reactivity and similarity in the amino acid sequence; it is not clear how far such correlation is linear since published data are not perfectly concordant. Of course, the immunological reactivity only directly involves 1–2% of the whole molecule, and only its external surface; however, amino acid substitutions in any part of the molecule may affect the overall steric arrangement and hence the steric compatibility for the determinants to react with the corresponding antibodies (so-called "carrier specificity"). (5) The present knowledge about structure and reactivity of determinants does not allow cladistic analyses to be conducted on serological

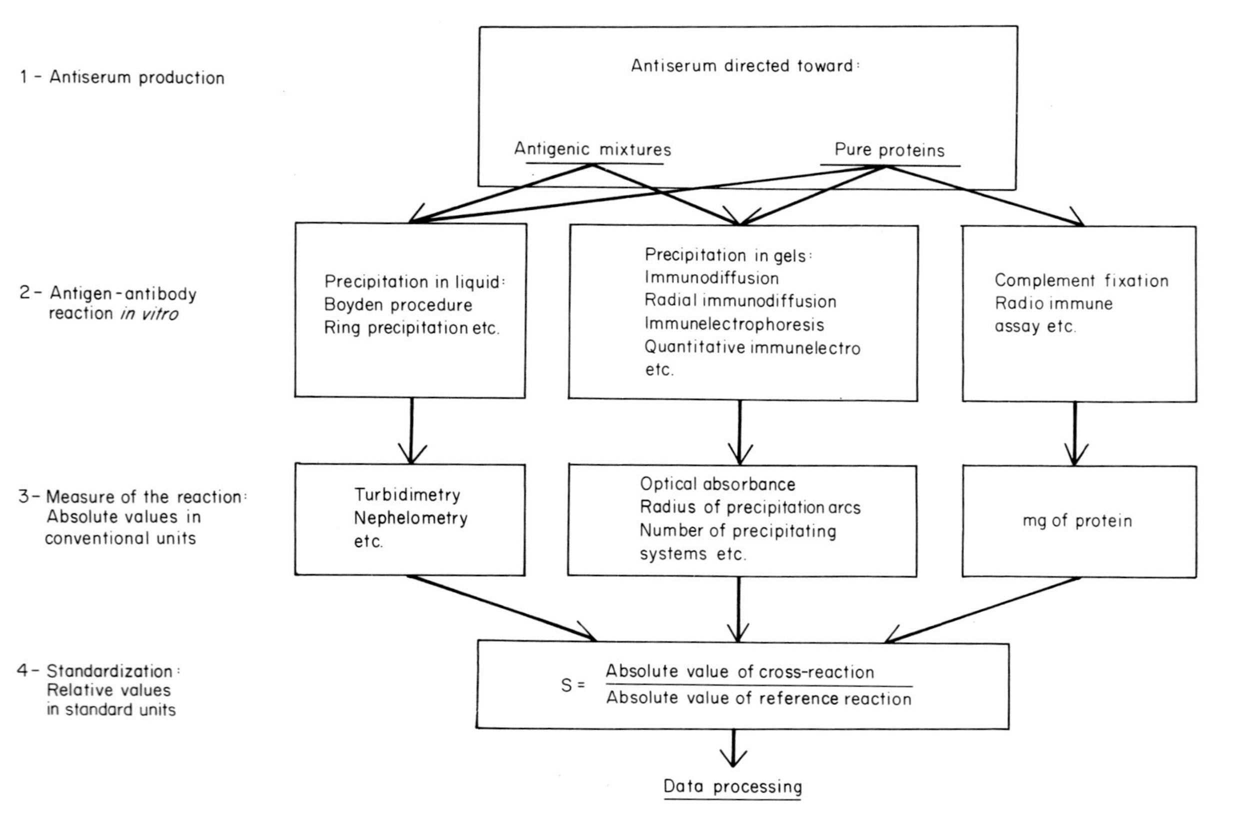

Fig. 1. Logical sequence of steps in serological taxonomy.

characters in practical sero-systematical work. The systematist must limit himself, at present, to assess phenetic relations, although a number of evolutionary inferences can be drawn from them.

GATHERING SEROLOGICAL DATA

Four steps can be logically distinguished in the sero-taxonomic praxis (Fig. 1): antiserum production; antigen–antibody reaction *in vitro;* measure of the reaction; standardization and processing.

Antisera can be raised against pure proteins or against mixtures of them. Also, monoclonal antibody populations directed toward a single determinant can be produced. Consequently, antigen–antibody reaction *in vitro* can be performed following numerous techniques, the choice of which depends on the material under study (pure antigens or mixtures). Comprehensive surveys of current techniques can be found in Williams and Chase (1971) and in Daussant (1975). If a precipitin reaction is carried out in a liquid, then the amount of protein involved in insoluble complexes can be measured optically by turbidimetry or nephelometry. A special development of these techniques is constituted by the Boyden (1932, 1954, 1964) procedure; by this technique the turbidity of a series of antigen–antibody proportions is measured, from antigen excess (where there is no precipitate, because antigens and antibodies form insoluble complexes) to optimum point (antigen fully saturated by antibodies to form insoluble aggregates) and to antigen extreme dilution (where again no insoluble complexes can form). A modification of this technique has been proposed by Kloz (1960, 1961) as "ring precipitation"; in this modification, overlaying the antigen on the antiserum makes the series of dilutions unnecessary. The techniques above are designed to be used both with pure antigens and with mixtures.

Precipitation in gels can also yield quantitative measures of complex antigenic systems, either in terms of number of precipitin arcs (especially possible by immunelectrophoresis) or in terms of overall optical density of the precipitate in immunodiffusion (Cristofolini, 1971). If pure proteins are being reacted, radial immunodiffusion or quantitative (bidimensional) immunelectrophoresis give indirect estimates of the reactants concentration. Direct estimates of the amount of protein involved in the serological reaction are provided by complement fixation technique as well as by radio-immuneassay.

The output of any of these methods is an absolute quantitative measure expressed in conventional units; the units are peculiar to the method employed. Such measures can be normalized by referring them to the absolute value of the corresponding self reaction (or reference reaction), that is the reaction of the tested antiserum with the corresponding antigen.*

Mainardi (1959) proposed that the distance between two taxa be computed as the geometrical mean of the measures obtained by the Boyden procedure:

$$D = \sqrt{\left(\frac{\text{Cross reaction } A \cdot B}{\text{Self reaction } A} \cdot \frac{\text{Cross reaction } B \cdot A}{\text{Self reaction } B}\right)}$$

He justified this formula arguing that the dilutions in the Boyden procedure follow a geometrical series. Sokal and Sneath (1963) proposed an improvement to this index by expressing it on a logarithmic scale to the base 2. On the other side, Kubo (1964) suggested that the amount of precipitated protein in a series of Boyden dilutions is an exponential function of the antigen concentration, and formulated a generalized empyrical equation to describe the relation between the turbidity and the antigen concentration; however, it is not known how general this type of relation is. Besides, it is not known how far the amount of precipitated protein is linearly proportional to the similarity in the primary structure: reports presently available are not fully consistent on this point. Facing this complex of unknown relations, it seems justified to assume the simplest formula, that is the original Boyden's (1932) principle of computing the similarity as the arithmetic mean:

$$S = \left(\frac{\text{Cross reaction } A \cdot B}{\text{Self reaction } A} + \frac{\text{Cross reaction } B \cdot A}{\text{Self reaction } B}\right) \cdot \frac{1}{2}$$

In its original version, this index was expressed as a measure of

*Several authors use the terms "homologous" and "heterologous" instead of "self" and "cross" reaction. Such terms should be avoided, as they have a different—and universally accepted—meaning in general biology, and may therefore generate confusion (see also Williams, 1964; Fairbrothers, 1967).

distance and therefore was transformed into a percentage and subtracted from 100, an expedient that seems useless. In its original form, or slightly modified, it has been used by many authors (see, for example, Lanzavecchia *et al.*, 1965; Lee and Fairbrothers, 1969; Cristofolini and Poldini, 1972). The similarity expressed by this index is a figure that varies linearly from 0 to 1, where 0 indicates complete dissimilarity and 1 identity. The principle, the range of variation and the progression are the same as those of Jaccard's (1908) Similarity Index, which is often used in numerical taxonomy and presents therefore several advantages for computational purposes. The Index should be regarded as a relative measure of similarity for comparisons between taxa which have been subjected to the same experimental procedure, and applies not only to Boyden procedure, but to any kind of measures of serological correspondence.

ESTABLISHING SEROLOGICAL DISTANCES

Whenever several antigenic systems are reacted against the same antiserum, the degree of similarity (or of distance) of each of them to the reference OTU can be plotted on an axis (Fig. 2a). Of course, it is possible to measure the distance within any pair of OTUs along the same axis. The question is whether this elementary computation is correct or not: if it is, then it is possible to add a second axis, representing the distance from a second antiserum (Fig. 2b), and compute the euclidean distance within any pair of OTUs in the plane defined by the orthogonal axes. More antisera can be added to the system, and the euclidean distance can be similarly computed in the hyperspace defined by the axes corresponding to the antisera.

First it should be noted that the axes do not represent quantitative characters but reference sera, that is sets of quantitative characters. Each point along any axis corresponds to the sum of several unknown characters, each of them being present in an unknown amount. This is true even if a serum to a single protein is being used: indeed, in this case several determinants and several antibody populations are involved, and the individual contribution of each of them to the overall reaction remains unknown. Properly speaking, the character in systematic serology is the determinant site rather than the whole protein or the complex of proteins that are represented

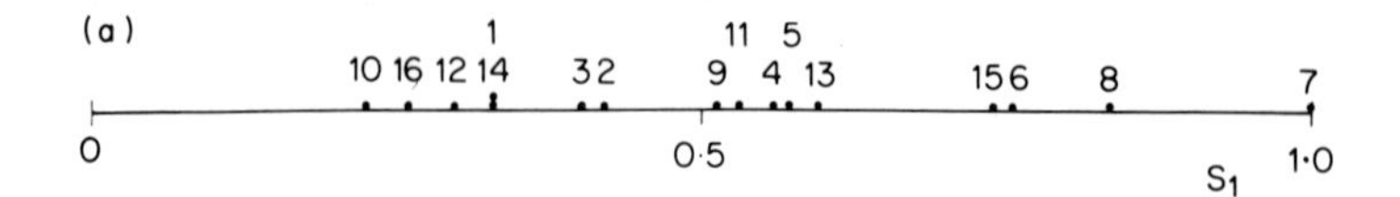

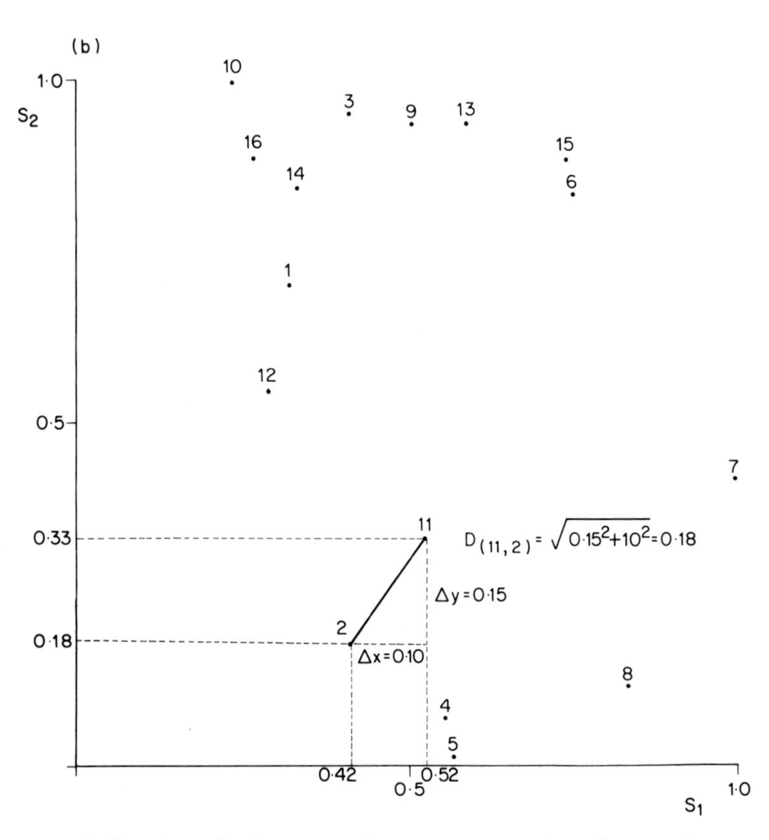

Fig. 2. (a) Ordination of 16 species of Genisteae according their similarity to an antiserum to *Genista sagittalis.* (b) Ordination of the same species according their similarity to *Genista sagittalis* and to *Genista tridentata,* and computation of the serological distance between two species. Discussion in text. S = (cross reaction)/(self reaction), S1 = similarity to *G. sagittalis,* S2 = similarity to *G. tridentata,* 1 = Adenocarpus complicatus, 2 = Calicotome villosa, 3 = Cytisus pseudoprocumbens, 4 = C. scoparius, 5 = C. sessilifolius, 6 = Cytisanthus radiatus, 7 = Genista sagittalis, 8 = G. sericea, 9 = G. sylvestris, 10 = G. tridentata, 11 = Laburnum anagyroides, 12 = Lembotropis nigricans, 13 = Retama sphaerocarpa, 14 = Spartium junceum, 15 = Teline monspessulana, 16 = Ulex europaeus.

by the axes. In spite of this fact, ordering species on the axes seems correct, if it is acceptable to treat a sum of quantitative characters as a single all-comprehensive quantitative character. This procedure only involves a loss in analytical information.

As a second observation, special attention should be paid to the possible presence of redundancies, a very common problem in numerical taxonomy. If an OTU A is similar under some respect to an OTU B, there will be a number of determinants common to both OTUs. If other OTUs, C, D, etc., are ordered according to their similarity to A and to B, those determinants are considered twice. The danger of a classification based upon such redundant characters is overcome by checking first the correlation between axes. A significant correlation proves the presence of a set of redundant characters.

The main problem, however, lies in the fact that the measure of interspecific distance is affected by a double uncertainty: first, no information is available about those proteins that do not react with the reference sera; secondly, it is not known whether the reacting determinants are the same in the different antigens. The first point is resolved only if an antiserum is produced for each species, so that the system includes as many axes as there are points. With regard to the second question, it is apparent that two OTUs which display the same degree of similarity toward a reference antiserum may be perfectly equal or completely different. In Fig. 2a *Laburnum anagyroides* shows a similarity $S = 0.52$ with the serum to *Genista sagittalis*, whereas *Calicotome villosa* has $S = 0.42$ with the same serum. However, the real similarity between the two species (with respect only to those characters that are detected by a serum to *Genista sagittalis*) is in the range 0–0.42, as shown in Fig. 3. Its real measure within this range cannot be defined. In general, if the similarity of two OTUs with a reference antiserum is $S = 1$, the probability that they are immunologically equivalent is $P = 1$; if the similarity with the reference antiserum is $S = 0$, their mutual similarity is completely indetermined; in the intermediate range, the probability that the real similarity corresponds to the apparent similarity decreases exponentially with decreasing similarity to the reference antiserum. The distance between OTUs having a low similarity level with the reference antiserum is of little significance.

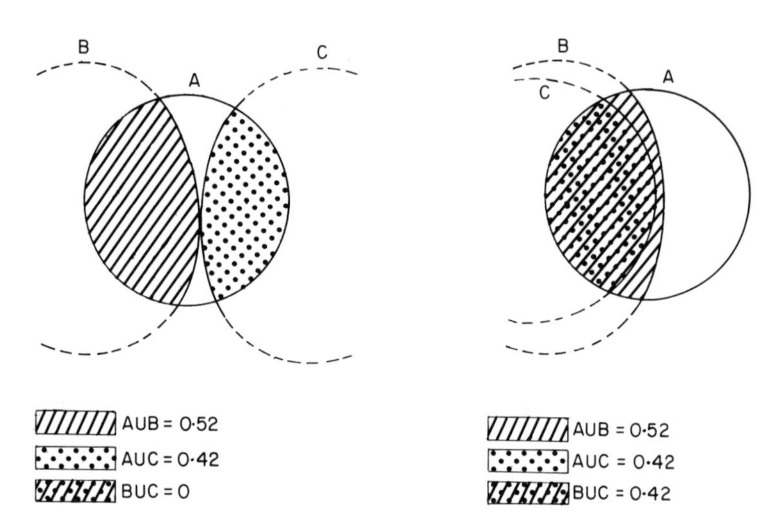

Fig. 3. Possible mutual similarity between two species, whose similarity to a reference antiserum is known. Data from Fig. 2a. A: Antiserum to *Genista sagittalis*, B: *Laburnum anagyroides*, C: *Calicotome villosa*.

Figure 4 presents an example of misclassification due to this type of error. A number of legume genera have been tested against five antisera (data from Cristofolini, in press); the precipitin reaction in gel has been classified in four conventional levels (absent, weak, middle, strong) and the interspecific correlation coefficients have been computed. The correlation coefficient has in this case been chosen in preference to the index of similarity; the latter can be used when two taxa show the occurrence or non-occurrence of the same characters: in many immunological reactions this is not the case. What is observed is just the joint occurrence of a reaction against the same antiserum: this does not mean that the same characters are present. From the matrix of correlation coefficients a dendrogram of phenetic similarity has been constructed by average linkage clustering (unweighted pair-group method, Sneath and Sokal, 1973). The systematic unsoundness of the resulting classification needs no special comment; the reason for it lies in the uncertainty about the reciprocal position of the species: the similarity to the reference antisera generally being low (because the taxa under study are distributed throughout the range of Leguminosae),

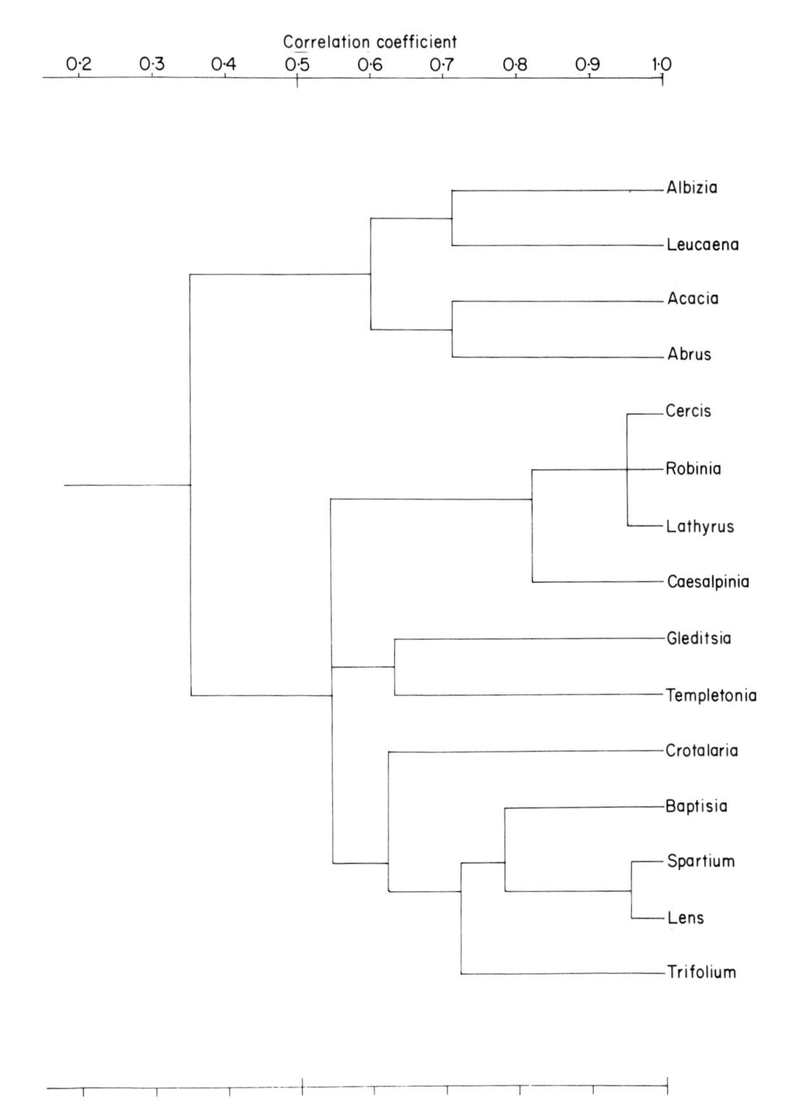

Fig. 4. Clustering of Legume genera based on a quantitative evaluation of their reaction with five antisera (*Acacia, Albizia, Gleditsia, Templetonia* and *Baptisia*). Discussion in text.

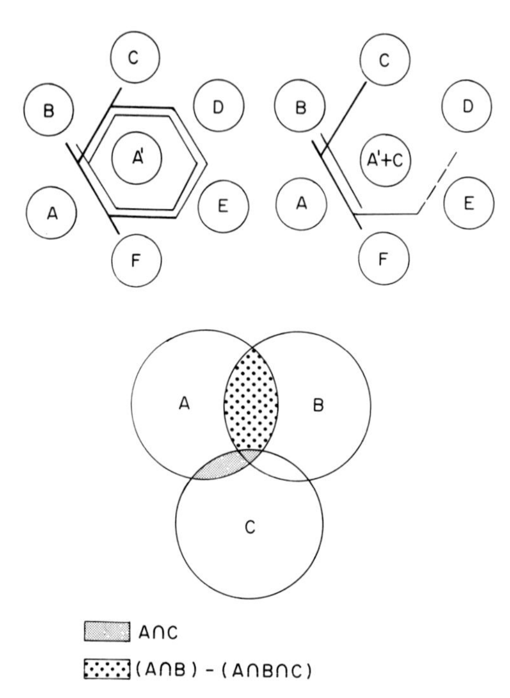

Fig. 5. Logical background of pre-absorption procedure. The amount of precipitate with unabsorbed antiserum (left cluster) corresponds to $A \cap B$, $A \cap C$, etc. The amount of precipitate with absorbed antiserum (right cluster) corresponds to $(A \cap B) - (A \cap B \cap C)$ etc. By pre-absorbing successively an antiserum A with two antigens B and C, the real distance between B and C measured on A can be computed as

$$D(B, C) = \{[A - (A \cap B)] \cap C\} U \{[A - (A \cap C)] \cap B\}$$

it happens that species which display the same level of similarity to the same reference antiserum are often completely different, although their mutual correlation is high. The limited number of antisera is responsible for the heavy distortion.

Of course, one could object that antisera to all taxa under examination should have been used, but it should be remembered that real systematic work is not conducted under ideal conditions. In practical serobotanical work, the condition of having all antisera cannot be fulfilled, unless the systematic study is restricted to those

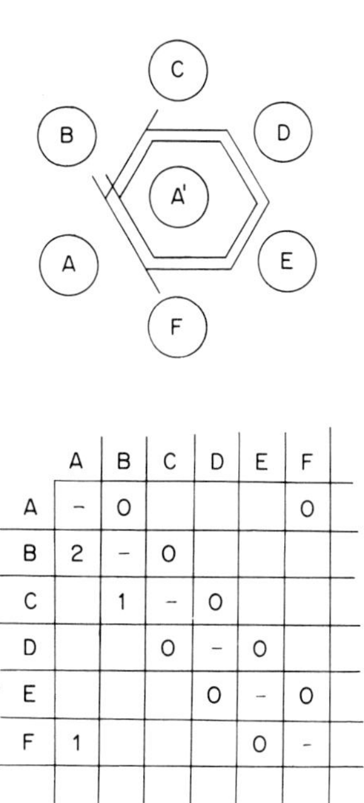

	A	B	C	D	E	F
A	–	0				0
B	2	–	0			
C		1	–	0		
D			0	–	0	
E				0	–	0
F	1				0	–

Fig. 6. Interpretation of immunodiffusion plate and codification in a double entry table. A separate matrix is required for any antiserum. The matrix is completed when all pairs of antigens have been compared directly (side by side). Discussion in text.

taxa of which one can get enough material to produce a good antiserum.

USING QUALITATIVE DATA

To overcome such limitations some authors prefer using pre-absorption (e.g. Jensen, 1968; Lester, 1979; Pearce and Lester, 1979). Basically, pre-absorption (or pre-saturation) consists of first reacting an antiserum to an OTU A with the antigenic mixture of another OTU B, so that all antibodies that are precipitated by B are taken from the system. The remaining $(A-B)$ complex is then reacted with any other antigenic mixture C. The principle of this procedure is represented in Fig. 5. Of course it is correct pre-saturating a serum with an antigenic mixture, and not vice-versa, because most antigens are polyspecific.

The pre-saturation technique removes the obstacle due to non-correspondence between reacting systems, although the measure of distance it yields is still affected by the ignorance about the subset of non-reacting systems. This ignorance, again, is effectively surmounted only if there are as many sera as there are OTUs being examined, and if all redundancies are removed.

There is another method for establishing correspondence between reacting systems of different species, i.e. using the qualitative information contained in the immune reaction in gels. Corresponding proteins of different species can be identified, as well as differential ones, as schematized in Fig. 6. The reaction of an immunodiffusion plate is coded in a double entry table. In each case the presence and the number of fractions which characterise the OTU of the column and are absent from the OTU of the row are recorded. For example, if a fraction is present in the OTU *B* and not in OTU *C*, then "1" is recorded at the crossing of column *B* and row *C*, whereas "O" is recorded at the crossing of column *C* and row *B*.

The sum of the scores of each pair of symmetrical cases is a rough measure of distance within the corresponding pair of OTUs with respect to the reference serum. On the other hand, each column represents the number of characters a given OTU shares with the reference serum, that are absent from other OTUs, i.e. the sum of scores in the column is an indirect measure of similarity to the reference serum. For the same reason, the sum of scores of the row is an indirect measure of dissimilarity.

If only few sera have been raised, the interspecific distance can be computed comparing each pair of columns and each pair of rows. This can be achieved by a similarity coefficient (e.g. Jaccard's coefficient, as only positive characters are relevant), by computing an euclidean distance (e.g. the taxonomic distance, Sokal and Sneath, 1963), or the correlation coefficient. The last method, used by Cristofolini and Feoli Chiapella (1977) should be preferred for the reasons mentioned above, when not all antisera are considered and the identification of reacting systems is consequently incomplete. Moreover, by computing the correlation coefficient, its significance level can be defined; this is a major advantage in comparison with other methods.

A more refined separation of proteins can be achieved if an antigenic mixture is subjected to electrophoresis, and after then an antiserum is layed on one side and a reference antigen on the other side of the electrophoretic pattern (Osserman, 1960; Williams and Chase, 1971), so that it can be checked if there are antigenic fractions to differentiate the two mixtures, to define how many they are and to recognize, for each of them, if there is a partial identity or a complete difference (Fig. 7). This technique involves all information offered by pre-absorption, and furthermore it offers detailed qualitative information. It largely overcomes the normal immunodiffusion, although it requires more time and material.

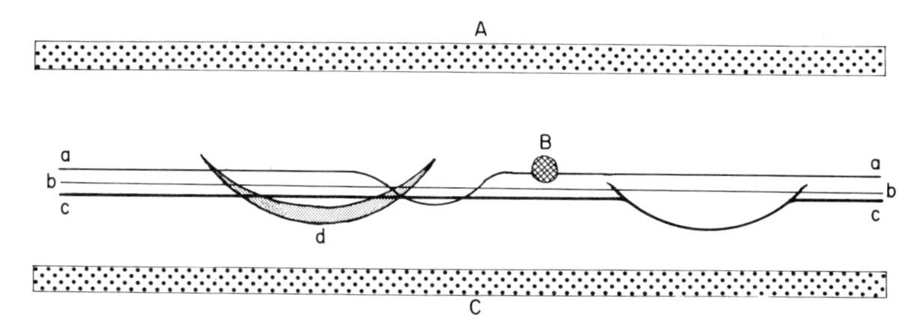

Fig. 7. Identification of immunological correspondence of proteins by means of immuno-electrophoresis with a reference antigen (Osserman technique). A: Reference antigen (*Templetonia australis*). B: Electrophorized antigen (*Spartium junceum*), C: Antiserum (*Baptisia australis*). Immune precipitates: *a*: identical in A and B, *b*: present in A, absent from B, *c*: partially identical in A and B, *d*: present in B, absent from A.

A number of the same legume genera clustered in Fig. 4 have been reprocessed by this method. A preliminary processing involved two antisera (*Baptisia australis* and *Acacia retinodes;* analytical data in Cristofolini, in press); the individual fractions have been identified and their occurrence in each species has been recorded. The similarity of each species to both antisera has been plotted (Fig. 8) in a rough ordination on two axes. Most of the information content of the OTUs under study cannot be displayed, since only two reference sera are employed: as a consequence, several OTUs are crowded in the lower left corner of the diagram, the region of low similarity to both reference sera. Nevertheless, Mimosoideae and

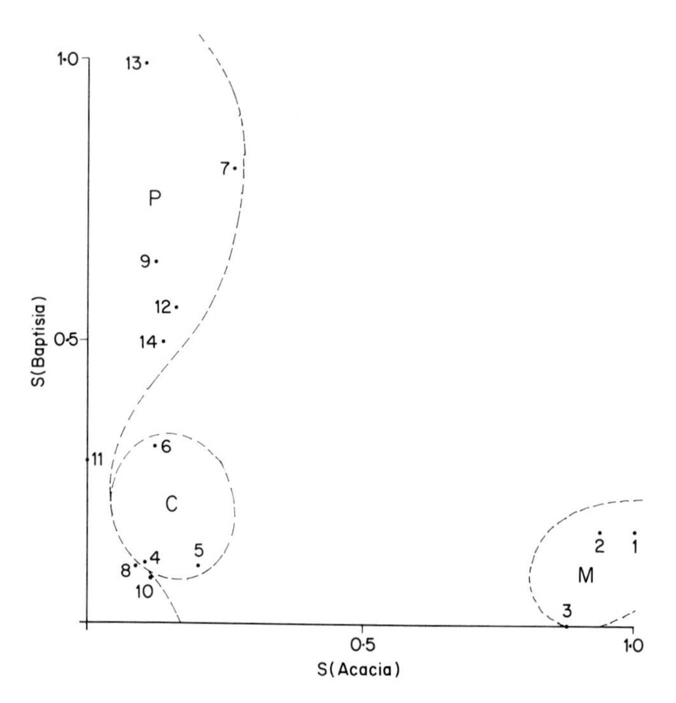

Fig. 8. Ordination of legume genera according to their similarity to two reference sera (anti-*Acacia* and anti-*Baptisia*). S = (number of immune-fractions detected in the cross reaction)/(number of immune-fractions detected in the self-reaction). M = Mimosoideae, C = Caesalpinioideae, P = Papilionoideae 1 = *Acacia* 2 = *Leucaena* 3 = *Albizia* 4 = *Cercis* 5 = *Caesalpinia* 6 = *Gleditsia* 7 = *Robinia* 8 = *Astragalus* 9 = *Lens* 10 = *Trifolium* 11 = *Crotalaria* 12 = *Templetonia* 13 = *Baptisia* 14 = *Spartium*

Caesalpinioideae appear very homogeneous, whilst Papilionoideae, the largest subfamily, is much more diversified: they are not differentiated by anti-*Acacia* serum, but they are by anti-*Baptisia.* The axes are fully independent, as one can establish visually, by observing the distribution of the OTUs – and in fact the correlation between x- and y-coordinates is $r^2 = 0.166$ – that is to say that there are no redundant characters.

As the number of reference sera increases, several ordination techniques can be employed. Principal component analysis has been

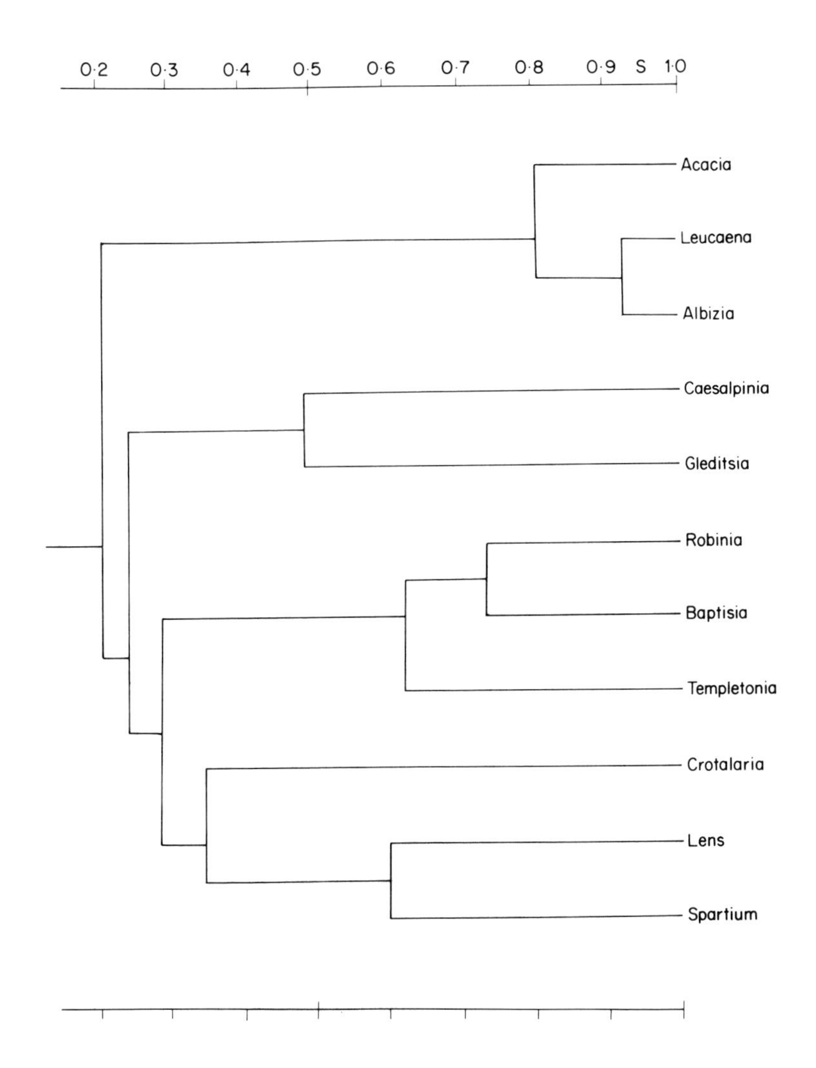

Fig. 9. Clustering by average linkage of Legume genera. Reference antisera to *Acacia retinodes* and *Baptisia australis*. S = (number of immune fractions common to two taxa)/(total number of fractions).

used by Cristofolini and Feoli Chiapella (1977) on data based upon 15 antisera, and by Peace and Lester (1979) on data based upon one serum.

Usually, ordinations are the best way of displaying the pattern of diversity, for they introduce a minimum of distortion. On the other hand, any classification requires hierarchical systems, where OTUs are arranged by categories. A classification has been attempted for the same genera ordered in Fig. 8. The intergeneric correlation coefficients based on presence or absence of the identified immune fractions have been computed. From the matrix of similarity coefficients, the OTUs have been clustered by average linkage (Fig. 9). In spite of the incompleteness of the data (a total of eight immune fractions have been detected by anti-*Acacia* serum and a total of six by anti-*Baptisia*) the subfamilies cluster separately; within the Papilionoideae the splitting of Genisteae *sensu* Taubert (1894) agrees remarkably with modern views on the systematics of the tribe (Polhill, 1979). In comparison with the dendrogram of Fig. 4, it is evident that introducing qualitative information has resulted in a great improvement of the classification in spite of the narrower number of reference sera.

In my opinion, serological reactivity allows one to detect a large set of characters of great systematic value, so that it can give a major contribution to systematics. However, qualitative information should possibly be joined to rough quantitative measures. Furthermore, it should not be overlooked that several antibody systems must be used in any analysis, and that an appropriate numerical processing is necessary in order to handle data and to draw out the results. In this respect, there is a great need for a further development of the theory of serological data processing.

REFERENCES

Boyden, A. (1932). Precipitin tests as a basis for quantitative phylogeny. *Proc. Soc. Exp. Biol. Med.* **29**, 955–957.

Boyden, A. (1954). The measurement of significance of serological correspondence among proteins. *In* "Serological Approaches to Studies of Protein Structure and Metabolism" (W. C. Cole, ed.). Rutgers University Press, New Brunswick.

Boyden, A. (1964). Perspectives in systematic serology. *In* "Taxonomic Biochemistry and Serology" (C. A. Leone, ed.). Ronald Press, New York.

Cristofolini, G. (1971). Contributo sierodiagnostico alla sistematica di Euphorbia triflora Schott, Nym. & K. *Giorn. Bot. Ital.* **105**, 145–156.

Cristofolini, G. (in press). Serological systematics of the Leguminosae. *In* "Advances in Legume Systematics" (R. M. Polhill and P. H. Raven, eds).

Cristofolini, G. and Feoli Chiapella, L. (1977). Serological systematics of the tribe Genisteae, Fabaceae. *Taxon* **26**, 45–56.

Cristofolini, G. and Poldini, L. (1972). Ricerche morfologiche e sierodiagnostiche su una nuova entitá di Cytisus emeriflorus scoperta nelle Alpi Carniche. *Giorn. Bot. Ital.* **106**, 277–279.

Daussant, J. (1975). Immunochemical investigations of plant proteins. *In* "The Chemistry and Biochemistry of Plant Proteins" (J. B. Harborne and C. F. Van Sumere, eds), pp. 31–69. Academic Press, London and New York.

Fairbrothers, D. E. (1967). Chemosystematics with emphasis on systematic serology. *In* "Modern Methods in Plant Taxonomy" (V. H. Heywood, ed.), pp. 141–174. Academic Press, London and New York.

Hurrell, J. G. R., Nicola, N. A., Broughton, W. J., Dilworth, M. J., Minasian, E. and Leach, S. J. (1976). Comparative structural and immunochemical properties of leghaemoglobins. *Eur. J. Biochem.* **66**, 389–399.

Hurrell, J. G. R., Thulborn, K. R., Broughton, W. J., Dilworth, M. J. and Leach, S. J. (1977). Leghaemoglobins: immunochemistry and phylogenetic relationships. *FEBS Letters* **84**, 244–246.

Jaccard, P. (1908). Nouvelles récherches sur la distribution florale. *Bull. Soc. Vaud Sci. Nat.* **44**, 223–270.

Jensen, U. (1968). Serologische Beiträge zur Systematik der Ranunculaceae. *Bot. Jb.* **88**, 204–268.

Kabat, E. A. and Mayer, M. M. (1961). "Experimental Immunochemistry". 2nd edn. Thomas Publications, Springfield, Illinois.

Kloz, J. (1960). The quantitative ring (layering) precipitation reaction. *Folia Biologica* **6**, 319–325.

Kloz, J. (1961). The use of the quantitative precipitation reaction for determining the intensity of cross reactions with special reference to quantitative serology. *Biol. Plant.* **3**, 217–227.

Kubo, K. (1964). An empyrical equation for the precipitin reaction system and its application to systematic serology. *Serol. Mus. Bull.* **31**, 1–3.

Lanzavecchia, G., Parisi, V. and Saita, A. (1965). Osservazioni comparative sulle proteine del tuorlo di alcuni anfibi. *Accad. Naz. Kincei, Rendic. Sci. Fis. Mat. Nat.* 8, **38**, 732–736.

Lee, D. W. and Fairbrothers, D. E. (1969). A serological and disc electrophoretic study of North American Typha. *Brittonia* **21**, 227–243.

Lester, R. N. (1979). The use of protein characters in the taxonomy of Solanum and other Solanaceae. *In* "The Biology and Taxonomy of the Solanaceae" (J. G. Hawkes *et al.*, eds), pp. 285–304. Linn. Soc. Symp. Ser. 7.

Mainardi, D. (1959). Immunological distances among some gallinaceous birds. *Nature* **184**, 913–914.

Mez, C. and Ziegenspeck, H. (1925). Serodiagnostischer (Koenigsberger) Stammbaum des Pflanzenreiches, 1924. *Bot. Arch.* **13**, 483–486.

Osserman, E. F. (1960). A modified technique of immunelectrophoresis facilitating the identification of specific precipitin arcs. *J. Immunol.* **84**, 93–97.

Pearce, K. and Lester, R. N. (1979). Chemotaxonomy of the cultivated eggplant. A new look at the taxonomic relationships of Solanum melongena L. *In*

"The Biology and Taxonomy of the Solanaceae" (J. G. Hawkes *et al.*, eds), pp. 615–628. Linn. Soc. Symp. Ser. 7.

Polhill, R. M. (1976). Genisteae (Adans.) Benth. and related tribes (Leguminosae). Bot. Syst. 1, 143–368.

Prager, E. M. and Wilson, A. C. (1971a). The dependence of immunological cross-reactivity upon sequence resemblance among lysozymes. I. Microcomplement fixation studies. *J. Biol. Chem.* **246**, 5978–5989.

Prager, E. M. and Wilson, A. C. (1971b). The dependence of immunological cross-reactivity upon sequence resemblance among lysozymes. II. Comparison of precipitin and micro-complement fixation results. *J. Biol. Chem.* **246**, 7010–7017.

Sneath, P. H. A. and Sokal, R. R. (1973). "Numerical Taxonomy". Freeman and Co., San Francisco.

Sokal, R. R. and Sneath, P. H. A. (1963). "Principles of Numerical Taxonomy". Freeman and Co., San Francisco.

Taubert, P. (1894). "Leguminosae". *In* "Die Natuerlichen Pflanzenfamilien" (A. Engler and K. Prantl, eds), III, 3, pp. 70–388. Engelmann, Leipzig.

Wallace, C. A. and Boulter, D. (1976). Immunological comparison of higher plant plastocyanins. *Phytochem.* **15**, 137–142.

Williams, C. A. (1964). "Immunochemical Analysis of Serum Proteins of the Primates: a Study in Molecular Evolution". Academic Press, New York and London.

Williams, C. A. and Chase, M. W. (1971). "Methods in Immunology and Immunochemistry". Vol. III. Academic Press, New York and London.

14 The Status of Immunological Distance Data in the Construction of Phylogenetic Classifications: a Critique

A. E. FRIDAY

University Museum of Zoology, Downing Street, Cambridge CB2 3EJ, England

Abstract: In the reconstruction of evolutionary history the usefulness of data derived from immunological cross-reactions involving proteins is generally viewed as depending on a correlation between such data and amino acid sequence difference. The available evidence suggests that this correlation is at best very approximate.

Phylogenetic analysis of immunological distance data by currently available techniques converts the data to "additive" structure. This structure represents an entirely divergent process without distortion; non-divergent changes are dealt with by default, implying the assumption that such changes are few and randomly distributed. Because of the claim that molecular evolution is mostly divergent, phylogenies have been supported on the extent to which observed data approach additive structure. Such an argument, based on the self-consistency of the data, can be misleading, and more direct evidence from amino acid sequences indicates that a substantial amount of non-divergent change is involved in molecular evolution. Phylogenetic methods in which non-divergent change is explicitly analysed require more information about the evolutionary process than we can provide at present.

In these circumstances immunological data do not have the same status as cladistically-coded character state data derived from morphological comparisons,

Systematics Association Special Volume No. 16, "Chemosystematics: Principles and Practice", edited by F. A. Bisby, J. G. Vaughan and C. A. Wright, 1980, pp. 289–304, Academic Press, London and New York.

because we lack biological criteria for evaluating competing solutions involving different distributions of non-divergent change.

INTRODUCTION

There are three sources of error involved in the use of immunological techniques to reveal similarities and differences between organisms. First, the error inherent in experimental procedures; secondly, the interpretation of raw immunological results in such a way that comparisons are correlated with the known or suspected nature of variation between the large molecules concerned; thirdly, the additional limitations of the techniques used to reveal taxonomic or phylogenetic structure in data already containing the first two sorts of error.

Immunological methods can reveal characters which are useful in comparing organisms. They have proven worth in this respect. However, it is when phylogenetic conclusions are drawn from immunological data that major difficulties occur, and it is this aspect that will be briefly examined here.

The first source of error, experimental, is not peculiar to immunological methods, of course. Most currently available methods for phylogenetic reconstruction have not, however, been constructed to incorporate estimates of error, and it may be unrealistic to expend much effort on evaluating competing solutions where the differences between them fall within the bounds of experimental uncertainty. Because of variability between animals in their immune response, immunological data suffer particularly in this respect compared to, for example, amino acid sequence data.

INTERPRETATION OF IMMUNOLOGICAL CROSS-COMPARISONS

The second source of error, the transformation of raw immunological data, has been well aired but can hardly be regarded as resolved. Experimental techniques which employ antibody to measure similarities and differences between antigens from different species are very varied. Fundamentally, however, the behaviour of the antigen-antibody system is quantified to yield a set of interspecies distances. For the purposes of phylogenetic reconstruction these distances, whether derived from double-diffusion, immuno-electrophoresis or reaction in liquid medium, are usually regarded as equal to, or (more realistically) proportional to the number of molecular events separating the relevant parts of the genomes concerned. The validity of such a relationship has been explicitly investigated by few workers: most have been content to make the assumption.

Evidence for a high correlation between their immunological index of dissimilarity (I.D.) and sequence differences between proteins was given by Prager and Wilson (1971) and Prager *et al.* (1972) using experimental results based on bird egg white lysozymes. The extent of the correlation was accepted by Nei (1975), who observed, "Using lysozymes instead of albumin, Prager and Wilson (1971) have shown that log I.D. is linearly related to the proportion of different amino acids between the two sequences compared", although he adds, "The reason why log I.D. should be a linear function of the proportion of different amino acids is not known." Read (1975) has suggested two different descriptive equations which fit the relation between immunological cross-reaction and protein sequence difference better than that originally put forward. Only the acquisition of more experimental evidence can resolve these issues.

Although there is now information about the structural basis of immunological reactivity from a wide variety of proteins, the muscle protein myoglobin has been studied to the extent that it is claimed that all the regions involved in the direct reaction with antibody have been localized. The results from these investigations cast doubt on the simplicity of relations between immunological cross-reaction and sequence difference. Atassi (1975) concludes that general rules are difficult to draw, but because of his work in mapping antigenic reactive regions it has been possible to further examine the implications for immunological comparisons, using the many available amino acid sequences for mammalian myoglobins.

Romero-Herrera *et al.* (1978) found that the myoglobins of man and horse differ at 18 positions (out of a total number of 153 comparable sites), while the myoglobins of sperm whale and horse differ at 19 positions. Goat antisera to horse myoglobin are reported to react with the human protein to the extent of 30% of the homologous cross-reaction, yet the same antisera do not react at all with the sperm whale myoglobin (see Atassi *et al.*, 1970, for details of the immunological procedures).

There may be disproportionate immunochemical effects of substitutions fixed in common by relatively distantly related organisms; related, that is, in terms of recency of common ancestry. Nisonoff *et al.* (1970) give a striking example of this taken from immunochemical studies on cytochrome *c*.

It was pointed out by Romero-Herrera *et al.* (1978) that if the parts of a molecule which are involved in the induction of the precursors of antibody-forming cells are not the same as those parts which finally interact directly with the evoked antibodies, then, even with the three-dimensional structure of myoglobin and the painstaking mapping work of Atassi and his colleagues, firm conclusions are difficult to draw on theoretical grounds. The three-way nature of immunological comparison (immunizing antigen; foreign antigen; antibody) probably contains more information about differences between organisms than we are currently capable of analysing, because the animal providing the antibody is taking more than an impartial interest in the immunizing substances.

All this serves to illustrate that immunological comparisons between proteins are not merely "poor man's sequencing"; there are other considerations peculiar to these approaches which have not yet been fully explored. The attempt to force a correlation between amino acid differences and immunological cross-reaction may be diverting us from dealing with immunological data so that we can make use of the properties of the three-way comparison.

DATA STRUCTURE: SOME DESIRABLE PROPERTIES

It is not, in any case, crucial to the value of immunological methods in the reconstruction of evolutionary history that they yield data which behave as, or can be converted into, amino acid distances. Such a relation would, however, be of use in comparing conclusions from the two techniques and also if it aided us in formulating a model for the evolutionary process in antigens in which probabilities could be assigned to the changes which occur. However, the idea that most fixed change at the molecular level has no significance as regards the process of natural selection, raising the possibility that molecular evolution can be fairly readily modelled (by a Poisson or negative binomial relation, say), has caused considerable argument (reviewed by Wilson *et al.*, 1977).

For appropriate techniques of analysis it matters little whether or not rates of change in whatever is being used to measure distance are constant. Indirectly, rate can matter because low or high rates of change may cause the data to have different properties and so be best analysed by different techniques. The relationship between the minimum evolution and maximum likelihood methods mentioned below is crucially dependent on rate of change per character, for example (Felsenstein, 1973).

For the types of method now being used to analyse immunological distance data it may be sufficient for the measure to have other properties. If we assume that evolutionary relationships can be represented geometrically by branching networks or trees (trees have a "root", networks are undirected in the discussion below, but others, especially graph theorists, use the terms in different ways sometimes), then to derive easily a geometric representation from some distance data it is desirable that the measure of difference should have at least *metric* structure.

The concept of a metric can be found explained at greater length in Copson (1968), Jardine and Sibson (1971), Sneath and Sokal (1973), Beyer *et al.* (1974) and Sattath and Tversky (1977). In the simplest terms, a metric is a function which obeys certain mathematical conditions, and the resulting measure then has features which make it particularly appropriate for the expression and representation of distance between species or other entities. There are many kinds of metric, and also many measures which are not, by definition, true metrics yet which have desirable properties for the manipulation of distance in specialized contexts.

For a set of species, to the pairs of which a non-negative real number can be assigned as a distance, then the conditions to be met are:

1. $d(x, y) > 0$, for $x \neq y$
2. $d(x, y) = 0$, only when $x = y$
3. $d(x, y) = d(y, x)$, for all x, y
4. $d(x, z) \leqslant d(x, y) + d(y, z)$, for all x, y, z

As Beyer *et al.* (1974) observe, the last property, the so-called "triangle inequality", is "essential for relating meaningful topological notions to properties defined by distance".

Non-metric data often cause difficulties when networks are to be constructed from distance matrices. It is simple to test data for the triangle inequality, and massive deviations from metric structure often indicate that inappropriate distance measures are being used. Some data transformations may also result in non-metric distances; for example, amino acid distances calculated between homologous proteins of the same length are metric, but the corresponding minimum mutation distances need not be, although deviations are usually small. The transformations commonly used for amino acid sequence data are usually intended to compensate for genetic changes implicit in, but masked by, the raw number of differences. Correction for multiple events or translation into number of nucleotide differences depends on knowledge of the total number of residues in the protein (sometimes, more precisely, on the proportion of the sites known to vary), or on the actual chemical identity of the amino acids concerned. This information is not usually available for immunological cross-comparisons.

An apparently even more desirable property for distance data in a phylogenetic context is that they be "additive". The concept of additivity is explained well by Waterman *et al.* (1977), who describe the "four-point condition", and refer to the original proofs concerned. Under the "four-point condition", for all sets of four species one searches for a way of labelling the four (call them w, x, y, z) such that:

$$\mathrm{d}(w, x) + \mathrm{d}(y, z) = \mathrm{d}(x, z) + \mathrm{d}(w, y) \geqslant \mathrm{d}(w, z) + \mathrm{d}(x, y)$$

Just as it is possible to check distance data for the triangle inequality, it is possible to check them for the "four-point condition", although computationally the easiest way to do this would appear to be to attempt to construct an additive network from them. This is because additive data have a most valuable property: *such data define one, and only one, metric network.* A proof of this property is given by Waterman *et al.* (1977). Strictly speaking, additive data are pseudo-metric, because it is possible for species to be joined by a link of zero length on a network (Buneman, 1971).

METHODS OF PHYLOGENETIC RECONSTRUCTION

It is sad that data are rarely additive in practice: if they were we could proceed directly to a unique network pattern. Waterman *et al.* (1977) supplement their proof by a simple method for reconstructing the implicit network pattern from an additive dissimilarity matrix, and extend this method to non-additive matrices. Their justification for this extension to the method is that "it works in the case of additive dissimilarity matrices", a justification which will be examined below. The extended method makes use of linear programming, for which universally available computer routines exist. These linear algebra techniques had previously been advocated by the same group (Beyer *et al.*, 1974) for the analysis of amino-acid sequence data, and essentially involve simultaneous equations in which the branch lengths of a test network pattern are the unknown variables to be solved.

Experience with this method confirms its theoretical similarity to the Wagner method of Farris (1972), and the two techniques often produce identical solutions in practice. Farris originally wrote the distance Wagner technique with immunological distance data in mind. He assumed for this purpose that the immunological distances were at least approximately proportional to amino acid distances, but even true amino acid distances do not conform to the ideal metric for the distance Wagner equations. The ideal metric is the so-called "Manhattan" metric, and reference should be made to Farris' account for a description of its properties. The Farris method is, however, robust in operation, and these deviations from ideal data structure do not usually greatly affect the accuracy of its solutions with test data.

It might be thought that, because we cannot distinguish ancestral and derived character states from immunological distance data, species will necessarily be closely linked on a reconstructed network when they share many unmodified ancestral features. However, as long as some divergent change is involved (a requirement for any phylogenetic analysis), the data can have additive structure no matter how many shared unmodified ancestral character states are present. Such data will, in the absence of non-divergent change, lead to the "true" network pattern when analysed by a Farris Wagner type of procedure, but other techniques may not achieve a correct solution. This limitation on technique has recently been examined by Baverstock *et al.* (1979), in the context of electrophoretic data.

Nei (1975) has criticized the Wagner approach because it does not allow the distance between two species on a reconstructed network to be less than the distance between them in the original data matrix. A consequence of this is that if experimental error is inherent in the measurements it is treated as significant. The linear programming techniques could be used to investigate the stability of networks under varying amounts of error, but no published study has yet appeared which makes use of this possibility. Including error bounds greatly increases the complexity of network reconstruction. Cavalli-Sforza and Edwards (1967) explicitly avoided this complication whilst noting that if errors in the interspecies distances are known not to be independently distributed, or to have unequal variances, the corresponding variance–covariance matrix should be incorporated in the estimation procedure. There remains the need for such developments in current methods. Cavalli-Sforza and Edwards, using gene frequency data, placed emphasis on the nature of the character space appropriate to their data and, therefore, viewed the problem of quoting standard deviations for their estimates as best approached by an examination of the consequences of small displacements of the populations in the character space.

MEASURING "GOODNESS-OF-FIT"

The issue of experimental error is important because it has lead to the evaluation of several methods in the light of Nei's criticism of the Wagner technique. Prager and Wilson (1978) examined the Unweighted Pair Group (see Nei, 1975), Farris Wagner and Fitch and Margoliash (1967) approaches. To evaluate the phylogenies produced by these methods from test data, they used the "goodness-of-fit" criterion. Under this criterion the best network is that whose output (network) distances between species match most closely the corresponding input (data) distances. A variety of measures of goodness-of-fit or stress are in common use; Prager and Wilson chose to compare the "standard deviation" and their F value. The standard deviation measure is given by:

$$\sqrt{\left(\sum_{j=1}^{n}\left[\frac{(i_j - o_j)\,100}{i_j}\right]^2 \bigg/ \left[\frac{N(N-1)}{2} - 1\right]\right)}$$

where i is an input distance, o an output distance, N is the number of species and n the total number of comparisons (that is, $N(N-1)/2$). Prager and Wilson's F value is given by $100\,f/I$, where f is the sum of absolute values of differences between corresponding input and output distances, and I is the sum of all input values. Essentially this is the Farris f value converted to a percentage.

Prager and Wilson discuss the various merits of the goodness-of-fit statistics, advocating the principal use of their F value.

As far as the performance of the various procedures went, these authors found that the Farris method was competitive with the Fitch and Margoliash method when the goodness-of-fit measures had low values for given data. This is the same as saying that the methods converge when the data are near additive structure, which is not surprising.

Given that immunological distances contain experimental error, Prager and Wilson advocate the Fitch and Margoliash method for such data, with the Farris method as second choice and the Unweighted Pair Group method rejected because of its excessive dependence on constancy of rate of change.

Whereas the Farris Wagner and linear algebra methods (as described) are minimum evolution methods, in which the total network length is minimized, the Fitch and Margoliash method as used by Prager and Wilson is a means of imposing a best-fitting additive matrix upon the input distances. This emphasizes the formal similarity of the Fitch and Margoliash method to the Least Squares Additive approach of Cavalli-Sforza and Edwards (1967) and to the procedure of Buneman (1971). In fact, with appropriate choice of the function to be minimized the linear algebra methods can be used to fit best additive matrices to data; Beyer *et al.* (1974) have used the following function for this purpose:

$$\sum_{ij} (o_{ij} - i_{ij})/i_{ij}$$

This function was used together with the constraint that all estimated branch lengths on the network should be $\geqslant 0$, but the constraint could be relaxed and the absolute differences between input and output distances taken. Dividing by the original input distance gives a statistic with different properties again to the two examined by Prager and Wilson, and indeed a whole family of functions could be used as goodness-of-fit statistics. I have elsewhere (Friday, 1977) advocated the *delta* series of statistics used by Jardine and Sibson (1971) because they enable one to assess the balance of stress in an analysis; that is, whether distortion is attributed largely to a few entries in the matrix or is spread more evenly. In practice one often comes across different solutions which can be separated in this fashion, although their relative merit is a more contentious issue.

The matter of negative branch lengths in phylogenetic reconstructions has been widely discussed. Cavalli-Sforza and Edwards (1967) interpreted the presence of a negative branch length in a Least Squares Additive solution as indicating the need to change the network form, although it was not clear to them that a solution with no negative branches might always be found. Kidd and Sgaramella-Zonta (1971) reinforced the observation that one may get negative branch lengths from the imposition of an incorrect solution on additive data; even for real data their analysis appears to support rejection of negative branch lengths on empirical grounds. Both the original linear algebra and Farris Wagner methods are explicitly designed to prevent the generation of negative branch lengths, although, as noted by Prager and Wilson (1978), small negative lengths may be encountered with the Farris method. I have found these only for large sets of data; they are presumably a feature of the accumulated approximations used in the Wagner distance equations.

It is difficult to see what negative branch lengths might mean in terms of an evolutionary process, but one can base a defence for their presence on the statistical nature of the goodness-of-fit approach. Often an iterative method may pass through solutions with negative branch lengths to reach a more satisfactory final solution (Cook and Hewett-Emmett, 1974).

Kidd and Sgaramella-Zonta (1971) were not able satisfactorily to resolve the relationship between minimum evolution and least squares methods, but conjectured that the network with the lowest $\Sigma\,(\text{error})^2$ value among those with no negative segments is the network with the minimum length. This led them to advocate the use of a combined method searching for the best-fitting (least squares) solution with minimum length. It should be pointed out that the Farris method does attempt to obtain a good additive fit to the data at the same time as pursuing its prime objective of a minimum length solution, and the linear algebra methods can be engineered to work in the same way.

Can either the minimum evolution or best-fit additive approaches be rigorously justified? Edwards (1971), the begetter of the method of minimum evolution, is under no illusions about this approach: "I now know the conjecture on which it was based to be false", and

Buneman (1971) has observed of the additive approach that it will usually produce a detailed tree even from distances which do not at all resemble an additive tree metric!

Furthermore, Beyer *et al.* (1974) demonstrate that the minimum evolution approach may produce a variety of solutions at the same total length but with different branch length assignments. They claimed that the use of their version of the goodness-of-fit criterion avoided this for the best-fit additive approach and led to a unique assignment of branch lengths, but this is not the case for all goodness-of-fit criteria.

THE PROBLEM OF NON-DIVERGENT CHANGE

Perhaps the most telling criticism of the minimum evolution and best-fit additive methods comes from their comparison with the method of maximum likelihood. For gene frequency data, Thompson (1973) found both methods lacking in comparison with a likelihood estimation; the minimum evolution method being somewhat more justifiable than the Least Squares Additive. The maximum likelihood approach was also advocated by Felsenstein (1973), and promises rich returns (including estimates of branching times and the position of the "root") if only a model can be framed in which probabilities can be assigned to events. Likelihood methods are particularly well suited to comparing the relative support for competing hypotheses.

Farris (1973) and Beyer *et al.* (1974) have both argued that committing ourselves to a detailed hypothesis about the nature of evolutionary change is inappropriate when we are trying to determine how that change took place. Except for gene frequency data, it may prove impracticable to carry out an analysis to estimate both the progress and pattern of evolutionary change at a single stroke, which would be required of a full likelihood approach. In any case, as Felsenstein (1978) has recognized, convincing models for evolution are difficult to think of. Nevertheless, minimum evolution and best-fit additive methods do not always give the same answers as likelihood methods in those situations in which they can be tested against each other; indeed, the restrictions on the data to make the methods agree are quite severe (Felsenstein, 1973).

If data are subject to error, then, this may result in non-additive (or even non-metric) distances, and influence our choice of estimating procedure, albeit a choice from methods which lack rigorous justification.

Even in the absence of error, non-divergent change also causes non-additive distances. With robust methods, such as the Wagner procedure, moderate and localized non-divergent change can be recognized and coped with. Larger proportions cause an incorrect network structure to be assigned. This can be demonstrated by using data derived from simulations in which a true network structure is known. Some of the stress introduced into the data by the analysis may represent non-divergent events being revealed.

Current procedures fail on test data when non-divergent change is considerable. At a certain level of such change it may be more parsimonious or less stressful to assign a different network pattern from the true one. Peacock and Boulter (1975) have investigated this problem using amino acid sequences.

The "justification" for the method of Waterman *et al.* (1977), quoted above, reveals forcibly that these procedures fail on data with many non-divergent changes because such change is not part of the underlying model which forms the basis of the analysis. If we are forced to use methods which deal with this problem by default, we cannot expect to resolve the distribution of non-divergent changes between competing network patterns with any conviction.

To model non-divergent change we need to understand much more about the reasons for change, either resolving the selective reasons or creating a model in which such events are expected with a known probability. From immunological distance data we cannot recognize and examine competing distributions of non-divergent change in biological terms. The inability to translate the immunological data in a functionally intelligible way suggests that such data are intrinsically inferior to cladistically-coded character state data, where we can at least attempt to understand function and to assign the polarity of change on the basis of this understanding. Interspecies distances always represent a relatively unsatisfactory approach to phylogenetic problems; to quote Buneman (1974), "It is not surprising that by avoiding a DC [dissimilarity coefficient], one can build trees which give much better descriptions of the data. Finding trees, and possibly clusters, from raw attribute data is something that deserves further investigation." Phylogenetic methods have had intense study in recent years, but methods of scoring immunological comparisons have not kept pace.

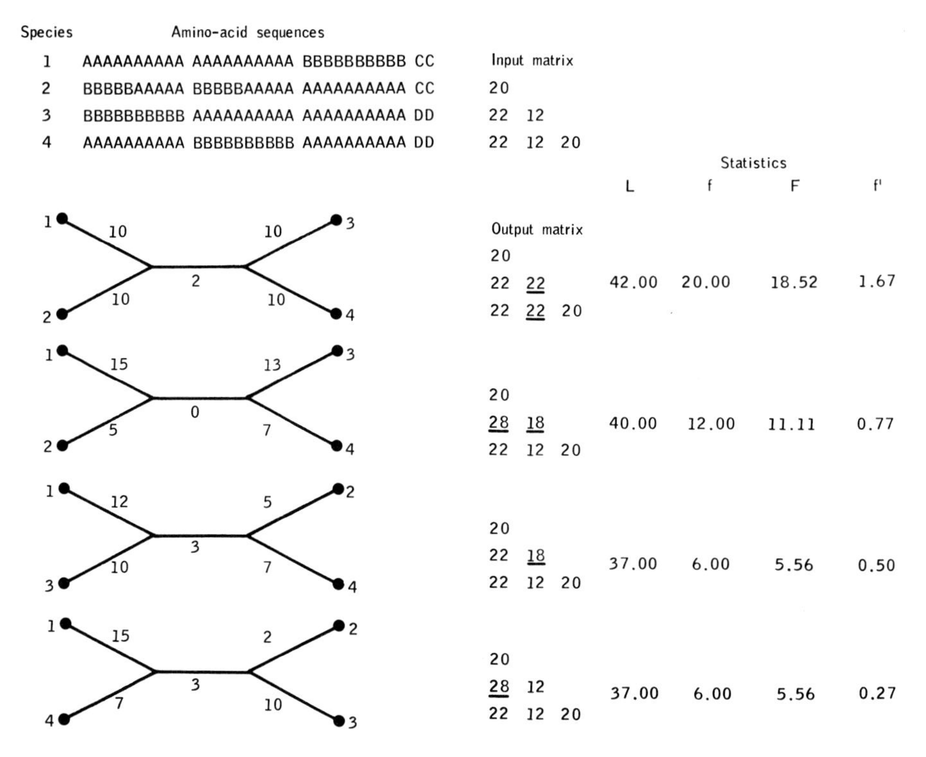

Fig. 1. The set of hypothetical amino acid sequences at the top of the figure yields an input distance matrix of differences between the four species; this matrix is shown on the right of the sequences.

Of the four networks, the first shows the known course of change. The remaining ones are estimates obtained for each of the three possible network patterns dichotomously linking four species. Since there is no experimental error in this case, linear algebra methods were used to obtain integral branch lengths none of which could be negative.

For each network a matrix of interspecies distances has been calculated by summing the branch lengths between the species. To the side of each matrix are given the total length (L) of the network concerned and three distortion statistics, f (Farris, 1972), F (Prager and Wilson, 1978) and f' (Beyer *et al.*, 1974; called by them F_2). The distortion statistics are explained in detail in the text.

Because of the extensive parallel change shown by species 2 with species 3 and 4, the networks showing less distortion are nearest the input matrix but farthest from the known pattern of change. If we had not known the original pattern of events, then accepting a solution with the wrong pattern would enable us to "support" a reconstruction which grossly underestimates the number of non-divergent changes.

Without additional information we cannot break the circularity of such an analysis. If it could be shown from other evidence that the process of evolution was entirely, or very substantially, divergent, the analysis of distance data on an additive model would be more justifiable.

THE LOGICAL CIRCULARITY OF "SELF-CONSISTENCY"

The additive model breaks down in the presence of considerable non-divergent change, but perhaps it is possible to recognize and avoid sets of data where much non-divergent change is revealed on analysis. In other words, can we rely on the self-consistency of the data to support a case for mostly divergent change in the process which generated those data? Cronin and Sarich (1975) note the fact that "protein and nucleic acid evolution are basically divergent processes". In keeping with this assumption, and supposing that non-divergent changes, where they occur, are randomly distributed, Cronin and Sarich take an additive best-fit approach. Their aim is to minimize the differences between the cladogram they construct and the actual one. By this tactic they avoid the temptation to test their conclusions other than internally.

Relying on an additive best-fit approach to corroborate the mainly divergent nature of one's data can, however, be seriously misleading. As we have seen, at high levels of non-divergent change the assignment of an incorrect network pattern will improve the goodness-of-fit and reinforce the impression that "the extent to which the differences we measure among modern species are indeed uniquely apportionable into a phylogeny" (Cronin and Sarich, 1975) is a reliable criterion.

This is not, of course, to say that Cronin and Sarich will necessarily be wrong in any given analysis of their data, merely that their approach is self-reinforcing and logically incapable of revealing considerable non-divergent change where this has occurred. This is illustrated in the example data of Fig. 1.

For character state data, Gaffney (1979) has pointed out that one cannot logically advance an argument for parallelism before a hypothesis of relationship is even tested. Gaffney's contention that the only objective evidence we can have for the existence of non-divergent change is the appearance of character contradictions is an important one. From attempts at phylogenetic analysis of amino acid sequence data there are claims (Peacock and Boulter, 1975; Romero-Herrera *et al.*, 1978) that considerable amounts of non-divergent change must be a feature of molecular evolution. In recognition of this, and in keeping with the logic advanced by Gaffney, there has been a promising start with the application of compatibility analysis (see Estabrook *et al.*, 1977, and the references therein) to amino acid sequence data (Fitch, 1977; Boulter *et al.*, 1979). Compatibility methods test for character contradictions. Where different characters cannot be manipulated to support an identical phylogeny, non-divergent change may well be the cause of this. It is not appropriate to attempt here a critique of the logic of support for competing phylogenetic schemes advanced by the compatibility school. It is sufficient to point out that compatibility methods are currently the best means we have of revealing non-divergent change, and they cannot be used on immunological distance data.

REFERENCES

Atassi, M. Z. (1975) Antigenic structure of myoglobin: the complete immunochemical anatomy of a protein and conclusions relating to antigenic structure of proteins. *Immunochemistry* **12**, 423–438.

Atassi, M. Z., Tarlowski, D. P. and Paull, J. H. (1970). Immunochemistry of sperm whale myoglobin. VII. Correlation of immunological cross-reaction of eight myoglobins with structural similarity and its dependence on conformation. *Biochim. biophys. Acta* **221**, 623–635.

Baverstock, P. R., Cole, S. R., Richardson, B. J. and Watts, C. H. S. (1979). Electrophoresis and cladistics. *Syst. Zool.* **28**, 214–219.

Beyer, W. A. Stein, M. L., Smith, T. F. and Ulam, S. M. (1974). A molecular sequence metric and evolutionary trees. *Math. Biosci.* **19**, 9–25.

Boulter, D., Peacock, D., Guise, A., Gleaves, J. T. and Estabrook, G. (1979). Relationships between the partial amino acid sequences of plastocyanin from members of ten families of flowering plants. *Phytochemistry* **18**, 603–608.

Buneman, P. (1971). The recovery of trees from measures of dissimilarity. *In* "Mathematics in the Archaeological and Historical Sciences" (F. R. Hodson, D. G. Kendall and P. Tatu, eds), pp. 387–395. University Press, Edinburgh.

Cavalli-Sforza, L. L. and Edwards, A. W. F. (1967). Phylogenetic analysis: models and estimation procedures. *Evolution* **21**, 550–570.

Cook, C. N. and Hewett-Emmett, D. (1974). The uses of protein sequence data in systematics. *In* "Prosimian Biology" (R. D. Martin, G. A. Doyle and A. C. Walker, eds), pp. 937–958. Duckworth, London.

Copson, E. T. (1968). "Metric Spaces". University Press, Cambridge.

Cronin, J. E. and Sarich, V. M. (1975). Molecular systematics of the New World monkeys. *J. hum. Evol.* **4**, 357–375.

Edwards, A. W. F. (1971). Mathematic approaches to the study of human evolution. *In* "Mathematics in the Archaeological and Historical Sciences" (F. R. Hodson, D. G. Kendall and P. Tatu, eds), pp. 347–355. University Press, Edinburgh.

Estabrook, G. F., Strauch, J. G. and Fiala, K. L. (1977). An application of compatibility analysis to the Blackiths' data on orthopteroid insects. *Syst. Zool.* **26**, 269–276.

Farris, J. S. (1972) Estimating phylogenetic trees from distance matrices. *Am. Nat.* **106**, 645–668.

Farris, J. S. (1973). A probability model for inferring evolutionary trees. *Syst. Zool.* **22**, 250–256.

Felsenstein, J. (1973). Maximum likelihood and minimum-step methods for estimating evolutionary trees from data on discrete characters. *Syst. Zool.* **22**, 240–249.

Felsenstein, J. (1978). Cases in which parsimony or compatibility methods will be positively misleading. *Syst. Zool.* **27**, 401–410.

Fitch, W. M. (1977). On the problem of discovering the most parsimonious tree. *Am. Nat.* **111**, 223–257.

Fitch, W. M. and Margoliash, E. (1967). Constructions of phylogenetic trees. *Science, N.Y.* **155**, 279–284.

Friday, A. E. (1977). Evolution by numbers. *In* "Myoglobin" (A. G. Schneck and C. Vandercasserie, eds), pp. 142–166. Editions de l'Universite, Brussels.

Gaffney, E. S. (1979). An introduction to the logic of phylogeny reconstruction. *In* "Phylogenetic Analysis and Paleontology" (J. Cracraft and N. Eldredge, eds), pp. 79–111. Columbia University Press, New York.

Jardine, N. and Sibson, R. (1971). "Mathematical Taxonomy". Wiley, London.

Kidd, K. K. and Sgaramella-Zonta, L. (1971). Phylogenetic analysis: concepts and methods. *Am. J. hum. Genet.* **23**, 235–252.

Nei, M. (1975). "Molecular Population Genetics and Evolution". Amsterdam, North-Holland.

Nisonoff, A., Reichlin, M. and Margoliash, E. (1970). Immunological activity of cytochrome *c*. II. Localization of a major antigenic determinant of human cytochrome *c*. *J. biol. Chem.* **245**, 940–946.

Peacock, D. and Boulter, D. (1975). Use of amino acid sequence data in phylogeny and evaluation of methods using computer simulation. *J. mol. Biol.* **95**, 513–527.

Prager, E. M., Arnheim, N., Mross, G. A. and Wilson, A. C. (1972). Amino acid sequence studies on bobwhite quail egg white lysozyme. *J. biol. Chem.* **247**, 2905–2916.

Prager, E. M. and Wilson, A. C. (1971). The dependence of immunological cross-reactivity upon sequence resemblance among lysozymes. I. Microcomplement fixation studies. *J. biol. Chem.* **246**, 5978–5989.

Prager, E. M. and Wilson, A. C. (1978). Comparison of phylogenetic trees for proteins and nucleic acids: empirical evaluation of alternative matrix methods. *J. mol. Evol.* **11**, 129–142.

Read, D. W. (1975). Primate phylogeny, neutral mutations, and "molecular clocks". *Syst. Zool.* **24**, 209–221.

Romero-Herrera, A. E., Lehman, H., Joysey, K. A. and Friday, A. E. (1978). On the evolution of myoglobin. *Phil. Trans. R. Soc. B.* **283**, 61–163.

Sattath, S. and Tversky, A. (1977). Additive evolutionary trees. *Psychometrika* **42**, 319–345.

Sneath, P. H. A. and Sokal, R. R. (1973). "Numerical Taxonomy". Freeman, San Francisco.

Thompson, E. A. (1975). "Human Evolutionary Trees". University Press,

Waterman, M. S., Smith, T. F., Singh, M. and Beyer, W. A. (1977). Additive evolutionary trees. *J. theor. Biol.* **64**, 199–213.

15 | Classification from Chemical Data

J. A. HARRIS

Bedford College, Regent's Park,
London NW1 4NS, England

and

F. A. BISBY

Biology Department, Building 44,
Southampton University, Southampton S09 5NH, England

INTRODUCTION

During the late 1960s and early 1970s there was a wide-spread assumption that *taxonomic data from chemical analyses of plants and animals would suggest classifications similar to those based on anatomical and morphological data.* It has at times been invoked both in phenetic work aimed at the recognition and delimitation of taxa based on overall resemblance and in phylogenetic work aimed at arranging taxa in phylogenetic sequences. We shall confine ourselves to the former.

The assumption is a special case of the hypothesis of taxonomic congruence (Farris 1971), here applied to chemical and morphological data sets and to pairs of chemical data sets, and referred to as the hypothesis of chemotaxonomic congruence (HCC). Despite its imprecision, the HCC is fundamental to most chemotaxonomic

Systematics Association Special Volume No. 16, "Chemosystematics: Principles and Practice", edited by F. A. Bisby, J. G. Vaughan and C. A. Wright, 1980, pp. 305–327, Academic Press, London and New York.

studies up to the present. Although some theoretical support did come from the non-specificity hypothesis of Sokal and Sneath (1963), the principal support comes from practical experience. Mayr (1964) writes "newer methods and characters only rarely lead to drastic changes in the recognition and arrangement of taxa, providing the original arranging had been done by a biologically thinking taxonomist". Likewise Sneath (1971) reports, "No obvious difference between the patterns of morphological and chemotaxonomic characters in higher organisms have yet been reported." Seigler (1974) is even more adamant, "All the characters of a plant must be related and consistent."

1. Doubts about the HCC

Confidence in the HCC is now waning. We suggest three factors which contribute doubts as to whether the hypothesis holds.

(a) Counter examples. Counter examples did appear in the 1960s and have continued to appear, albeit at a low frequency. Simon and Goodall (1968) investigated relationships of *Medicago* using polyphenolic compounds but found little agreement with the accepted orthodox classification. Similarly, Brown (1967) found wing pigment characters of nymphalid butterflies that suggested taxonomic groupings contradictory to those suggested by morphological, breeding and behavioural studies.

(b) Under-reporting of counter examples. Many cases of partial or total disagreement are dismissed as peculiarities, failures or "useless" results and the data or patterns in them are not published. In this context it is interesting to examine the crude conundrum:

(1) If chemical data fit the morphological pattern, they are taxonomically meaningful and publishable (but, as they agree, the classification remains unchanged and the exercise was a waste of time!).

(2) If chemical data disagree with the morphological pattern they are not taxonomically meaningful or publishable (and again the exercise was a waste of time!).

Are chemotaxonomists really confined to the uneasy path between these extremes, to locating data of partial agreement and using it to tinker with the detail of the morphological classification?

(c) Doubts about comparisons. Doubts about the significance of many comparisons reported centre on exactly what is being compared and how. Many of the generalized statements are based on comparisons made "by eye" between new chemical data and one or several orthodox classifications. The commonest practices are either to record "agreement" when just some characters in the new data agree or to select just these few characters that agree and add them on to the morphological data. Such "agreement" between the characters and one of the morphological classifications may then be used to establish its superiority (Bernasconi *et al.*, 1965; Farnsworth and Trojanek, 1973).

This type of reporting, which is still carried out today, has what we consider to be three shortcomings. First, classification of the new chemical data, the production of the orthodox classification and the comparison are done "by eye", that is by inspection, without recourse to the explicit methods now available. Many orthodox taxonomists can produce excellent classifications by inspection and others have "nearly explicit" methods which they use, albeit in their heads rather than on paper or in a computer. It is not our intention to criticize their normal products. It is only in the context of assessing congruence that we insist that explicit methods must be used to remove at least some of the many variables from the classifications to be compared. Often the chemical analysis and the orthodox taxonomic analysis have been done by quite different people. If we eliminate this source of variation by having one worker make both classifications we come to a different problem: orthodox taxonomists have good memories and are trained to pick out correlated characters. As a consequence, whichever classification is made second is biased by knowledge of the first. Often this occurs subconsciously because of the use of names based on the morphological classification to label chemical samples.

Secondly, the agreement reported often involves only a selection of the chemical characters. Our assumption, that chemical and morphological patterns might agree, should surely apply to the whole set of chemotaxonomic characters and the whole set of morphological characters, whereas many supposed examples apply to subsets or single chemical characters compared with the whole morphological set. Clearly the observation that the occurrence of the amino acid canavanine corresponds with the morphological subdivision of the Leguminosae into subfamilies is very different from the observation that in the tribe Vicieae the morphological pattern and overall amino acid patterns agree in delimiting two large genera *Lathyrus* and *Vicia*.

Thirdly, it is possible for some chemical characters to agree with one morphological classification and for others to agree with another. Therefore the observation of a selection of characters from one type of chemical character (where many others remain to be observed) that agree with one of several morphological classification though interesting is in no way conclusive.

2. *Requirements for Comparisons*

If we set two requirements for congruence tests:

(a) that the two data sets be analysed by the same explicit classification procedure and

(b) that the two results be compared by an explicit technique, then surprisingly we eliminate, with very few exceptions, all the examples in the literature. Basford *et al.* (1968) came close to meeting these requirements when they found a high level of congruence between numerical classifications of Coleoptera based on protein-characters and an orthodox classification converted to numerical form for comparison. One of the most interesting examples that does meet the requirements is that of Mickevich and Johnson (1976) in which they demonstrate the incongruence of patterns based on morphological and allozyme data on species of *Menidia*.

In our opinion cases where there is severe incongruence between chemical and morphological data do exist and such cases do pose both fundamental and practical problems for taxonomists. We illustrate these below with the interesting, if extreme case of the *Cytisus/Genista* complex in the Leguminosae.

INCONGRUENCE IN THE *CYTISUS/GENISTA* COMPLEX

The *Cytisus/Genista* complex in the tribe Genisteae of the family Leguminosae contains about 195 species if the marginal genera *Adenocarpus* and *Argyrolobium* are omitted. They are mostly yellow flowered shrubs found as a major component of shrubby vegetation in Europe, but with a few species found in the Canaries, Asia Minor and N. Africa. *Ulex europaeus* (Gorse) and *Cytisus scoparius* (Broom) are well-known members in N.W. Europe, whilst *Laburnum anagyroides* and *Spartium junceum* are widespread in Central and in Southern Europe respectively.

The complex has a history of unstable classification caused primarily by a reticulate and poorly differentiated pattern of morphological variation. Plants in the complex are rich in flavonoids and alkaloids and we may suppose that the early chemotaxonomic work by Harborne (1969) and Faugeras and Paris (1971) was aimed at using chemical data to provide clarification and stability. Four different morphological classifications are still in use today: those of Bentham and Hooker (1865), Rothmaler (1944), Hutchinson (1964) and Polhill (1976). Rothmaler's system provides the basis for the arrangement in Flora Europaea (Tutin *et al.*, 1968) and, because this divides the complex into the largest number of subdivisions, it is the nomenclature of this system which will be used below. To some extent the small genera of Flora Europaea can be united in different combinations to produce the other systems.

One of us (JAH) collected comparative data for a selection of species using ten different sources – observations on morphology, scanning electron microscopy of pollen surfaces, floral fragrance, UV reflectance of petals, electrophoresis of seed proteins, serology of seed proteins, seed alkaloids, seed free amino acids, flavonoids and chromosome numbers. All but the second in this list yielded variations which could be used as taxonomic characters, that is which (a) were wholly or nearly invariant within species, (b) varied between some of the species and (c) were similar between some of the species.

The number of species investigated for each type of information was limited by the availability of seeds and of living material. Thus for the UV reflectance data, which could be collected from live or pressed material, 91 species were recorded, whereas morphological data, based on living material, was limited to 48 species.

Table I. Ten Sources of Taxonomic Information

		No. species	No. characters
1.	Morphology	48	59
2.	Pollen grain morphology	59	–
3.	Floral fragrance	40	18
4.	Ultraviolet patterns of flowers	62	3
5.	Electrophoretic mobility of seed proteins	41	20
6.	Serological properties of seed proteins	42	4
7.	Seed alkaloids	53	9
8.	Free amino acids in the seeds	48	7
9.	Leaf flavonoids	90	14
10.	Chromosome numbers	89	7

The identity of all material was carefully checked. All too often good chemical work has lost credibility through mistakes in identification. Many of the seeds used in these chemical analyses were supplied by European botanic gardens, often with very little information as to the origin of the parent material. All such supplies were grown to flowering plants and their identity verified. About 10% of seed batches received proved to be wrongly labelled.

1. Nine Sources of Information

Detailed description of methods, data and results are available in Harris (1980).

(a) Morphological data. The principal reasons for collecting a set of comparative morphological data was to produce a morphological classification by exactly the same methods and for exactly the same living material as was used for the other types of data. This could then be used for comparison with the other results. The 59 morphological characters used were those described in Harris (1980), and differ from those used in Bisby and Nicholls (1977) and Bisby (1980). Secondly, it was of interest to compare the result with the existing morphological classifications.

(b) Floral fragrance. The volatile compounds in crushed petals were extracted in *N*-pentane and analysed by gas liquid chromatography. Eighteen peaks on the chromatogram were used as characters.

(c) UV reflectance of petals. The plain yellow petals of most species in the *Cytisus/Genista* complex bear marks and lines which do not reflect UV and other areas that do. The resulting appearance to a bee is of a two coloured flower sometimes with guide lines.

The UV reflectance patterns were detected by photographing mature untriggered flowers through a filter that transmits only UV. An alternative which yields similar patterns is to observe the fluorescence of pressed petals on herbarium sheets when illuminated with UV. Several distinct patterns on the different petals were recorded as three characters amongst the species.

(d) Electrophoresis of seed proteins. Proteins in mature seeds were extracted into a phosphate buffer and analysed by electrophoretic and serological methods. Poly-acrylamide gel disc-electrophoresis was used to separate the proteins according to their charge, size and shape. Once the gels had been stained those protein bands clearly common to several extracts were recorded as 20 characters.

(e) Serology of seed proteins. Both double diffusion and immuno-electrophoretic techniques were investigated. It was the results of the immuno-electrophoretic work which were used to produce four characters.

(f) Seed alkaloids. Alkaloids were extracted from mature seeds into chloroform, separated by thin layer electophoresis on prepared cellulose sheets and then identified by mass spectroscopy. The presences of 9 alkaloids were then recorded for each species studied.

(g) Free amino acids in seeds. These were extracted into ethanol and then separated by high voltage electophoresis. The amino acids have so far only been tentatively identified from Rf values, colour reactions and comparisons with known standards. The distribution of seven amino acids amongst the species was recorded.

(h) Flavonoids. Leaf flavonoid data was obtained from work done by Harborne (1969). Herbarium material used in his study is known to have been identified authoratitively by Gibbs. Fourteen substances were recorded.

(i) Chromosome numbers. Records of chromosome numbers were obtained from the literature and used as the basis for a classification using seven characters.

2. Taximetric Methods

The explicit taximetric method used to analyse and form a classification from each set of data was the single link cluster analysis technique using graph theory described by Wirth *et al.* (1966). The classifications produced by this method are strictly phenetic, that is, they are based on overall resemblance: there is no attempt to assess their evolutionary history which would result in a phylogenetic classification. The results can be drawn as a dendrogram or as a series of linkage diagrams. The linkage diagrams were found to be more useful in that they showed the exact links by which clusters form and then link together in each classification.

Figure 1 shows one of these linkage diagrams. Each symbol represents a species. Each pair of symbols joined by a line, known as a link, is a pair of species with a similarity recorded as more than the threshold similarity for which the diagram has been drawn. Thus in Fig. 1 the symbols linked by lines represent species with similarity greater than 0.66 and the symbols listed as single member clusters do not have a similarity of 0.66 with any other species in the study. A group of symbols with one or more links is known as a cluster. Symbols within a double circle represent species with identical data. Species within single circles have more than three links with other species in the circle.

For the sake of visual comparisons different symbols have been allocated to the species of each of the genera recognized in Flora Europaea (Tutin *et al.* 1968) as shown in Table II. Thus the reader can see that in Figs. 1 and 2 the classification produced from our morphological data by this technique shows a strong resemblance to the Flora Europaea system: for instance the species of *Chamaecytisus* (open circles) link together, as do the species of

Table II. Two generic Classifications of Genisteae

Generic symbols	No. of spp.	Tutin et al. (1968)	Polhill (1976)		
	200	Lupinus	Lupins		
	70	Argyrolobium	Argyrolobium		
	15	Adenocarpus	Adenocarpus		
▬	2	Laburnum	Laburnum		CYTISUS – GENISTA COMPLEX
▭	1	Hespero-laburnum	Hespero-laburnum		CYTISUS – GENISTA COMPLEX
◊	1	Podocytisus	Podocytisus		CYTISUS – GENISTA COMPLEX
●	31	Cytisus	Cytisus	'Cytisus Group'	CYTISUS – GENISTA COMPLEX
✖	1	Petteria	Cytisus	'Cytisus Group'	CYTISUS – GENISTA COMPLEX
⚲	1	Argyrocytisus	Cytisus	'Cytisus Group'	CYTISUS – GENISTA COMPLEX
○	42	Chamaecytisus	Cytisus	'Cytisus Group'	CYTISUS – GENISTA COMPLEX
△	2	Chronanthus	Cytisus	'Cytisus Group'	CYTISUS – GENISTA COMPLEX
⊖	2	Lembotropis	Cytisus	'Cytisus Group'	CYTISUS – GENISTA COMPLEX
⧓	2	Calicotome	Calicotome	'Cytisus Group'	CYTISUS – GENISTA COMPLEX
◆	1	Erinacea	Erinacea	'Cytisus Group'	CYTISUS – GENISTA COMPLEX
✦	1	Spartium	Spartium		CYTISUS – GENISTA COMPLEX
⋈	3	Gonocytisus	Gonocytisus		CYTISUS – GENISTA COMPLEX
●–●	4	Retama (Lygos).	Retama		CYTISUS – GENISTA COMPLEX
■	76	Genista	Genista	'Genista' Group	CYTISUS – GENISTA COMPLEX
+	10	Teline	Genista	'Genista' Group	CYTISUS – GENISTA COMPLEX
□	2	Chamaespartium	Genista	'Genista' Group	CYTISUS – GENISTA COMPLEX
▲	3	Echinospartum	Genista	'Genista' Group	CYTISUS – GENISTA COMPLEX
⬢	8	Ulex	Ulex	'Genista' Group	CYTISUS – GENISTA COMPLEX
⬡	2	Stauracanthus	Ulex	'Genista' Group	CYTISUS – GENISTA COMPLEX
	195				

Teline (crosses). Support for Polhill's system (see Table II) might appear as links between *Cytisus* and *Chamaecytisus* symbols or between *Teline* and *Genista* symbols.

3. The Resulting Classifications

Figures 3–10 show a linkage diagram from the results for each of the remaining data sets. In each case the reader can see that the symbols are mixed and consequently the arrangement cuts across the Flora Europaea system. Indeed the mixtures of species found as clusters is different in each case. Not one of the data sets produces clusters that resemble the large groupings in Hutchinson's or Polhill's systems. Detailed versions of these diagrams showing the identity of each species are in Harris (1980).

4. Measures of Congruence

Comparisons were made between each classification and every other classification. For each pair of classifications the comparison was made on just those species present in both. The method used was to calculate the product moment correlation coefficient between equivalent elements of the similarity matrices used in the cluster analyses. Thus the estimates of congruence fall within the range + 1 to − 1, with + 1 being perfect congruence.

It was found that all the values (see Table III) were low, with the highest value 0.23 between the alkaloid and the amino acid data. If $n(n-1)/2$ is used for the degrees of freedom, none was significantly different from zero. Perhaps of special interest was the fact that even the two classifications based on seed protein information were not congruent. This has been reported by other workers and Pickering (1971) suggests an explanation. Polyacrylamide gel electrophoresis and serology provide information on different characteristics of the proteins and these characteristics may evolve independently and yield incongruence. We conclude that the eight data sets, seven of them chemical sets, are incongruent with each other and with the morphological data.

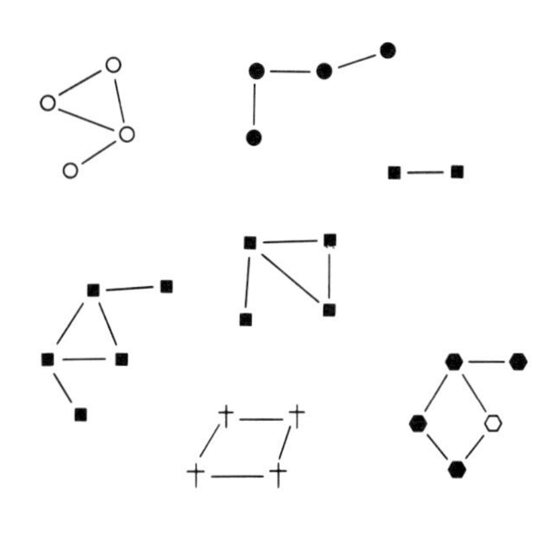

Single member clusters, 20

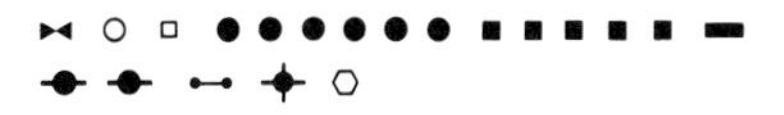

Fig. 1 (left). Morphological data. Threshold similarity 0.66, Level 9. 48 species Genisteae, 59 characters.

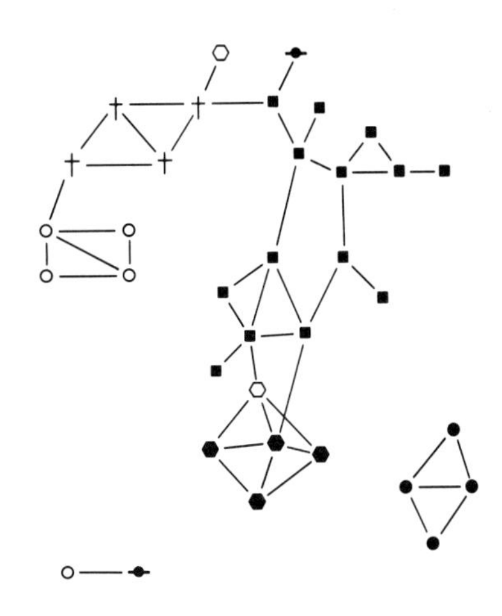

Single member clusters, 13

Fig. 2 (right). Morphological data. Threshold similarity 0.61, Level 11. 48 species Genisteae.

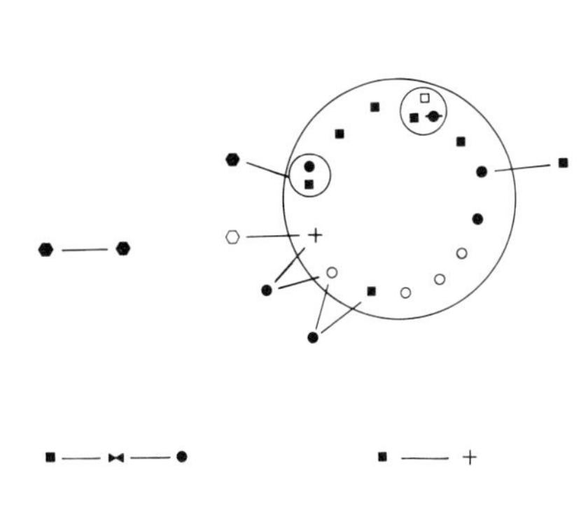

Single member clusters

Fig. 3 (left). Floral fragrance data. Threshold similarity 0.78, 40 species Genisteae, 18 characters.

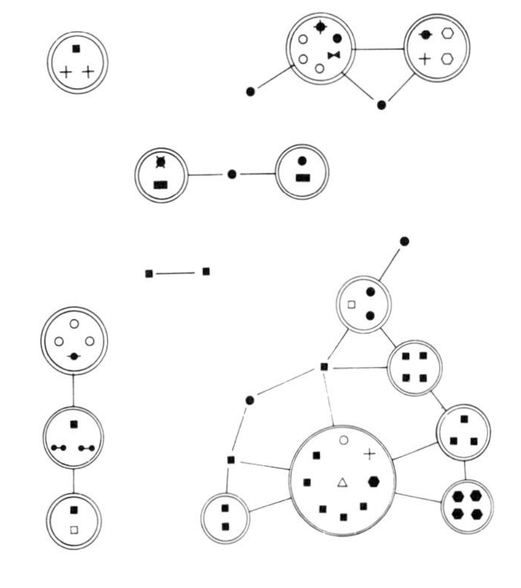

Single member clusters

Fig. 4 (right). Ultraviolet patterns of flowers. Threshold similarity 0.98. 62 species Genisteae, 3 characters.

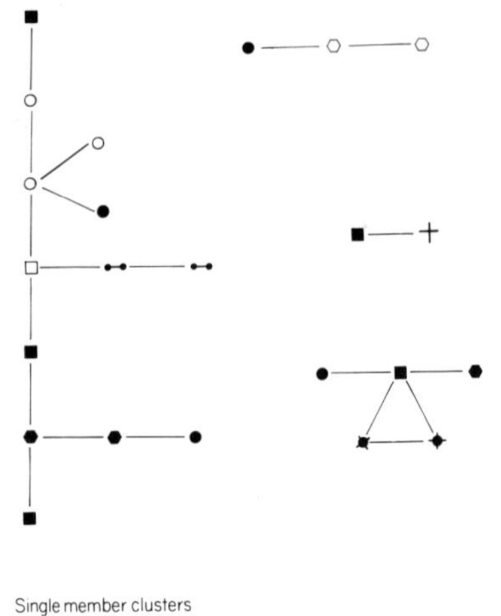

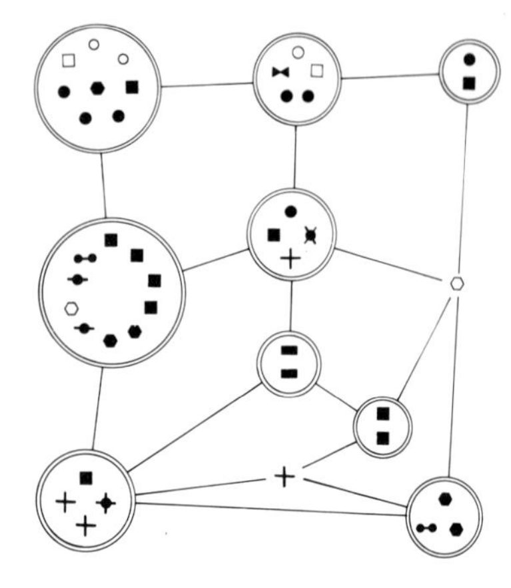

Fig. 5 (left). Electrophoretic mobility of seed proteins. Threshold similarity 0.84. 41 species Genisteae, 20 characters.

Fig. 6 (right). Serological properties of seed proteins. Threshold similarity 0.75. 42 species Genisteae, 4 characters.

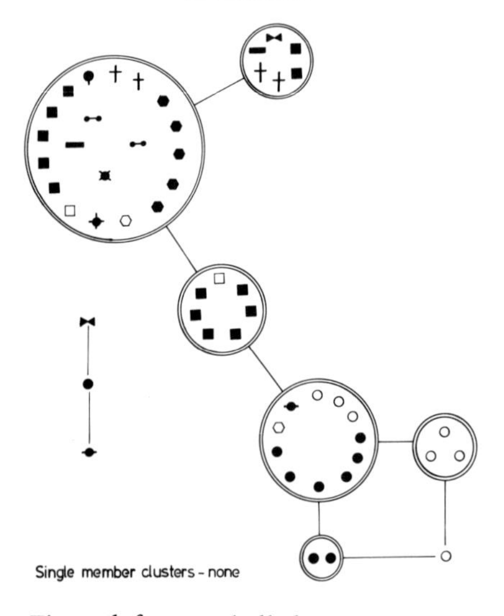

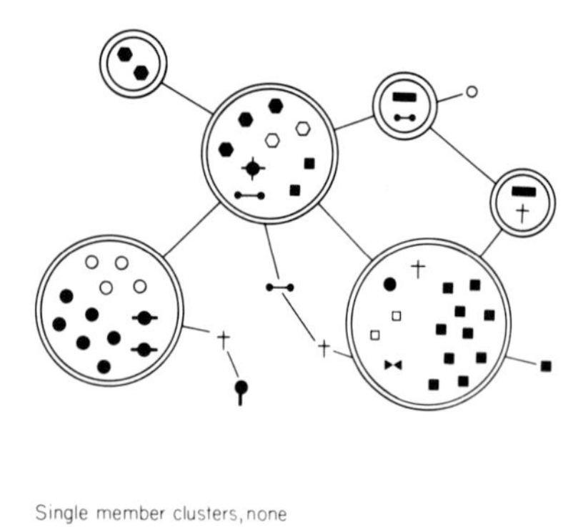

Fig. 7 (left). Seed alkaloids. Threshold similarity 0.86. 53 species Genisteae, 9 characters.

Fig. 8 (right). Free amino acids in the seeds. Threshold similarity 0.85, Level 1. 48 species Genisteae, 7 characters.

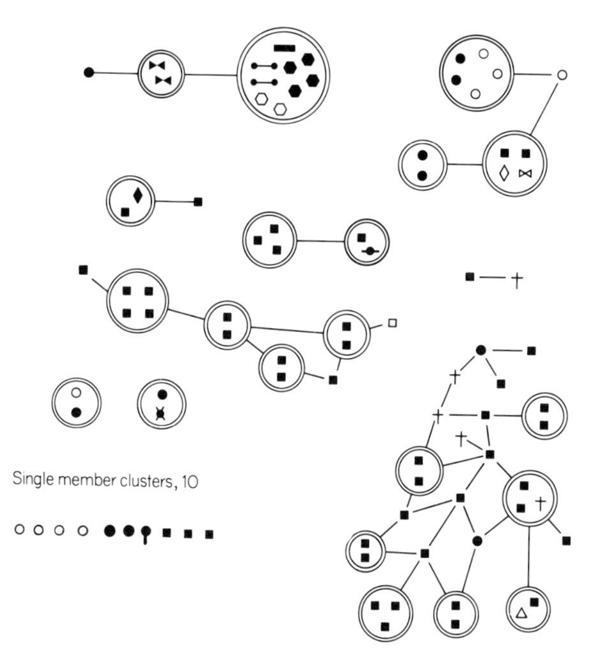

Fig. 9. Leaf flavonoids. Threshold similarity 0.93, Level 1. 90 species Genisteae, 14 characters.

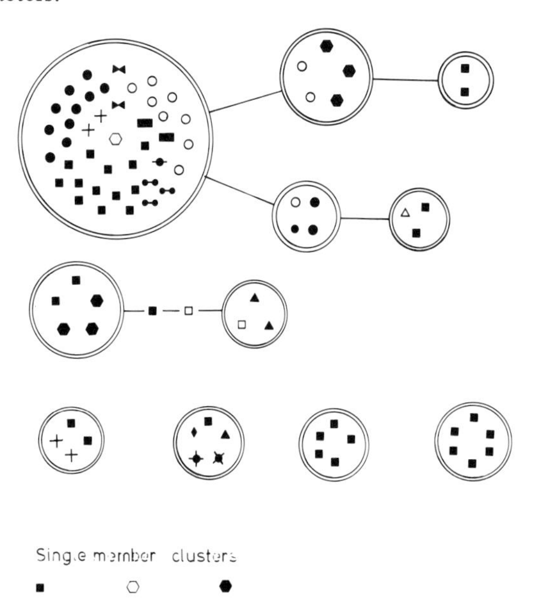

Fig. 10. Chromosome numbers. Threshold similarity 0.56. 89 species Genisteae, 7 characters.

Table III. Congruence between Classifications

values of product moment correlation coefficient
+ 1 complete congruence
− 1 no congruence

M	Morph								
FF	Fl. fragrance	−0.2							
UV	UV. fls.	+0.13	+0.00						
E	El. seed pr.	−0.01	−0.12	−0.02					
S	Serol. seed pr.	+0.09	+0.16	+0.04	+0.04				
A	Seed alk.	+0.11	+0.11	+0.03	+0.03	+0.15			
AA	Free a.a, seed	+0.07	+0.06	+0.02	+0.03	+0.03	+0.23		
F	Leaf flav.	+0.13	−0.03	+0.14	+0.12	+0.08	+0.20	+0.17	
C	Chr. nos.	+0.08	−0.02	−0.08	+0.03	−0.19	−0.24	−0.05	−0.04
		M	FF	UV	E	S	A	AA	F

DEALING WITH INCONGRUENCE

We suspect that, as more investigations of the type described are performed, so more examples of incongruence between classifications based on "new" types of information and classifications based on morphology will appear. If such incongruence does prove to be a frequent occurrence then one will no longer be able to rely on a morphologically based classification to predict which species will be chemically, cytologically (etc.) most similar. The question therefore arises of how to cope with this problem of incongruence within the framework of taxonomy. There are at least five approaches worthy of further investigation:

1. Pooling All Available Characters

Lockhart (1964), Crovello (1969) and Sneath and Sokal (1973) advocate that a classification should be the product of all available characters distributed as widely and evenly as possible over the organisms studied. The argument against this approach is that, if character sets formulated from different types of taxonomic information are suggesting different relationships between species, how can one justify merging these character sets when they will obviously be working in different directions within the classification? It is difficult to envisage how the resulting classification could hope to reflect any "real" relationships between species.

The above argument does, however, conflict with the traditional idea of morphological classification being based on many morphological characters because classifications based on logical subsets of different morphological characters have been shown to be as incongruent with one another as classifications based on the individual character sets of the other types of data. This was, certainly, the case in this investigation. For example, the extremely low correlation value of + 0.17 was obtained for both the comparison of the classification based on characters of floral morphology with the classification based on characters of vegetative morphology, and the comparison of the classification based on leaf flavonoid characters with the classification based on seed amino acid characters.

Table IV. Congruence between Morphological Classifications

Floral		
Vegetative	+ 0.17	
Seed and Pod	+ 0.20	+ 0.17
	F	V

Floral morphology v's (Vegetative + Seed and Pod)	+ 0.24
Vegetative morphology v's (Floral + Seed and Pod)	+ 0.22
Seed and Pod morphology v's (Vegetative + Floral)	+ 0.24
29 random morphological characters v's 30 others	+ 0.48

Other workers that have reported disagreement between numerically derived phenetic classifications based on subsets of morphological characters include Johnson and Holm (1968) and Clifford and Lavarack (1974). The former, working with the genus *Sarcostemma,* also obtained a correlation value of + 0.17 from the comparison (using similarity matrices) of a classification based on floral characters with one based on vegetative characters. Clifford and Lavarack, working with the family Orchidaceae, obtained a correspondence value of 50% from the comparison (using truncated dendrograms) of a classification based on vegetative characters with a classification based on reproductive characters. Accepting this, Clifford and Lavarack still advised that these two sets of characters should be merged together to obtain a classification of Orchidaceae. One may, however, question the validity of their approach. They had separately compared the classification based on vegetative characters and the classification based on reproductive characters with the classification based on both vegetative and reproductive characters and found that the two comparisons revealed similar levels of correspondence (78% and 64% respectively). They therefore concluded that although neither was particularly reliable, both sets of characters were of equal efficiency in predicting the classification based on total data and so both should be employed in a general classification. Yet one might expect a high correlation value from comparing a classification based on one set of characters, with a classification based on that set plus another equal sized set of characters because the former set has contributed a substantial proportion of the information on which the overall classification is based. For example when we compared the classification based on the 28 vegetative and seed pod characters with the classification based on both these and the 31 floral characters we obtained a correlation coefficient value of + 0.70. But, whatever their reasoning, Clifford and Lavarack found that by merging all the available morphological characters they produced a classification that they considered reflected the most acceptable taxonomic arrangement. Likewise in this study the numerical classification based on combined morphological characters produced an arrangement of species that corresponded closely with the Flora Europaea system. If it is acceptable to obtain a classification by pooling incongruent sets of morphological characters then perhaps it should also be acceptable to aim at producing a classification based on the combined character sets formulated from as many different sources of taxonomic data as possible, regardless of the level of congruence between them. In this study the classification based on "all available characters" was found to contain groupings of species that were similar to those of the orthodox classifications. The stability of this was, however, strongly suspected. It was felt that the resultant groupings of species had only occurred because the classification was reflecting the relationships indicated by the one set of characters that formed the largest component of the total character set, i.e. the 59 morphological characters.

Both Clifford and Stephenson (1975) and Sneath and Sokal (1973) believe that as many characters should be merged together as is necessary to produce a stable classification whereby the addition of further characters does not alter the relationships indicated between species. Neither of these pairs of workers have considered whether other so-called "stable" classifications based on completely different sets of characters would themselves be congruent.

Furthermore, even if one overall, taxonomically stable classification did materialize for a particular set of taxa, the question would arise as to what the production of this type of classification (which would undoubtedly necessitate the collection of hundreds of characters) actually achieves. Could such a classification be used to store, retrieve or predict the relationship of species according to any one type of taxonomic information? The results obtained in this study indicate that this would not be possible. It was therefore concluded that the pursuit of this type of overall classification would not offer a practical solution to the problem of incongruence.

2. *Character Selection*

It has already been mentioned in the introduction that a common outcome of chemotaxonomic studies has been the subjective selection of those characters formulated from one particular source of chemical data which show a correlation with a current morphological classification. In some instances a few characters have been selected (for example, the distribution of several alkaloids) in others, only one (for example the distribution of one particular alkaloid). In all cases the outcome is the same: an alternative classification of the taxa being studied is approved just because one or a few "supporting" characters can be selected from the characters obtained for one particular source of chemical data.

The obvious problem with this approach is that one has not considered whether other characters obtained from the same or, even more importantly, different sources of taxonomic data (chemical or otherwise) support the same or some other classification

(Burtt, 1964). For instance, within the *Cytisus-Genista* complex one could pick our characters from one taxonomic data set that would support a feature of Polhill's classification and characters from another that would support a contrasting feature of Hutchinson's classification. Clearly, one or a few characters agreeing with a classification is very different from the overall pattern in a data set (or sets) doing so. In a sense, the above approach is merely using the morphological data for recognizing the pattern, but including one or a few characters from some other type of taxonomic data for the secondary process of delimiting or identifying taxa.

A non-arbitrary method of selecting characters for classification production has been achieved by Bisby (1970) using a taximetric procedure called "character analysis". Bisby explains how this procedure provides a quantitative measure of the potential taxonomic value of each character by calculating the overall correlation between that character and all the other characters included in the study. The characters are listed in descending orders of these "information contribution" values and Bisby then takes only the best correlated characters to produce the final classification. When Bisby applied this method to the morphological characters of *Crotalaria* he showed how it led to a simplification and clarification of the resulting classification.

This method of *a posteriori* character selection has not yet been applied to the total character set collected for the *Cytisus-Genista* complex, but when it is it will be most interesting to see if any characters of the eight other taxonomic data sets are more highly correlated with the morphological characters than other morphological characters.

The most obvious problem with this method is that, as in all taxonomic studies, one is dependent on the available data. It is always possible that the introduction of a new set of data might produce a different set of most highly correlated characters and a different classification.

3. Classification Analysis

This approach follows from Farris' (1971) theoretical studies on "The Hypothesis of Nonspecificity and Taxonomic Congruence". Farris used hypothetical models to show that two sets of characters that are incongruent according to a phenetic clustering technique may appear congruent if evaluated by cladistic techniques and vice versa. He believes that this phenomenon occurs because the two types of tree constructing methods are perceiving incongruence in fundamentally different ways, they are reacting to two separate classes of evolutionary phenomena. The incongruence measured by phenetic and not cladistic techniques is said to correspond to the idea of mosaic evolution, whereas the incongruence indicated by cladistic techniques is said to correspond to the numerical cladistic notion of incompatibility as defined by Camin and Sokal (1965). (Thus, two characters sets will be incompatible if homoplasy (convergence, parallelism and reversal) is present in at least one of the sets of characters.) Farris explains that these two types of incongruence (respectively called mosaic incongruence and incompatibility incongruence) are not mutually exclusive, they may be compounded.

If only mosaic incongruence exists between two sets of characters then one might expect them to yield congruent classifications when evaluated by cladistic techniques. An example of this has actually been published from Mickevich and Johnson's (1976) studies on *Menidia*. The incongruence they found between the morphological and allozyme character sets diappeared when they switched from a phenetic to a cladistic analysis of the data. Mickevich and Johnson further corroborated Farris' theory by demonstrating statistically that mosaicism existed between the two sets of characters. (They did, however, also show that even in cases where detectable homoplasy represented a substantial proportion of total evolutionary change (approximately one third in both character sets) cladistic congruence still holds.)

In future studies it would be very interesting to perform a similar investigation with all the character sets obtained for the *Cytisus-Genista* complex. It is, however, appreciated that even if congruent cladistic trees were obtained, the problem of producing a stable taxonomic classification would still not be resolved because the two cladistically compatible trees will not necessarily suggest the same "cut off points" for groupings of species. For example, consider the cladistically congruent trees A and B in which the species most likely to be grouped together in a classification are enclosed in rings (Fig. 11).

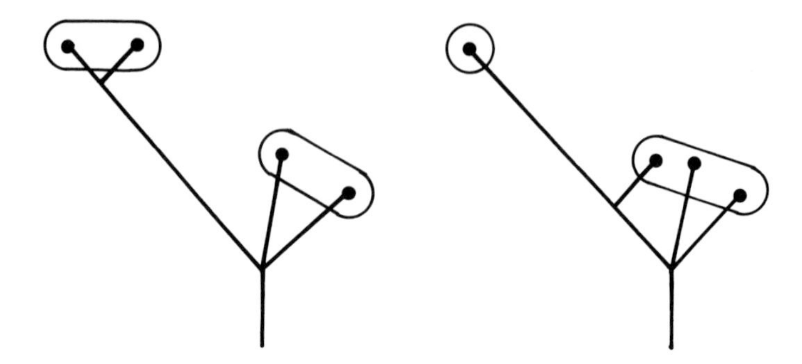

Fig. 11. Two cladistically congruent trees that might imply different classifications.

If both mosaic and incompatibility incongruence exist between the two character sets being compared then it seems probable that the classifications produced by both phenetic and cladistic techniques will be incongruent. Farris concludes that in order to achieve a stable general classification new numerical taxonomic techniques will have to be developed to minimize the influence on the classification of both these evolutionary phenomena that produce incongruence under current methods.

4. Special Purpose Classifications

A fourth approach for investigation involves the construction of separate classifications for each of the major new types of taxonomic data and then using each separately as a structure for providing a catalogue of the relevant data. Although this course of action might prove useful for workers in toxicology or pharmacognosy it was felt that putting it into practice would be fraught with difficulties. One would first have to decide which classes of compounds are likely to be most important and then, once the relevant individual classification had been produced, one would have to find a method of cross-referencing both the grouping labels and the contents of the group to the conventional morphological system and its formal nomenclature. For these reasons it was believed that, in the above form, this approach was impracticable, even if there were enough groups with both sufficient data available and sufficient incongruence. In a slightly different form, however, this leads on to the fifth approach for investigation, one which we consider will be of interest in the future.

5. Establishing a Monographic Data-base Service

This approach involves designing an alternative system which can store, analyse and disseminate taxonomic (that is comparative) data without being tied to a particular classification. If one likens taxonomic services to a comparative information service, then the question is whether one can provide the service without constraining the data to a fixed classification.

At Southampton we are experimenting with this approach in our Vicieae data-base project. The object of this is to collate chemical, morphological and other taxonomic data into a computer data-base, and then to experiment with providing a taxonomic information service from this. The products of the service will be catalogues, variation diagrams, informal classifications and keys available tailor-made to users for their chosen subsets of the taxa and for their chosen subsets of the characters. All taxa will be referenced by the correct scientific name applied to the taxa known from orthodox work on the morphological variation pattern.

CONCLUSIONS

1. Explicit taximetric methods of analysing the pattern in chemical data have opened up the whole field of comparing data sets.
2. Incongruence between data sets has been established in some taxa and has introduced a major problem in the seemingly simple taxonomic task of using chemical data in the recognition and delimitation of taxa.
3. We urge everyone to face the task of broadening the taxonomic information services to accommodate comparative chemical information.

REFERENCES

Aynilian, G. H., Farnsworth, N. R. and Trojanek, J. (1973). The use of alkaloids in determing the taxonomic position of *Vinca libanotica* (Apocynaceae).

In "Chemistry in Botanical Classification" (G. Bendz and J. Santesson, eds), pp. 189–204. Academic Press, New York and London.

Basford, N. L., Butler, J. E., Leone, C. A. and Rohlf, F. J. (1968). Immunologic comparisons of selected Coleoptera with analysis of relationships using numerical taxonomic methods. *Syst. Zool.* **17**, 388–406.

Bentham, G. and Hooker, J. D. (1865). Genera Plantarum. Vol. 1 (2).

Bernasconi, R., Gill, St. and Steinegger, E. (1965) Chemotaxonomic – phylogenetic classification of the *Genista* genus on the basis of alkaloid content. *Pharm. Acta. Helv.* **40**, 275–291.

Bisby, F. A. (1970). The evaluation and selection of characters in Angiosperm taxonomy: an example from *Crotalaria. New Phytol.* **69**, 1149–1160.

Bisby, F. A. (1980). Genisteae. *In* "Advances in Legume Systematics" (R. M. Polhill, ed.) Royal Botanic Gardens, Kew. In press.

Bisby, F. A. and Nicholls, K. W. (1977). Effects of varying character definition on classification of Genisteae (Leguminosae). *Bot. J. Linn. Soc.* **74**, 97–122.

Brown, K. S. (1967). Chemotaxonomy and chemomimicry: the case of 3-hydroxy nurenine. *Syst. Zool.* **16**. 213–216.

Burtt, B. L. (1964). Angiosperm Taxonomy in Practice. *In* "Phenetic and Phylogenetic Classification" (V. H. Heywood and J. Mc.Neill, eds), pp. 5–16. Systematics Association, London. (Publ. No. 6.)

Camin, J. H. and Sokal, R. R. (1965). A method for deducing branching sequences in phylogeny. *Evolution* **19**, 311–326.

Clifford, H. T. and Lavarack, P. S. (1974). The role of vegatative and reproductive attributes in the classification of the Orchidaceae. *Biol. J. Linn. Soc.* **6**, 97–110.

Clifford, H. T. and Stephenson, W. (1975). "An Introduction to Numerical Classification". Academic Press, New York and London.

Cronquist, A. (1973). Chemical plant taxonomy: a generalist's view of a promising speciality. *In* "Chemistry in Botanical Classification" (G. Bendz and J. Santesson, eds), pp. 29–40. Academic Press, New York and London.

Crovello, T. J. (1969). Effects of change of characters and number of characters in numerical taxonomy. *Am. Midl. Nat.* **81**, 68–86.

Farris, J. S. (1971). The hypothesis of non-specificity and taxonomic congruence. *Ann. Rev. Ecol. Syst.* **2**, 277–302.

Faugeras, G. and Paris, R. (1971). Nouvelles recherches phytochimiques sur les Papilionacées – Genistées d'Europe. *Boissiera* **19**, 201–218.

Harborne, J. B. (1969). Chemosystematics of the Leguminosae flavonoid and isoflavonoid patterns in the tribe Genisteae. *Phytochemistry* 8, 1449–1456.

Harris, J. A. (1980). Incongruence in taxonomic data of the *Cytisus – Genista* complex. Southampton University Ph.D. thesis.

Hutchinson, J. (1964). "The Genera of Flowering Plants". Vol. 1. Oxford University Press, Oxford.

Johnson, M. P. and Holm, R. W. (1968). Numerical taxonomic studies in the genus *Sarcostemma* R. Br. (Asclepiadaceae). *In* "Modern Methods in Plant Taxonomy" (V. H. Heywood, ed.), pp. 199–217. Academic Press, London and New York.

Lockhart W. R. (1964). Scoring of data and group-formation in quantitative taxonomy. *Devs. ind. Microbiol.* 5, 162–168.
Mabry, T. J. (1977). The order Centrospermae. *Ann. Mo. bot. gdn.* **64**, 210–220.
Mabry, T. J., Taylor A., and Turner B. L. (1963). Betacyanins and their distribution. *Phytochemistry* 2, 61–64.
Mickevich, M. F. and Johnson, M. S. (1976). Congruence between morphological and allozyme data in evolutionary inference and character evolution. *Syst. Zool.* 25, 260–270.
Pickering, J. L. and Fairbrothers, D. E. (1971). The use of serological data in a comparison of tribes in the Apioideae. *In* "Biology and Chemistry of the Umbelliferae" (V. H. Heywood, ed.), pp. 315–325. Academic Pres,, London and New York.
Polhill, R. M. (1976). Genisteae (Adans) Benth. and Related Tribes (Leguminosae). *Bot. Syst.* 1, 143–368.
Rothamaler, W. (1944). Die Gliederung det Gattung *Cytisus* L. *Feddes Repert.* 53, 137–150.
Seigler, D. S. (1974). Chemistry and taxonomy. *Chem. in Brit.* 10, 339–342.
Sneath, P. H. A. (1971). Numerical taxonomy: criticisms and critiques. *Biol. J. Linn. Soc.* 3, 147–157.
Sneath, P. H. A. and Sokal, R. R. (1973). "Numerical Taxonomy". Freeman, San Francisco.
Sokal, R. R. and Sneath, P. H. A. (1963). "Principles of Numerical Taxonomy". Freeman, San Francisco.
Tutin, T. G., Heywood, V. H., Burges, N. A., Moore, D. M., Valentine, D. H., Walters, S. M. and Webb, D. A. (1968). "Flora Europaea", Vol. 2. University Press, Cambridge.
Wirth, M., Estabrook, G. F. and Rogers, D. J. (1966). A graph theory model for systematic biology, with an example for the Oncidiinae (Orchidaceae). *Syst. Zool.* 15, 59–69.

ACKNOWLEDGEMENTS

We are grateful to the curators of many herbaria and botanic gardens for loans of material and gifts of seed, and to the Science Research Council for a studentship.

16 | Micromolecular Systematics: Principles and Practice

O. R. GOTTLIEB

Instituto de Quimica, Universidade de São Paulo, C.P. 20780, 05508 São Paulo, SP, Brasil

Abstract: Four chemosystematic principles are explained and exemplified. Each refers to the direction in which an evolutionary series of micromolecular markers has to be read for systematic purposes.

INTRODUCTION

A natural classification should consider the organizational levels of organisms (Fig. 1)(Takhtajan, 1973) in the sequence from genotype to phenotype as criteria for progressively lower hierarchic levels in an integrated system. Such a target is not yet within reach and at present we still have to choose one particular level as a classificatory criterion. If human well-being and even survival are motivating factors of one's endeavour, micromolecules must be selected for classificatory purposes, in view of their relevance towards the understanding of the interaction of living organisms on earth.

Why then have micromolecules so far been used merely as auxiliary markers in substantially morphological classifications? The most significant answer to this question concerns doubts about what constitutes a primitive or an advanced feature (Harborne *et al.*, 1976; Harborne, 1977). Consider a sequence of metabolites (Fig. 2) whose biosynthetic relationship we are able to predict. Now, if a taxon

---

Systematics Association Special Volume No. 16, "Chemosystematics: Principles and Practice", edited by F. A. Bisby, J. G. Vaughan and C. A. Wright, 1980, pp. 329–352, Academic Press, London and New York.

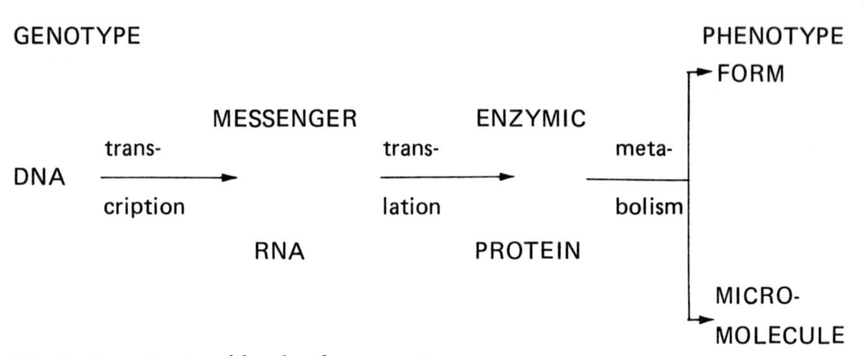

Fig. 1. Organizational levels of an organism.

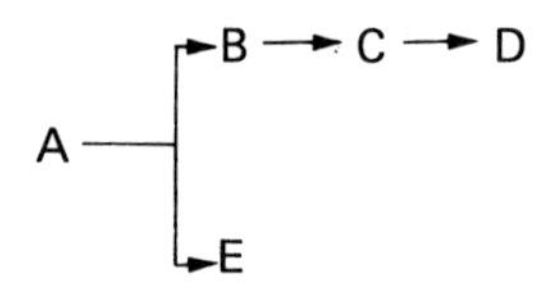

Fig. 2. Biosynthetic sequence of metabolites.

contains C, one will not be able to say if it evolved from a taxon accumulating B (by expansion of the reaction sequence), from a taxon accumulating D (by reduction of the reaction sequence) or from a taxon accumulating E (by concomitant blocking of the pathway leading to E and enhancement of the pathway B → D). This, however, is essential knowledge. It will never be possible to use micromolecular characters as a basis for phyletic classification until it becomes clear when and where to expect expansion, reduction or replacement of a reaction sequence, indeed until some basic principles foretell the direction in which an evolutionary series has to be read (Cagnin *et al.*, 1977; Gottlieb, 1978).

FIRST PRINCIPLE

With respect to chemical composition, plants may be divided broadly, i.e. without consideration of some borderline cases, into non-lignous (Thallobionta) and lignous (Embryobionta, Tracheophyta) categories. Within the latter category, for the primary precursors of the shikimate pathway, from which biogenetic groups characterizing taxa of high hierarchic level are derived, evolution proceeds by blocking of reaction steps.

Application of this First Principle to alkaloid evolution and angiosperm systematics (Gomes and Gottlieb, 1980) would require phenylalanine derived compounds to have appeared initially (Fig. 3). Indeed, benzylisoquinolines (toxicity index 3 (Levin and York, 1978)) occur in Magnolianae (*sensu* Dahlgren, 1975), generally considered the most primitive angiosperms (Cronquist, 1968), as well as in biosynthetically modified form (toxicity index 4) in Ranunculanae (Rezende *et al.*, 1975), whereas other phenylalanine derived compounds characterize Caryophyllanae on one hand and Lilianae on the other (Fig. 4). The sporadic occurrence of benzylisoquinoline alkaloids in many other superorders is taken as indication of the antiquity of the character.

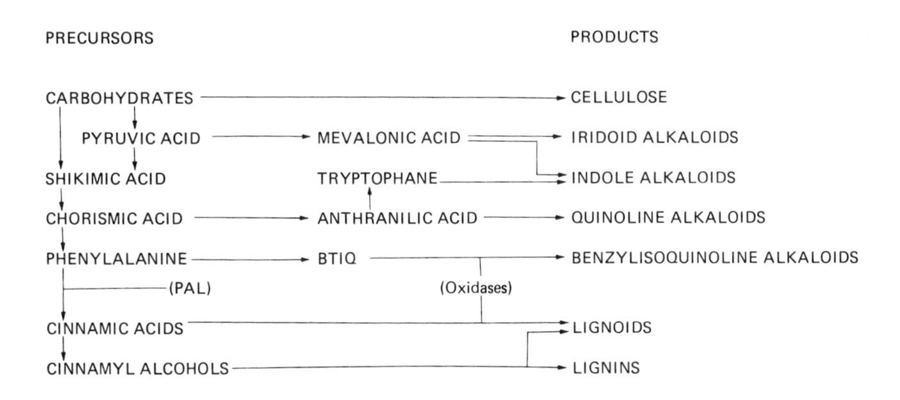

Fig. 3. Outline of the shikimate pathway.

By further reduction of the shikimate pathway chorismic acid becomes available for the production of anthranilic acid derived alkaloids. The characteristic is concentrated in Rutanae and extends to other superorders, with exception of the ones specifically mentioned above in connection with phenylalanine derived alkaloids. And again, it is postulated that the sporadic appearance of this character may reveal inheritance and thus help to date the origin of the groups concerned.

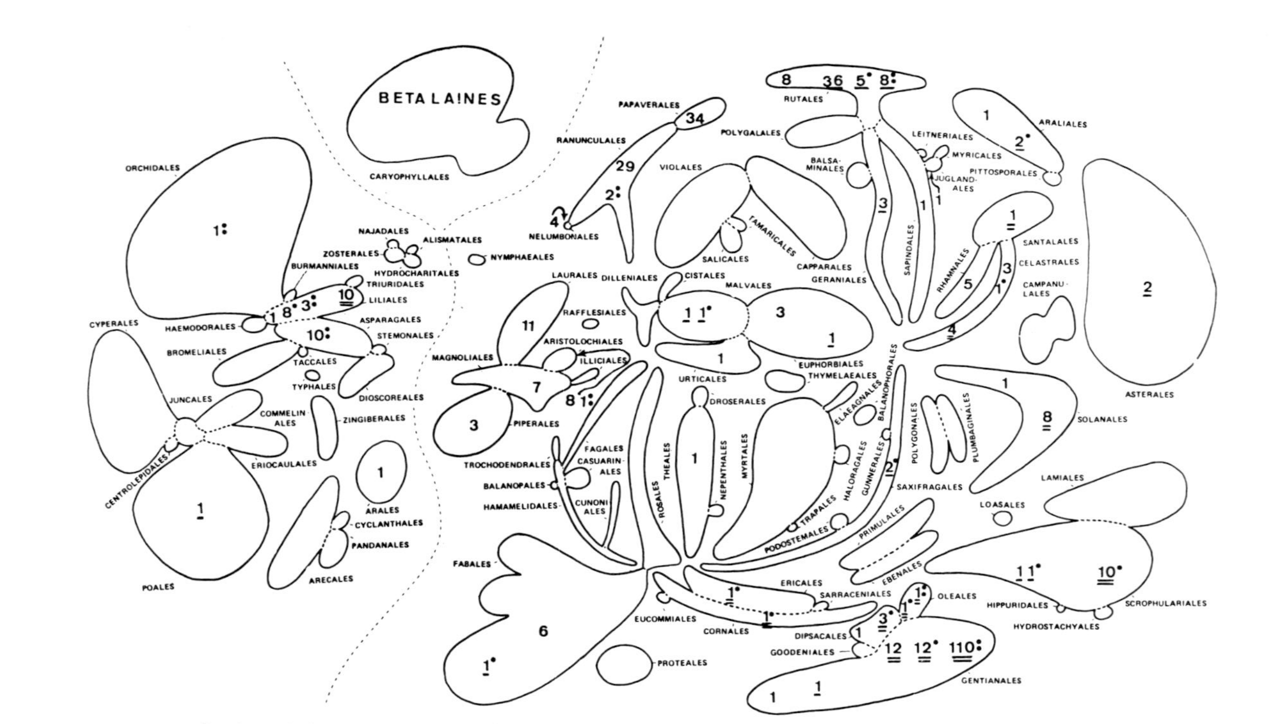

Fig. 4. Occurrence of phenylalanine derived alkaloids, indicated by the number of benzylisoquinoline, phenylethylisoquinoline (*) and benzyl-phenylethyl-amine (⁑) type skeletons; occurrence of anthranilic acid derived alkaloids (underlined), indicated by the number of quinoline, quinazoline (*) and acridine (⁑) type skeletons; occurrence of mevalonic acid derived alkaloids (double underlining), indicated by the number of steroid, iridoid (*) and indolo-iridoid (⁑) type skeletons.

In contrast to phenylalanine derived alkaloids, anthranilic acid derivatives frequently incorporate small terpenoid units. Such units are ubiquitous in the equally anthranilate derived indole alkaloids (toxicity index 6). The full expression of mevalonate utilization, however, occurs due to the total suppression of the shikimate pathway in the biosynthesis of steroid (toxicity index 5) and iridoid alkaloids in which nitrogen is linked directly to terpenoid moieties. All these mevalonate derived alkaloidal types are concentrated in Solananae, Lamianae, Gentiananae and Cornanae.

These postulates concerning evolution of alkaloids in angiosperms are summarized by a phylogenetic tree (Fig. 5). The superposition on the general use of nicotinic acid, histidine, ornithine and lysine, an ancient heritage (Gomes and Gottlieb, 1979), of phenylalanine, anthranilic acid and mevalonate for alkaloid production in three stages is consistent with the classification of the angiosperms into three major subclasses, Magnoliidae, Rosidae and Asteridae (Cronquist, 1968), and with the First Principle of Micromolecular Systematics.

SECOND PRINCIPLE

In oppostion to the First Principle, which considers primary precursors, the Second Principle is concerned with biogenetic groups (BGs) of secondary metabolites (Rezende and Gottlieb, 1973; Rezende *et al.*, 1975; Cagnin *et al.*, 1977; Gottlieb, 1978; Cagnin and Gottlieb, 1978; Salatino and Gottlieb, 1980; Gomes *et al.*, in press). Among taxa of lignous plants of low hierarchic level, biochemical evolution of secondary metabolites involves gradual specialization (variation) by substitutional or/and skeletal diversification of compounds within biogenetic groups. As a corollary, it is admitted that variation of substitution, chiefly oxygenation, precedes variation of skeletons.

The use of quinolizidine alkaloids as systematic markers of the Papilionoideae (Salatino and Gottlieb, 1980), a subfamily of the Leguminosae, will illustrate this principle.

Quinolizidine alkaloids are formed according to cytisine (III →), sparteine (II.2.1 →), ormosanine (II.2.2. →), matrine (II.3 →) and tetrahydroanabasine (I.2 →) pathways. Their biogenetic map (Fig. 6), constructed according to rules set forth previously (Cagnin and Gottlieb, 1978), served as a basis for the quantification of structural parameters for skeletons and for oxygenations. The relative probability of occurrence (RPO_x values) for skeletons, the quotient of the number of reaction steps (cf. Fig. 6) and the frequency of distribution (Salatino and Gottlieb, 1980), are registered in Table I. Oxygenation which seems essential for the development of skeletons is already shown on the biogenetic map (Fig. 6). Others, however, which appear with increasing rarity, were assigned increasing point

Table I. Determination of RPO_x Values for Quinolizidine Skeletons in Papilionoideae

Codes*		Codes*	
I.2	2/9 = 0.2222	II.2.1.1.1	6/3 = 2.0000
I.2.1	3/2 = 1.5000	II	2/109 = 0.0183
II.3	3/15 = 0.2000	III	3/48 = 0.0612
II.3.1	4/14 = 0.2857	IV	4/45 = 0.0889
II.3.1.1	5/6 = 0.8333	IV.3	5/2 = 2.5000
II.3.1.1.1	6/1 = 6.000	IV.3.1	6/1 = 6.0000
II.2.2	4/13 = 0.3077	IV.2	5/2 = 2.5000
II.2.2.1	5/11 = 0.4545	V	5/19 = 0.2632
II.2.2.1.3	6/1 = 6.0000	V.3	6/1 = 6.0000
II.2.2.1.2	6/1 = 6.0000	V.2	6/1 = 6.0000
II.2.2.1.1	6/4 = 1.5000	VI	6/11 = 0.5454
II.2.2.1.1.1	6/1 = 7.0000	VII	7/8 = 0.8750
II.2.1	4/28 = 0.1905	VII.2	8/1 = 8.0000
II.2.1.1	5/6 = 0.8333	VIII	8/1 = 8.0000

*Cf. Fig. 6

values (Table II) for use in calculation of the RPO_y values for substitution of each compound, the quotient of the sum of points and the number of compounds with identical substitution patterns. Determination of the mean RPO values for skeletons and substitutions of contained compounds gave the evolutionary advancement parameters EAP_x and EAP_y for genera plotted on the affinity diagram (Fig. 7).

Inclusion of points into areas according to geographical occurrence of the represented genera show African genera giving way to European, Asiatic and American genera, with increasing distance from the origin of the plot. The Australian genera are seen to be chemically akin to the Asian group. The further away a geographic complex of genera moves from Africa, the flatter the representative areas become,

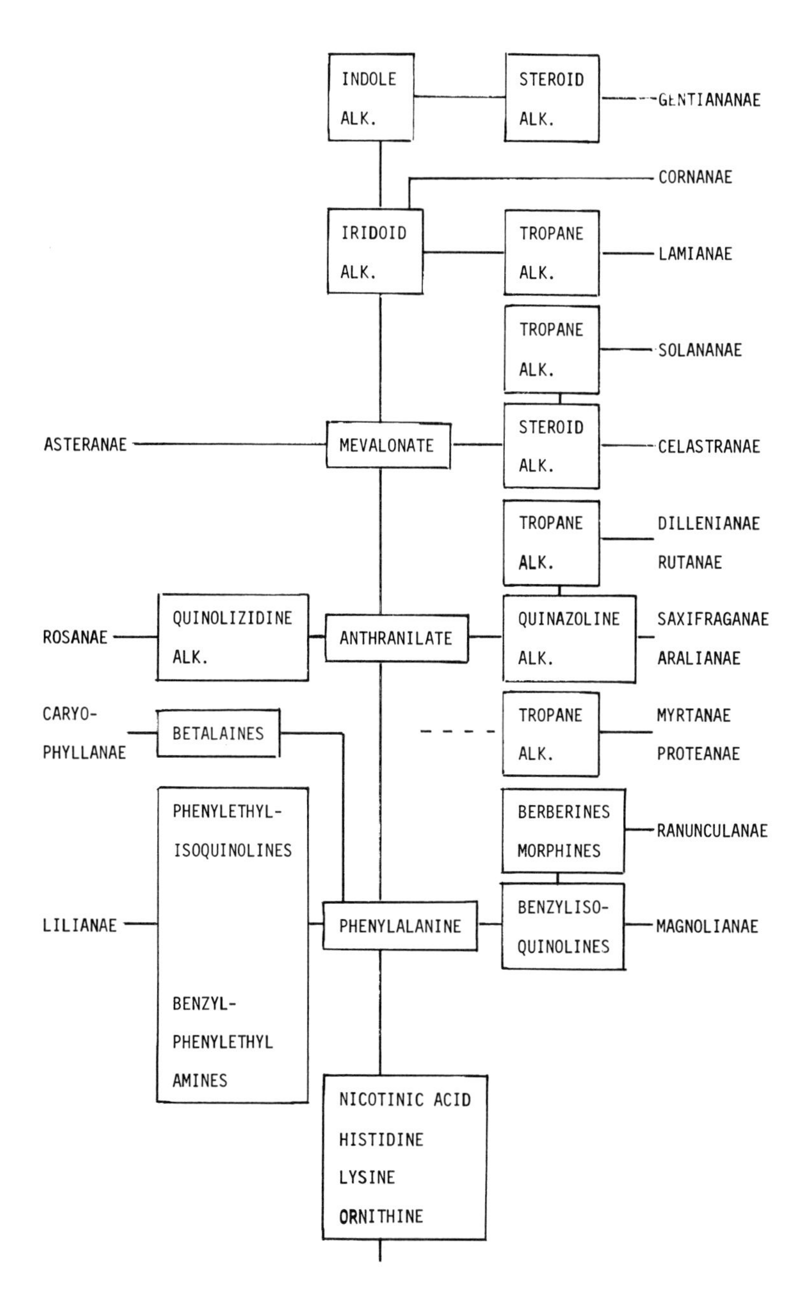

Fig. 5. Phylogenetic tree of angiosperm superorders (*sensu* Dahlgren, 1975) based on the evolution of metabolic pathways and on the wideness of distribution of alkaloids.

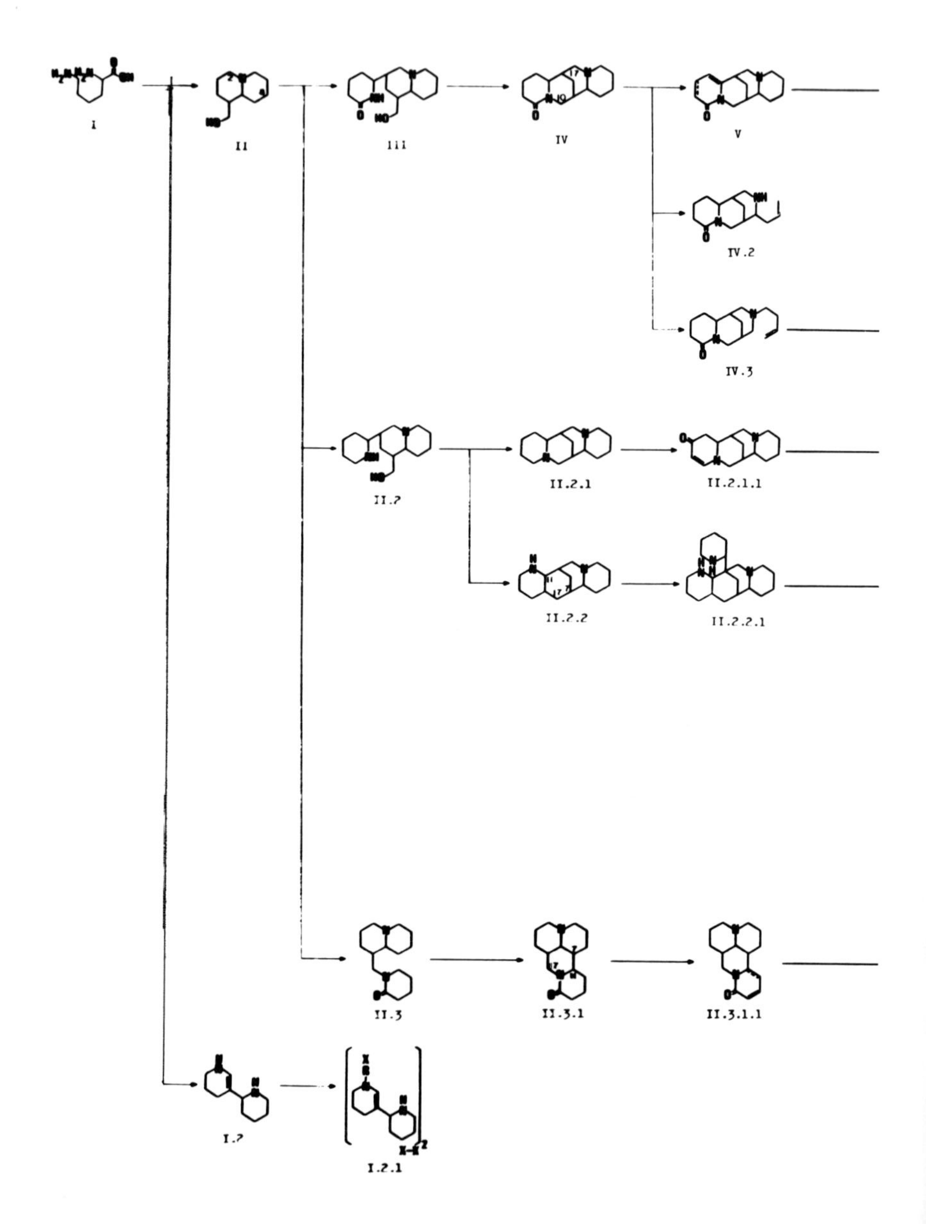

Fig. 6. Biogenetic map of quinolizidine skeletons from Papilionoideae. Codes refer to position of skeletons along the biogenetic pathways. Roman numerals are used for convenience, e.g. V = 1.1.1.1.1.

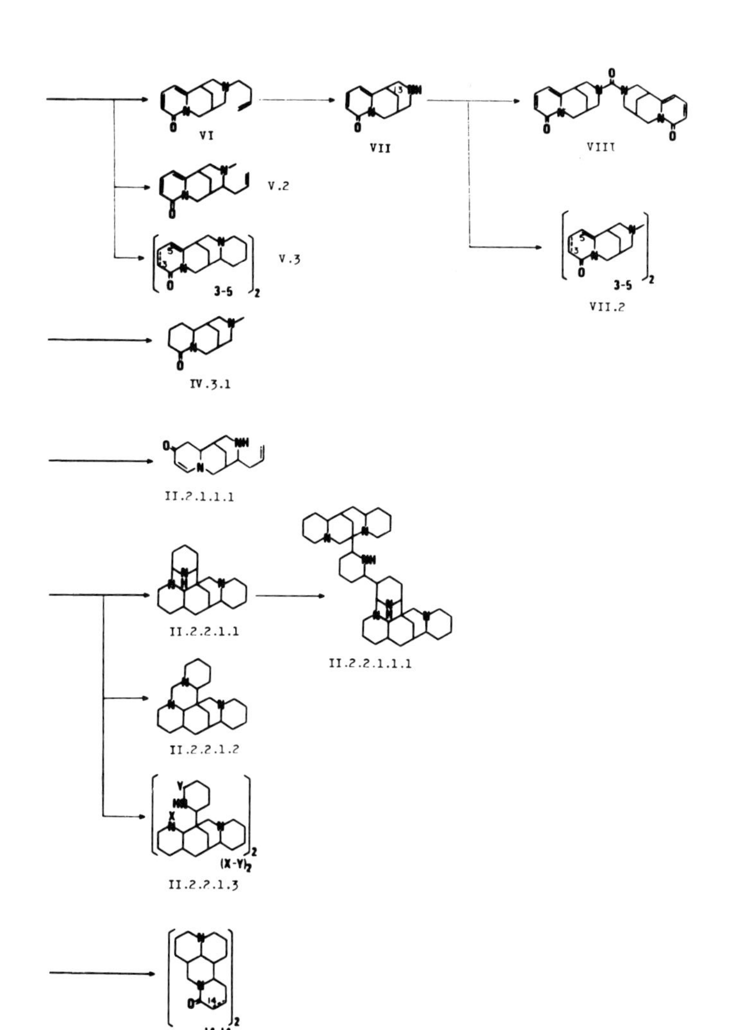

VI
VII
VIII
V.2
V.3
3-5
VII.2
IV.3.1
II.2.1.1.1
II.2.2.1.1
II.2.2.1.1.1
II.2.2.1.2
(X-Y)2
II.2.2.1.3
14-14
II.3.1.1.1

Table II. Rules for the Determination of RPO_y Values for Quinolizidine Substitution in Papilionoideae

C=O, C—OH or C=C	Point Value
α to N	1
γ to N	2
any other position	3

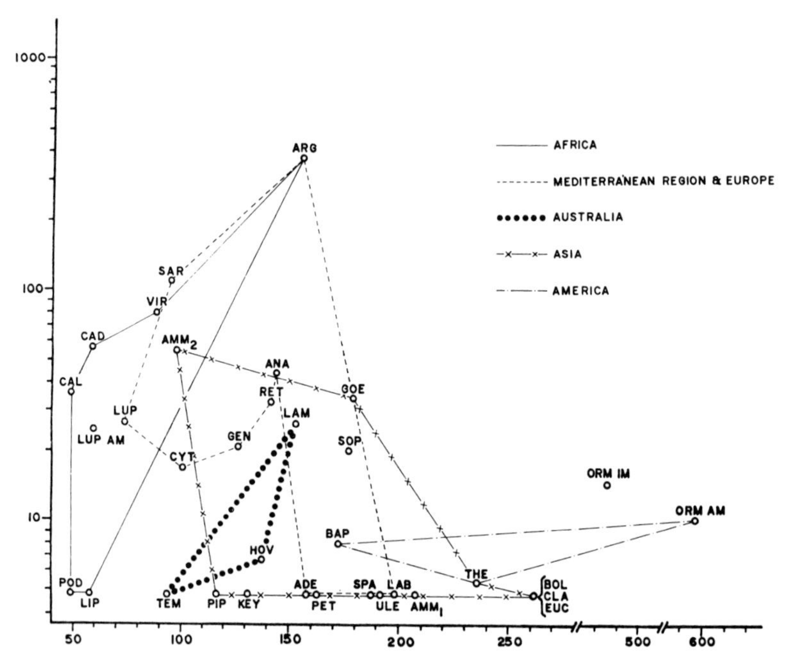

Fig. 7. Affinity diagram for Papilionoideae genera. Abscissa: 300. EAP_x. Ordinate: log 300. EAP_y. Excepting for BOL (South Africa), LUP-AM (South America), SOP (cosmopolitan) and ORM-IM (Indomalaysia), genera occurring in the indicated regions are related by lines. ADE, *Adenocarpus;* AMM_1, *Ammodendron;* AMM_2, *Ammothamnus;* ANA, *Anagyris;* ARG, *Argyrolobium*; BAP, *Baptsia*; BOL, *Bolusanthus*; CAD, *Cadia*; CAL, *Calpurnia*; CLA, *Cladrastis*; CYT, *Cytisus*; EUC, *Euchresta*; GOE, *Goebelia*; GEN, *Genista*; HOV, *Hovea*; KEY, *Keyserlingia*; LAB, *Laburnum*; LAM, *Lamprolobium*; LIP, *Liparia*; LUP-AM, *Lupinus* (America); LUP, *Lupinus* (Mediterranean); ORM-AM, *Ormosia* (America); ORM-IM, *Ormosia* (Indomalaysia); PET, *Petteria*; PIP, *Piptanthus*: POD, *Podalyria*; RET, *Retama*; SAR, *Sarothamnus*; SOP, *Sophora*; SPA, *Spartium*; TEM, *Templetonia;* THE, *Thermopsis;* ULE, *Ulex;* VIR, *Virgilia.*

signalling the gradual substitution of oxidative by skeletal variation. It is quite evident that conquest of new regions is conditioned by the acquisition of new biosynthetic pathways. Indeed, a detailed correlation of structural types and occurrence of their host genera suggests that quinolizidine containing Papilionoideae radiated from tropical Africa in three major directions (Fig. 8) along the southern temperate pre-cytisine route, the northern termperate pre-cytisine–cytisine route and the tropical ormosanine route with a temperate cytisine branch.

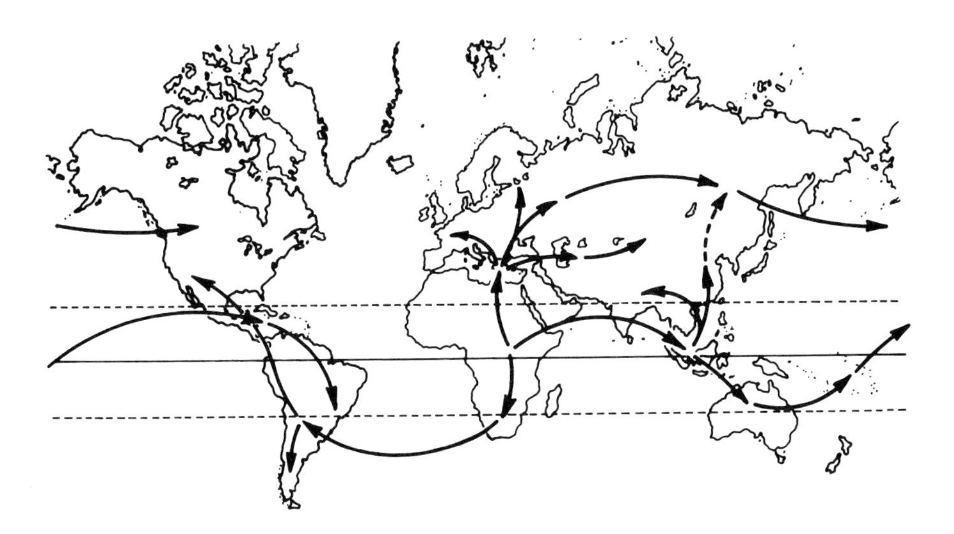

Fig. 8. Postulated radiation of Papilionoideae containing quinolizidine alkaloids.

Ecogeographic phytochemistry, a novel approach to micromolecular systematics with an exceptionally high predictive value, is conveniently studied by set theory. This is illustrated in Fig. 9 for coumarins, derived from umbelliferone (U), and quinoline alkaloids, derived from 4-hydroxy-2-quinolone (Q). Structural types of both classes of compounds are represented by codes which refer to their position in the respective biogenetic maps. For any biogenetic pathway only climactic types present in a taxon are registered. The obligatory presence of intermediate types makes their registry

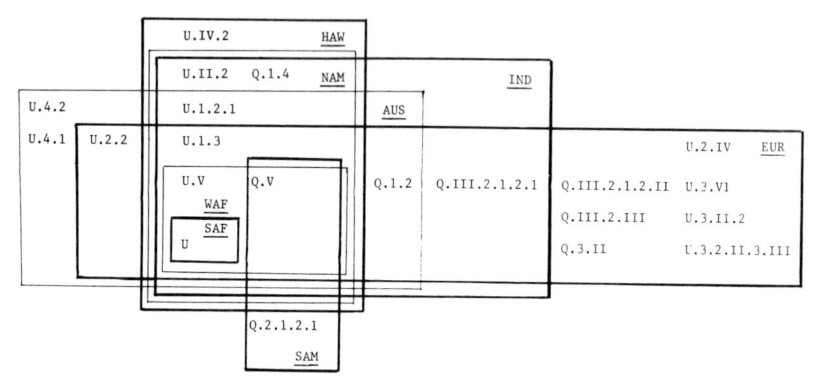

Fig. 9. Diversification of umbelliferone (U) and quinolone (Q) derived skeletal types of micromolecules upon radiation of their Rutoideae hosts. Codes refer to the position of skeletons along biogenetic pathways in Silva and Gottlieb (unpublished). Roman numerals are used for brevity, e.g. Q.III.2.1.2.II = Q.1.1.1.2.1.2.1.1. SAF, South Africa, Diosmeae; WAF, West Africa, Ruteae; EUR, Eurasia, Ruteae; SAM, South America, Cusparieae; IND, Indomalaysia, Zanthoxyleae-Lunasiinae; NAM, North America, Zanthoxyleae-Evodiinae; HAW, Hawaii, Zanthoxyleae-Evodiinae; AUS, Australia, Boronieae.

implicit. Clearly, simple U and Q derived compounds contained in south-west African Ruteae diversified along three routes: a temperate route through Eurasia with Ruteae and two tropical routes, one to South America with Cusparieae and one through Indomalaysia with Zanthoxyleae, subtribe Lunasiinae, to a bifurcation. From this point, one branch led over Polynesia to North America with Zanthoxyleae, subtribe Evodiinae, and the other to Australia with Boronieae (Silva and Gottlieb, unpublished).

The fundamental message derived from the geographic study of quinolizidine alkaloids in Papilionoideae concerns the ecological efficiency of the pyridones. Prior to the development of enzymic systems for the crucial biosynthetic sequence V → VI → VII (Fig. 8), seemingly many varied evolutionary attempts at enhancement of alkaloid toxicity occurred. The nature of these primitive chemical mutations was seen to diverge according to ecogeography. Does it also vary in consistence with morphology?

The alkaloid composition of a species can be characterized by the number of representatives of each type, expressed as a percentage of

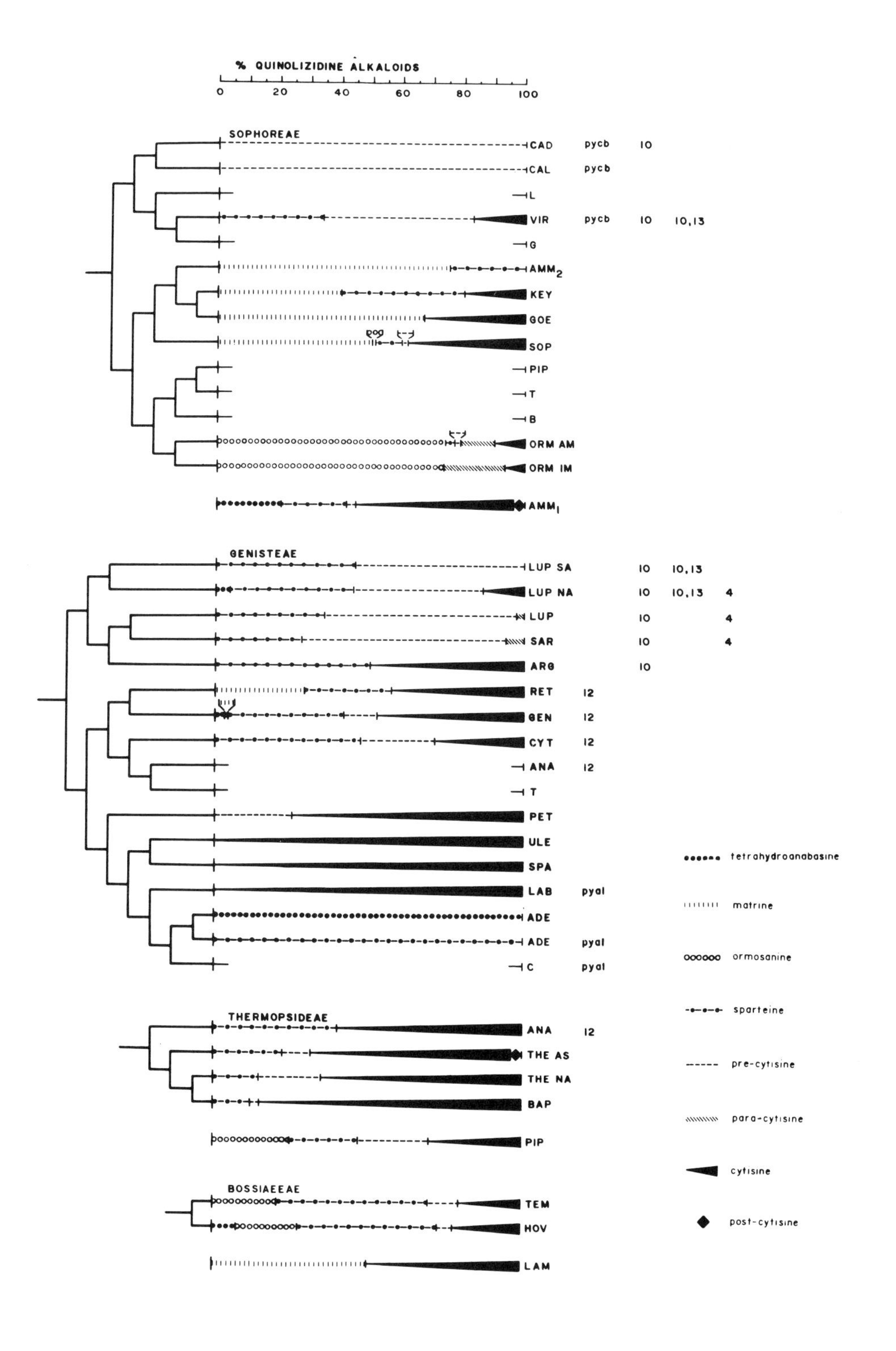

% QUINOLIZIDINE ALKALOIDS
0
20
40
60
80
100
SOPHOREAE
CAD pycb 10
CAL pycb
L
VIR pycb 10 10,13
G
AMM2
KEY
GOE
SOP
PIP
T
B
ORM AM
ORM IM
AMM1
GENISTEAE
LUP SA 10 10,13
LUP NA 10 10,13 4
LUP 10 4
SAR 10 4
ARG 10
RET 12
GEN 12
CYT 12
ANA 12
T
PET
ULE
SPA
LAB pyal
ADE
ADE pyal
C pyal
THERMOPSIDEAE
ANA 12
THE AS
THE NA
BAP
PIP
BOSSIAEEAE
TEM
HOV
LAM
tetrahydroanabasine
matrine
ormosanine
sparteine
pre-cytisine
para-cytisine
cytisine
post-cytisine

the total number of alkaloids reported for that species. The means of the percentages thus determined for each species within a genus are taken to characterize that genus. In Fig. 8, genera, grouped into tribes *sensu* Polhill (1976), are characterized graphically by this criterion of affinity. As a secondary criterion, special oxidation patterns of sparteine and pre-cytisine types are also considered. As shown by the phylogenetic tree, recording means of percentages for each genus within its chemical group (Fig. 11), the homogeneity of the proposed chemical – morphological (tribal) – geographical relations is practically complete.

THIRD PRINCIPLE

Although we have been able to rationalize phylogenesis considering separately primary and secondary metabolism, such a distinction is reasonably artificial. Indeed, consideration of single BGs for classificatory purposes will at best lead to partial results.

The precursors of all BGs are, of course, present in all plants where they provide essential macromolecules. It is thus plausible that chemical characters are polyphyletic. Although compounds may appear scattered throughout the plant kingdom, only if secondary metabolites are accumulated in certain plant groups will they serve a useful purpose in biochemical systematics. It may appear, *a priori*, that the distinction between haphazardly distributed metabolites and specially accumulated metabolites would be difficult or even impossible (Hegnauer, 1976). This is not the case, since accumulation of secondary metabolites from the chemosystematic standpoint refers not merely to quantitative aspects, but chiefly to versatility of structural variation of a biosynthetic theme.

One and the same species usually contains, against a background

---

Fig. 10. Proposed phylogenetic relations of Papilionoideae genera, classified into tribes, according to percentage of the indicated alkaloidal types (see text). Genera are represented by three letters (see Fig. 7). Single letters refer to tribes: L, Liparieae; G, Genisteae; T, Thermopsideae; B, Bossiaeeae; C, Crotalarieae; and indicate points of insertion of respective tribal dendrograms. Additional chemosystematically significant details, substitution by the pyrrolylcarboxyl (pycb) group, presence of pyrrolizidine alkaloids (pyal), oxygenations of pre-cytisine and sparteine types at positions 4, 10, 12 and 10,13, are registered on the right hand side of the diagram.

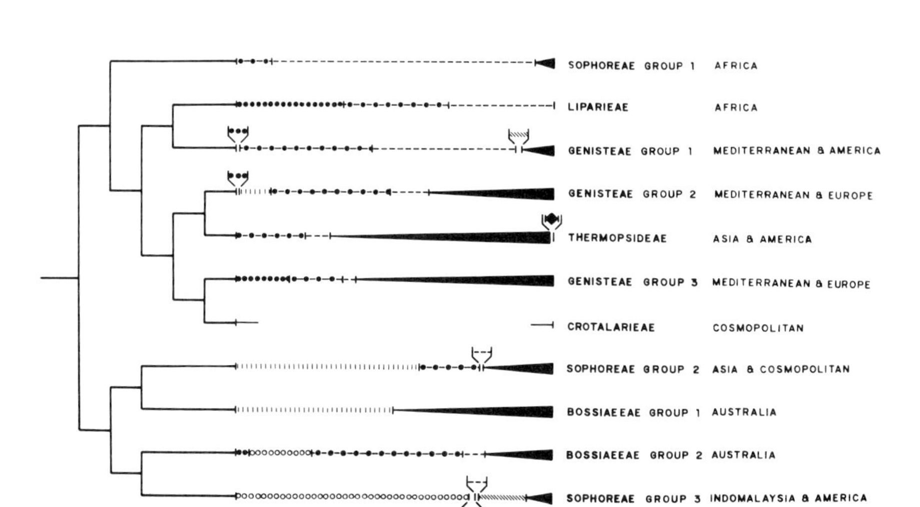

Fig. 11. Proposed phylogenetic relations of Papilionoideae tribes, according to percentage of the indicated alkaloidal types (see text and Fig. 10).

of chemosystematically irrelevant metabolites (unless considered in relation to ancestry), only a limited number of relevant BGs. Different BGs, supposedly with analogous functions, may nevertheless be produced by closely related plant taxa, possibly in response to environmental pressures. Enhancement of products along one biosynthetic route seems to trigger a regulating mechanism which blocks the formation of compounds along another.

From the standpoint of biochemical systematics, these arguments are embodied in the Third Principle: the homology of biosynthetic routes, not the substances produced, is a plausible indication of phylogeny (Brand, 1978); or, while resemblances of taxa with respect to structural variation of the metabolites of their chemosystematically meaningful BG is a reasonable indication of affinity, difference in chemical composition of taxa says nothing about their lack of affinity.

Let us consider the chemistry of the Lauraceae family as an example. In contrast with the structural variability of pyrones (Gottlieb, 1972) and neolignans (Gottlieb, 1977), alkaloids of the genus *Aniba* belong to the biosynthetically simplest benzyltetrahydroisoquinoline (BTIQ) type (Ferreira *et al.*, 1980), an indication of the early evolutionary divergence of the Anibineae (*Aniba*, *Licaria*) from the alkaloidal line (Rezende *et al.*, 1975) leading from the Perseae, with similar benzyltetrahydroisoquinolines, to the Cryptocaryeae, characterized by not less than seven different BTIQ-derived skeleta (Fig. 12)(Young and Gottlieb, unpublished).

The phenomenon is not limited to the Lauraceae. The accumulation of alkaloids in Ranunculales and Papaverales (Gomes and Gottlieb, 1978), as opposed to the strong chemical resemblance between *Aniba* and *Piper* species (Fernandes *et al.*, 1978), suggest that also in the subclass Magnoliidae as a whole, the basic chemistry of primitive Magnoliales suffered diversification of BTIQ-derived alkaloids (towards Ranunculales and Papaverales) versus diversification of phenylpropanoids (towards Piperales).

This balance between cinnamate- and phenylalanine-derived metabolites must be governed by some regulating mechanism, conceivably in connection with phenomena described by the First Principle. Thus, the interface of gymnosperm–angiosperm evolution was characterized by a decreasing trend of lignin production. Ferulic acid, which must have accumulated, should, as cinnamic acid (Havir and Hanson, 1968), inhibit L-phenylalanine ammonia-lyase (PAL) activity (Fig. 3). In response, enzymes were evolved in order to dispose of the excess phenylalanine generated. As beneficial as the resulting BTIQ production must have been to plant life (Swain, 1974), it is quite uneconomical in terms of nitrogen utilization. Elaboration of enzyme systems capable of metabolizing ferulic acid was thus of double advantage in producing alternative oxygenated deterrents as well as lifting or easing inhibition of PAL.

Such mainly oxidative systems may even have embodied a third advantage. Oxidative coupling is required in the synthesis of both propenylphenol (P) and/or allylphenol (A) derived neolignans and benzyltetrahydroisoquinoline derived bis-BTIQ, aporphine, morphine and other alkaloids (Gottlieb, 1972). In other words, BTIQs are primitive characters and similar, mutually exclusive, pathways lead to the advanced characters, either complex alkaloids or complex neolignans.

Be this as it may, a chemical classification of Magnoliidae based either on phenylpropanoids or alkaloids would clearly be of limited

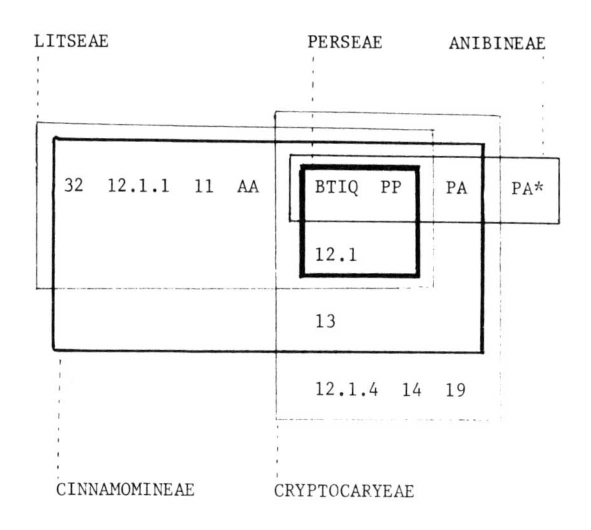

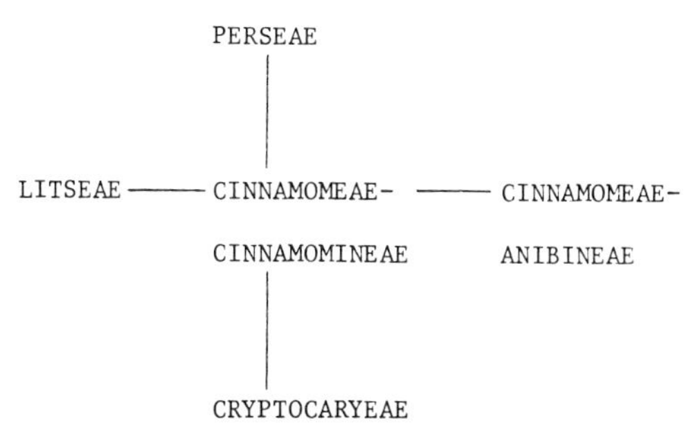

Fig. 12. Top: Composition of tribes of Lauraceae with respect to benzyltetrahydroisoquinoline (BTIQ) derived alkaloidal types (for meaning of codes see Rezende *et al.*, 1975), and to allylphenol (A) and/or propenylphenol (P) derived oxidative dimers. Outstanding structural diversity. Bottom: Morpholigical relationships among tribes of Lauraceae acc. to Kostermans (1957).

value. It should be realized that superposition of several BGs for systematic purposes is technically feasible. Following the development of flavonoids, isoflavonoids, quinolizidines and "uncommon" amino acids we arrive at a computed phylogenetic tree of Papilionoideae tribes (Fig. 13) (Gomes *et al.*, in press).

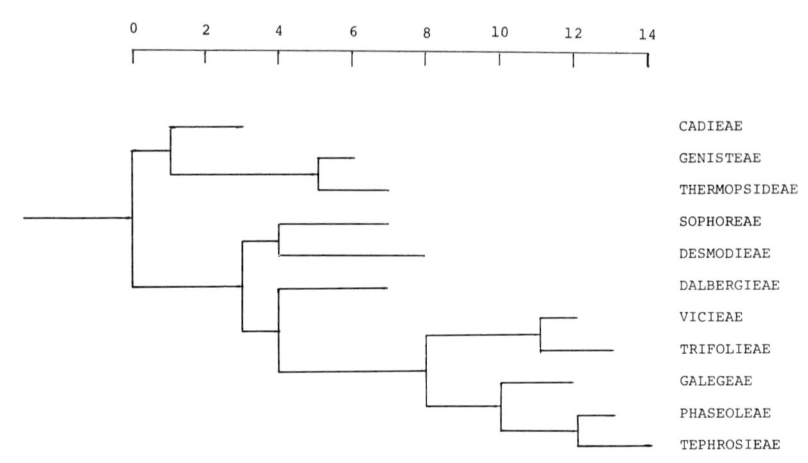

Fig. 13. Composite phylogenetic tree of Papilionoideae tribes based on distribution of flavonoids, isoflavonoids and bases (only tribes for which analyses of all three classes of compounds were reported are considered).

FOURTH PRINCIPLE

The processes by which genetic and ecologic information is translated into form and micromolecules are not yet understood. The significance of correlation or lack of correlation between morphology and chemistry, as given by the usual presence (+)/absence (−) data, consequently can not at present be evaluated. Such correlation, however, as can be observed induces many phytochemists to expect that morphological and chemical information for a taxon should always lead to coincident conclusions without bothering if and why this should happen at all.

Very tentatively we try to express an answer to this problem through a Fourth Principle: morphological and biochemical performances are linked by an equilibrium through control at the level of primary precursor supply. As already implicit by the discussion of the First and Third Principles above, it may well be that the equilibrium between the association of macromolecules and micromolecules within a BG is genetically held at a position where it would provide, given particular ecologic conditions, the best survival value.

This idea can be illustrated by a simple graphical model (Fig. 14, top) which appeared in our paper "the evolution of structural

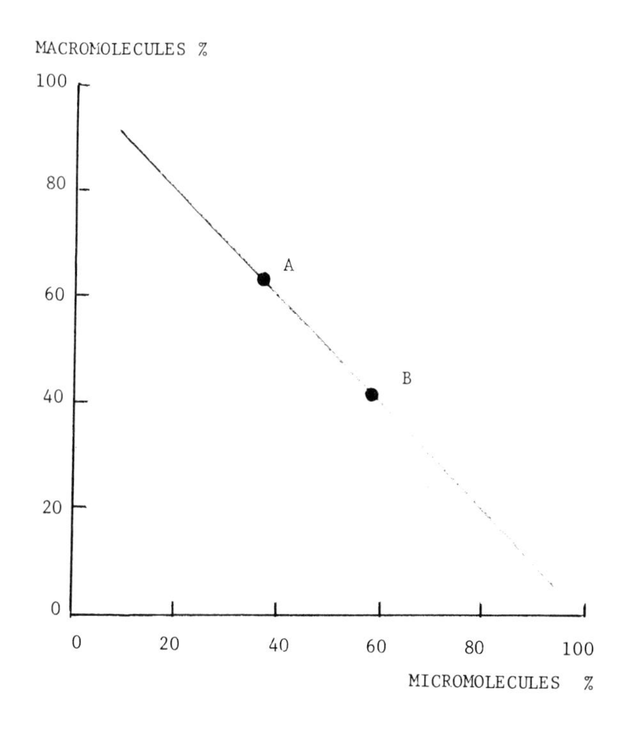
MACROMOLECULES %
100
80
60
40
20
0
A
B
0
20
40
60
80
100
MICROMOLECULES %

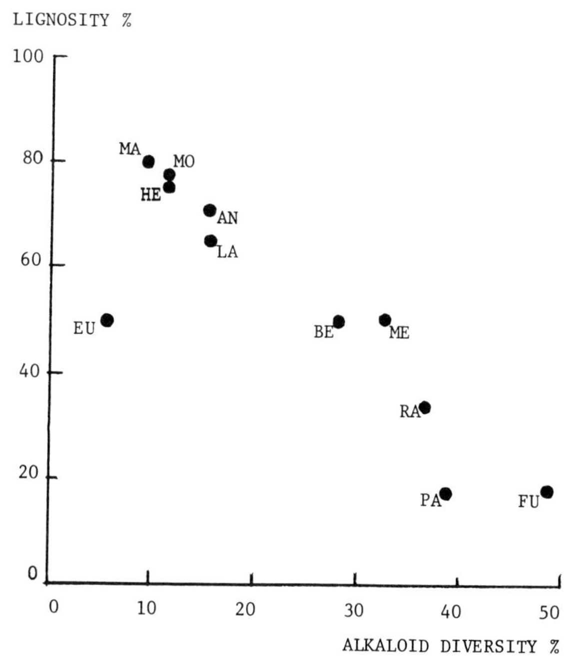
LIGNOSITY %
100
80
60
40
20
0
MA
MO
HE
AN
LA
EU
BE
ME
RA
PA
FU
0
10
20
30
40
50
ALKALOID DIVERSITY %

biopolymers and secondary metabolites is connected?" (Gomes and Gottlieb, 1978). Points represent the chemical composition of organisms. Imaginary organisms, made up either of 100% biopolymers or of 100% metabolites, would be represented by points on the extremes, respectively, of the x and y axes of coordinates whose origin symbolizes, of course, 0% biopolymers and 0% metabolites and hence the existence only of primary precursors.

Organisms made up entirely of macromolecules or micromolecules are both unviable (and it is for this reason that classification based either on morphology or on chemistry will always remain artificial, whatever the amount of supporting evidence). Real organisms are represented by points on the line linking the two extreme imaginary cases. At given ecologic conditions an organism A will be made up of a certain proportion of construction versus biologically active material. Now, if conditions change requiring, for example, more specialized defense, selection will take its course and produce an organism B, adapted to the new conditions.

So far we have tested this concept in very few instances. The plot of "lignosity" versus "alkaloid diversity" for Magnoliidae (Fig. 14, bottom) thus constitutes little more than a promising start to this approach to micromolecular systematics. Not only continuity but also disjunction between physical structure and chemical composition are seen to occur for Magnoliales, Ranunculales and Papaverales. Back extrapolation along this line leads to an ancestor ("lignosity"

Fig. 14. Top: Graphic illustration of postulated chemical equilibrium between structural biopolymers and secondary metabolites for two hypothetical related taxons A and B. Bottom: Correlation of "lignosity" versus "alkaloid diversity" for families of Magnoliidae. "Lignosity" was calculated by the expression [Σ(habit of genus × number of species of genus)]/[number of species of family]. Habits of genera were assigned values 1, 25, 50, 75 and 100, according to indications of Engler and Prantl (1897) and Hutchinson (1964), with respect to "herbs", "herbs or shrubs", "shrubs", "shrubs or trees" and "trees". Intermediate values, such as 62.5 for "trees predominantly shrubs", were also used. "Alkaloid diversity" refers to number of different skeletal types of benzylisoquinoline alkaloids, as defined in Rezende *et al.*, (1975). Magnoliales – MA, Magnoliaceae; MO, Monimiaceae; HE, Hernandiaceae; AN, Annonaceae; LA, Lauraceae; EU, Eupomatiaceae. Ranunculales – BE, Berberidaceae; ME, Menispermaceae; RA, Ranunculaceae. Papaverales – PA, Papaveraceae; FU, Fumariaceae.

100%, "alkaloid diversity" 0%) which may have also originated the remaining orders (Piperales, Nelumbonales, Aristolochiales), albeit by development of a non-alkaloidal chemistry (Third Principle). This is appreciated clearly in Fig. 15 (top), where 100% "lignosity" on the ordinate is placed at the origin of the plot. The relationships among the Magnoliidae are seen (Fig. 15, bottom) to be consistent

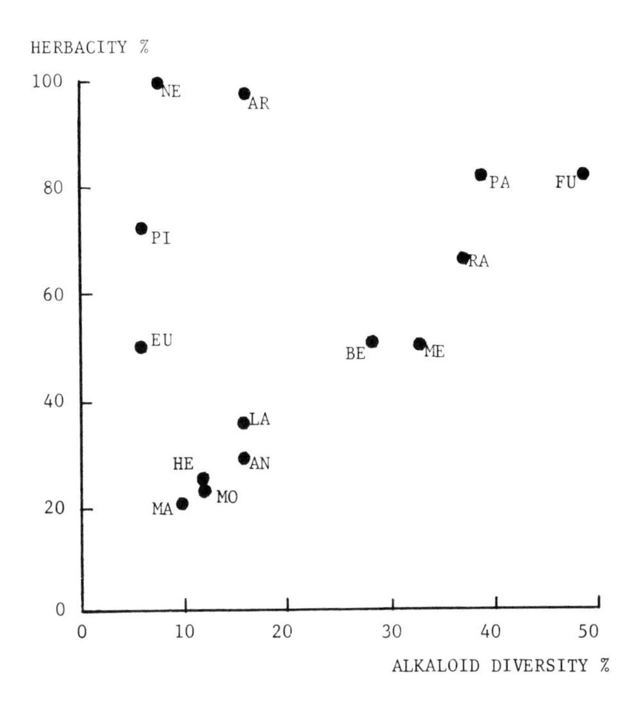

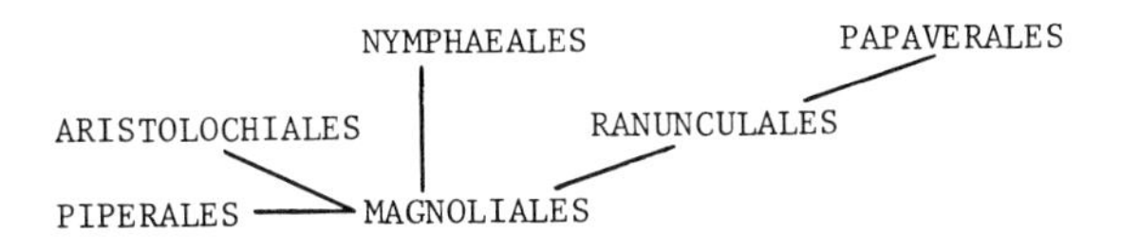

Fig. 15. Top: Correlation of "herbacity" (100 — "lignosity") versus "alkaloid diversity" for families of magnoliidae. Piperales – PI, Piperaceae. Nelumbonales – NE, Nelumbonaceae. Aristolochiales – AR, Aristolochiaceae. For other symbols see Fig. 14. Bottom: Morphological relationship among orders of Magnoliidae according to Cronquist (1968).

with the views of Cronquist (1968) and suggest that, if equal status is accorded to morphological and micromolecular systematics in accordance with Fig. 1, an integrated system of classification may be feasible for at least these two organizational levels of organisms.

CONCLUSION

It is suggested that four principles apply in studying the evolutionary direction of biosynthesis using the notation of Fig. 2.

1. *First Principle*

When B, C and D represent primary metabolites of the shikimate pathway, if a taxon accumulates derivatives of C it comes from a taxon characterized by derivatives of D.

2. *Second Principle*

When B → D represent a BG of secondary metabolites, if a taxon accumulates C it derived from a taxon characterized by B.

3. *Third Principle*

When B → D and E represent two homologous BGs, their mutually exclusive biosynthesis is governed by an equilibrium whose direction is controlled by blocking or change in the activity of an enzyme and may not imply genetic discontinuity.

4. *Fourth Principle*

When B → D and E represent respectively a BG of secondary metabolites and an association of biopolymers, their biosynthesis is connected through the genetically controlled supply of A.

REFERENCES

Brand, J. M. (1978). Fire ant venom alkaloids: their contribution to chemosystematics and biochemical evolution. *Biochem. Syst. Ecol.* 6, 337–340.
Cagnin, M. A. H. and Gottlieb, O. R. (1978). Isoflavonoids as systematic markers. *Biochem. Syst. Ecol.* 6, 225–238.
Cagnin, M. A. H., Gomes, C. M. R., Gottlieb, O. R., Marx, M. C., Rocha, A. I.

da, Silva, M. F. das G. F. da and Temperini, J. A. (1977). Biochemical systematics: methods and principles. *Plant. Syst. Evol.*, Suppl. 1, 53–76.

Cronquist, A. (1968). "The Evolution and Classification of Flowering Plants", 396 pp. Nelson, London.

Dahlgren, R. (1975). A system of classification of the Angiosperms to be used to demonstrate the distribution of characters. *Bot. Notiser* **128**, 119–147.

Engler, A. and Prantl, K. (1897). "Die Natürlichen Pflanzenfamilien", Vol. III, 3, 396 pp. Engelmann Verlag, Leipzig.

Fernandes, J. B., Gottlieb, O. R. and Xavier, L. M. (1978). Chemosystematic implications of flavonoids in *Aniba riparia. Biochem. Syst. Ecol.* 6, 55–58.

Ferreira, Z. S., Gottlieb, O. R. and Roque, N. F. (1980). Chemosystematic implications of benzyltetrahydroisoquinolines in *Aniba* species. *Biochem. Syst. Ecol.* 8, 51–54.

Geissman, T. A. and Crout, D. H. E. (1969). "Organic Chemistry of Secondary Plant Metabolism", p. 391. Freeman, Cooper, San Francisco.

Gomes, C. M. R. and Gottlieb, O. R. (1978). The evolution of structural biopolymers and secondary metabolites is connected? *Revta. brasil. Bot.* **1**, 41–45.

Gomes, C. M. R. and Gottlieb, O. R. (1980). Alkaloid evolution and Angiosperm systematics. *Biochem. Syst. Ecol.* 8, 81–87.

Gomes, C. M. R., Gottlieb, O. R., Gottlieb, R. C. and Salatino, A. (1978). Chemosystematics of the Papilionoideae. *In* "Advances in Legume Systematics" (R. M. Polhill and P. H. Raven, eds). The Royal Botanic Garden, Kew. In Press.

Gottlieb, O. R. (1972). Chemosystematics of the Lauraceae. *Phytochemistry* **11**, 1537–1570.

Gottlieb, O. R. (1977). Neolignans. *Fortschr. Chem. org. Naturst.* 35, 1–37.

Gottlieb, O. R. (1978). Biochemical systematics based on secondary metabolites: principles and methods. *Rev. Latinoamer. Quim.* 9, 138–147.

Harborne, J. B. (1977). Flavonoids and the evolution of the Angiosperms. *Biochem. Syst. Ecol.* 5, 7–22.

Harborne, J. B., Heywood, V. H. and King, L. (1976). Evolution of yellow flavonols in flowers of Anthemideae. *Biochem. Syst. Ecol.* 4, 1–4.

Havir, E. A. and Hanson, K. R. (1968). L-Phenylalanine ammonia-lyase II. Mechanism and kinetic properties of the enzyme from potato tubers. *Biochemistry* 7, 1904–1914.

Hegnauer, R. (1976). Accumulation of secondary products and its significance for biological systematics. *In* "Secondary Metabolism and Coevolution" (M. Luckner, K. Mothes, and L. Nover, eds), pp. 45–76. Deutsche Akademie der Naturforscher Leopoldina, Halle.

Hutchinson, J. (1964). "The Genera of Flowering Plants", Vol. I, 516 pp. Clarendon Press, Oxford.

Kostermans, A. J. E. H. (1957). Lauraceae. *Reinwardtia* 4 (2), 193–256.

Levin, D. A. and York, B. M. Jr (1978). The toxicity of plant alkaloids: an ecogeographic perspective. *Biochem. Syst. Ecol.* 6, 61–76.

Polhill, R. M. (1976). "List of genera and tribes of Papilionoideae (Faboideae)", 13 pp. The Royal Botanic Gardens, Kew.

Rezende, C. M. A. da M. and Gottlieb, O. R. (1973). Xanthones as systematics markers. *Biochem. Syst.* **1**, 111–118.

Rezende, C. M. A. da M., Gottlieb, O. R. and Marx, M. C. (1975). Benzyltetrahydroisoquinoline-derived alkaloids as systematic markers. *Biochem. Syst. Ecol.* **3**, 63–70.

Salatino, A. and Gottlieb, O. R. (1980). Quinolizidine alkaloids as systematic markers of the Papilionoideae. *Biochem. Syst. Ecol.* **8**, 133–146.

Swain, T. (1974). Biochemical evolution in plants. *In* "Comprehensive Biochemistry" (M. Florkin and E. H. Stotz, eds), vol. 29A, pp. 125–302. Elsevier, Amsterdam.

Takhtajan, A. (1973). The chemical approach to plant classification with special reference to higher taxa of Magnoliophyta. *In* "Chemistry in Botanical Classification" (G. Bendz and J. Santesson, eds), pp. 17–26.

17 | Hennig's Methods and Phytochemistry

C. J. HUMPHRIES

Department of Botany, British Museum (Natural History), Cromwell Road, London SW7 5BD, England

and

P. M. RICHARDSON

Department of Plant Sciences, Reading University, Whiteknights, Reading RG6 2AS, Berks, England

Abstract: The principles and methods of Hennig's phylogenetic systematics are briefly described. The value of chemical characters in cladistics is considered by using three separate examples to stress various theoretical points:

(a) The relationship of the Blue-green Algae with the Bacteria, Angio-aperms and Gymnosperms to demonstrate the concepts of parsimony and shared resemblance.

(b) The relationships of *Cicereae* tribe with others in the Leguminosae to illustrate the principles of phylogeny construction using chemical characters.

(c) The relationships of the families within the order Caryophyllidae to show the differences between traditional and analytical systematic methods.

Finally, it is generally concluded that chemosystematics is, on the whole, a vast data-gathering exercise with little recourse to theory. Consequently, until such time there is some agreement about its aims its efforts are likely to remain narrative, rather than analytical undertakings.

Systematics Association Special Volume No. 16, "Chemosystematics: Principles and Practice", edited by F. A. Bisby, J. G. Vaughan and C. A. Wright, 1980, pp. 353–378, Academic Press, London and New York.

INTRODUCTION

In the general field of systematics there are three fundamentally different approaches for solving problems of relationships between organisms. There are those essentially pragmatic methods, which adhere to methodological theory in attempts to improve understandings of relationships for practical purposes but in no way adopt the axioms of evolutionary theory as guidelines. Then, there are those methods which do adopt evolutionary principles in the development of a classifications but consider that evolutionary interpretation is a narrative rather than a analytical undertaking (Patterson, 1980; Rosen and Shuh, 1976). Finally, there are those analytical systematic methods developed during the 1950s and 1960s which have tried to extend and understand the principles of neo-Darwinism to reconstruct phylogenies.

Analytical methods are of three types: (i) Mathematical logic systems as developed, for example, in character compatibility methods (see Estabrook, this volume). (ii) Inductive evolutionary modelling techniques as initiated by Camin and Sokal (1965) and Felsenstein (1978). (iii) Philosophical parsimony methods as initiated by Hennig (1966) and elaborated in recent years by Kluge and Farris (1969), Farris (1970), Farris *et al.* (1970) and Moore *et al.* (1973).

Despite the active compilation of biochemical information in the last 20 years, small molecule chemical characters have only really ever been evaluated in a vague manner, typical of narrative evolutionary biology. We feel that this is due to a number of reasons but mainly for three: (i) chemosystematics is tied to outmoded classification schemes; (iii) there is a considerable lack of appreciation of the theoretical possibilities in an understanding of characters state distributions and the analysis of explicit relationships; (iii) it is generally believed that the accumulation of new data without recourse to theory will fill the gaps in our knowledge.

In this paper we would like to review these points by concentrating on the principles of one analytical approach, Hennigian Phylogenetic Systematics, and with the aid of simple examples consider its application to chemosystematics.

1. The Theory of Phylogenetic Reconstruction

The reasoning behind phylogenetic systematics is based on the idea that the processes of microevolution are the same today as they always were. In other words, observable hierarchies in the natural world are due to common descent through the process of speciation. For our purposes, speciation may be considered as the process whereby a phyletic line, formerly represented by one species, divides into two separate lines (Fig. 1). In the case where one new line is morphologically unchanged from the ancestral line, the process is often considered as the budding off of daughter species from a mother species which is itself considered to persist in time. However, since both lines have a genetic connection with the ancestral line, it is epistemologically best to consider that a mother species has divided into two daughter species (Bremer and Wanntorp, 1978). This is perhaps best represented by a rooted tree diagram (Fig. 1).

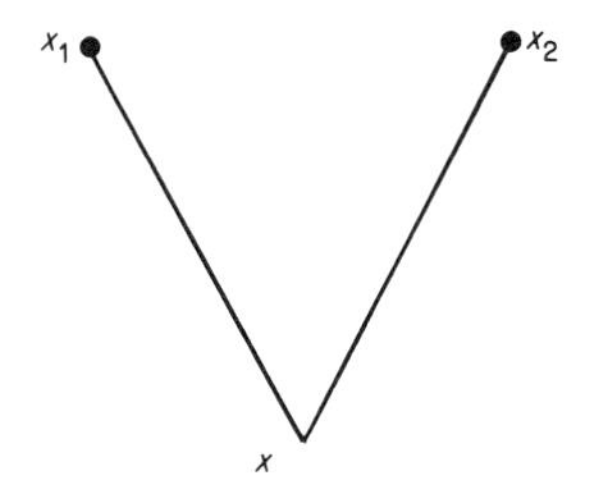

Fig. 1. A rooted tree diagram. In time species x divides into two daughter species x_1 and x_2.

It is worth pointing out that there is the possibility of multiple splitting and reticulate evolution by hybridization and polyploidy but it is not possible to go into details in this chapter. Phylogeneticists try to consider dichotomies or polychotomies as barriers or boundaries in a genealogical continuum.

With time, the process of evolution continues so that one species ultimately ends up as many different species. Each daughter species develops new features in one or more characters which separate them from each other and from the mother species. At each new species generation the unique characters of the first pair of daughter species will be inherited by the new daughter species, and so on. This stepwise process of anagenesis is what Darwin (1859) meant by "descent with modification". There are anagenetic theories in Botany which attempts to determine the ancester-descendant polarities of particular characters or organs. The Magnoioid theory for the origin of Angiosperms (Arber and Parkin, 1907) is one that determines the polarity of floral features in flowering plants by saying that *Magnolia* flowers are primitive. It is unfortunate for systematics that most narrative approaches are incomplete because they consider only, or over-emphasize anagenetic components. In a phylogeny there are two components, the second being cladogenesis, the actual splitting process. In Fig. 2 the species ABCDE are a hypothetical phylogeny comprised of an anagenetic component representing the historical thread of change and a cladogenetic component representing the process of diversification. If evolution had been entirely cladogenetic there would be a diverse array of extant forms with one branch connection (Fig. 2b) whilst, if it had been entirely anagenetic there would be one modern species with a long, stratigraphic continuum of ancestors (Fig. 2c) (after Hill and Crane, in preparation).

Any species within a phylogeny will, from any point in time, therefore possess a mixture of relatively primitive character states, i.e. character states the same as the mother species and a number of

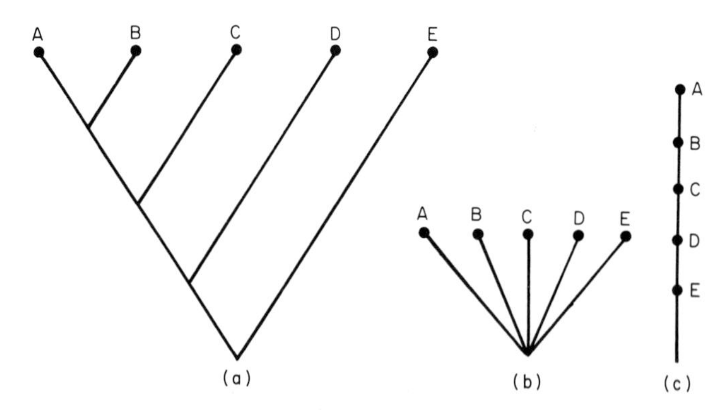

Fig. 2. Anagensis and Cladogenesis. See text for details.

relatively acquired or derived character states inherited from succeeding daughter species. This leads to the observation that within any group of species there will be a mosaic, or as Taktajhan (1959) prefers to call it, a heterobathmic distribution of relatively primitive and relatively derived character states. Hennig's method of phylogenetic reconstruction (1966) is one which tries to resolve the relative evolutionary orders of taxa and characters by having an understanding of heterobathmy and unique character inheritance from daughter species. The concept of a phylogenetic relationship is given in Fig. 3 which shows a cladogram of three recent species

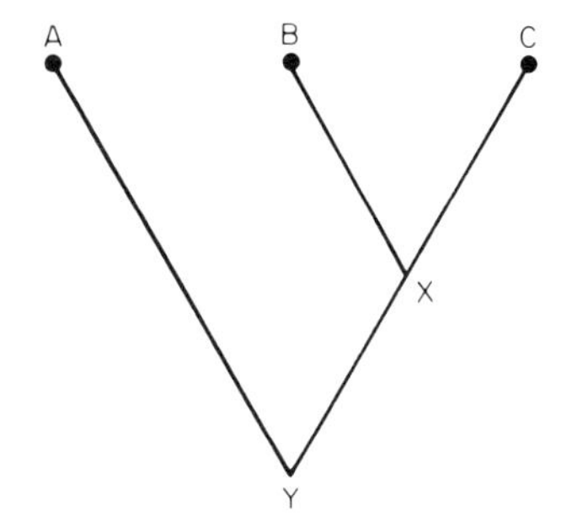

Fig. 3. A definition of a phylogenetic relationship. In three contemporary species ABC, BC share a more recent common ancestor X which is not shared by A (After Hennig, 1966).

A, B and C and two hypothetical ancestors X and Y. Species B and C share a more recent common ancestor X which is not shared by A. Characters originating from the ancestor Y are relatively primitive (plesiomorphic) and can be equally inherited by species A, B or C. Characters originating at X, or between X and Y can only be inherited by B and C and are, therefore, unique to the monophyletic group BC. The joint possession of the derived characters (synapomorphies) indicate a phylogenetic or cladistic relationship. It is possible, in addition to possessing shared derived characters, for a group of taxa to have shared primitive characters and/or shared independantly derived characters. Figure 4 shows that groups formed on the basis of shared primitive characters are paraphyletic. There are

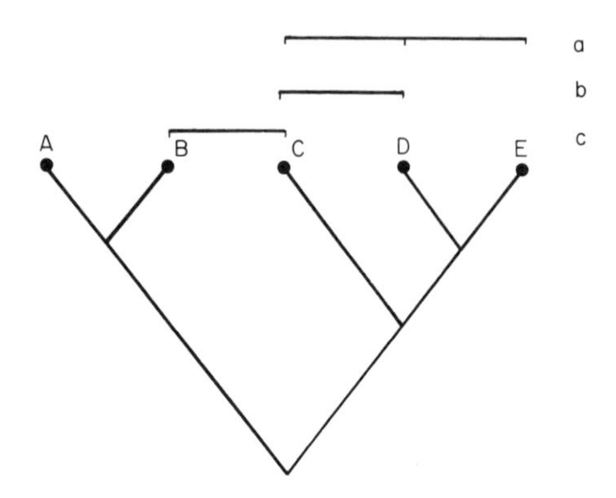

Fig. 4. The relationship between resemblance and group formation. (a) a monophyletic group formed on the basis of shared derived characters (CDE). (b) a paraphyletic group (CD) formed on the basis of shared primitive characters. (c) a polyphyletic group (BC) formed on the basis of shared independantly derived characters (After Hennig, 1966).

many paraphyletic groups in the angiosperms as indicated by such terms as "primitive stocks", "central cores" or "core groups", which in reality are unresolved phylogenies.

Groups formed on the basis of independantly derived characters or false synapomorphies are polyphyletic. Polyphyletic groupings or convergent groups are based on apparent similarity in non-homologous features. The difference between a paraphyletic group and a monophyletic group is only methodological, since it depends on whether groupings are arranged on shared primitive or shared derived homologies. Paraphyletic groupings are unacceptable for the purpose of phylogenetic discussion since they invariably exclude some taxa. In short, phylogenetic reconstruction relies on splitting up the concept of resemblance and forming monophyletic groups on the basis of shared derived character states.

2. The Method of Reconstruction

Since shared derived characters are used for the formation of monophyletic groups, the first step in a phylogenetic reconstruction is the identification of the relatively primitive (plesiomorphous) and the relatively derived (apomorphous) character states. In specialized groups there are several obvious apomorphies but there are usually two criteria for determining ancestral descendant polarities: (i) where there is a gradual transformation series it is possible to identify irreversible steps, e.g. organ loss and neotinous changes. (ii) Since all characters ultimately have the potential for forming a group, the most common approach for any particular level of universality is to examine particular character states in wider related groups. Wide-spread character states are probably primitive relative to the distribution of derived features which usually occur in smaller groups.

Finding reliable apomorphous states becomes increasingly difficult the smaller the group becomes, because the number of inherited derived characters gets less and less. When apomorphous steps have been identified the phylogeny is reconstructed in a piecemeal way by the successive identification of sister groups. The gradual building of the cladogram, or synapomorphy scheme, is a gradual linking together of sister groups. Thus in Fig. 5, B is the sister taxon to the group C, D and at a higher rank A is the sister taxon to the group B, C. D. As Hennig (1966) carefully pointed out, every species or group has a sister species or group, however remote, in the modern biota. The sister group concept is one which overcomes the absence of fossils and it is a concept which attempts to define possible branching points. Of course, when fossils are available they can be included in the cladogram by the same principles applied to modern taxa. The only guiding principles for the reconstruction of cladograms are those of homology and parsimony, so the most parsimonious cladograms are considered the best possible concepts of phylogenetic relationship. Reconstructions can be tested in a variety of ways. Vicariance distribution patterns are often the result of allopatric speciation, and it is common to find that vicariants are sister species. Congruence with cladograms produced on totally independent data sets, such as those produced on biochemical or cytological information, are usually indicative of a "good" hypothesis whilst incongruency can provide a falsification.

3. Chemical Characters

Within the phylogenetic systematic paradigm, genealogical relationships are expressed in cladograms which link together sister groups by the most parsimonious distribution of homologous character

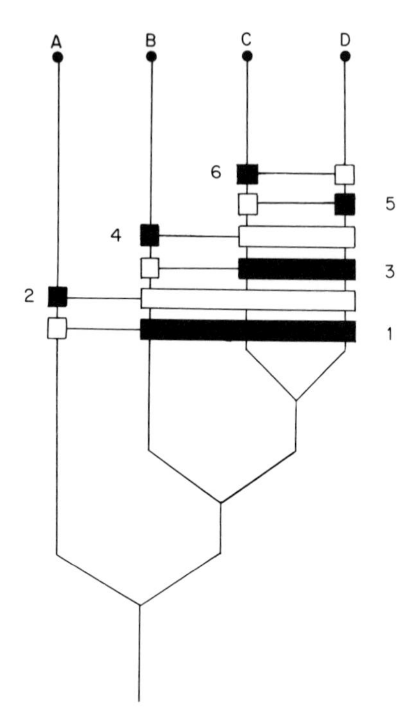

Fig. 5. Argumentation diagram (after Hennig, 1966). Black squares and triangles define derived character states, open squares and rectangles define relatively primitive character states. Characters 2, 4, 5, 6 are autapomorphic for A, B, D and C, characters 1 and 3 are synapomorphic for the groups BCD and CD respectively.

states. Homologies in small molecules are difficult to establish, since for the most part they are recorded simply as being either present or absent. In this situation there can be only two possible ancestral derived polarities, either prescence is primitive or derived. Absence can only be a derived feature if it is genuinely known to be due to loss. Much more difficult to establish is the situation whereby a chemical character state or a suite of chemical character states of restricted distribution is known to be a replacement of a more common system and therefore interpreted as a derived feature. Although ultimately knowledge of biogenetic pathways will be the best arbiter for polarity decisions, commonness, i.e. widespread distribution in related groups, is more often that not a good starting position for deciding on relatively primitive conditions (Szalay, 1977).

A third possible way for identifying homologies in chemical characters comes from an understanding of metabolic sequences. Apomorphies can easily be identified if particular compounds are associated with irreversible steps in a metabolite sequence or if they occupy terminal positions in a chain of reactions.

CHEMOSYSTEMATIC EXAMPLES

1. The Logic of ABC

If we accept the idea that three is the minimum number of taxa upon which a phylogenetic hypothesis can be made and that every taxon has a sister group in the modern biota, it follows that phylogenetic hypotheses can be made on any three taxa. A recent paper in *Nature* (Gardiner *et al.,* 1979) considered the phylogenetic relationships of the salmon, the lungfish and the cow to stress the concepts of cladism, so we do the same for the three taxon group; Angiospermae, Cyanophyta and Bacteria.

Phylogenetic Systematics has frequently been criticized by narrative phylogenists for its methodical rigidity in defining Monophyletic groups on the basis of shared derived characters (e.g. Sneath and Sokal, 1973). Since all hypotheses of relationship depend on some measure of similarity to form groups, the differences in approach depend on different views of homology, or in other words whether the pairs of similarities to be emphasized should be advanced or primitive characters. The cladogram in Fig. 6 is based on one biochemical character, chlorophyll A, a key pigment in photosynthesizing organisms, and one structural character, the presence of an organized nucleus. For over 100 years (Cohn, 1853) the Cyanophyta have been closely related to the Bacteria. For example, Stanier and Van Niel (1941; see Round, 1965) used the name Monera to group the Cyanophyta with Bacteria using three characteristics of (a) disorganized nuclei, (b) the absence of sexual reproduction and (c) the absence of plastids as evidence for close relationships; a grouping based on shared primitive features. In phylogenetic terms this hypothesis suggests that the Cyanophyta have a more recent common ancestry with the Bacteria as compared with Angiospermae, with the consequence that the Angiospermae must

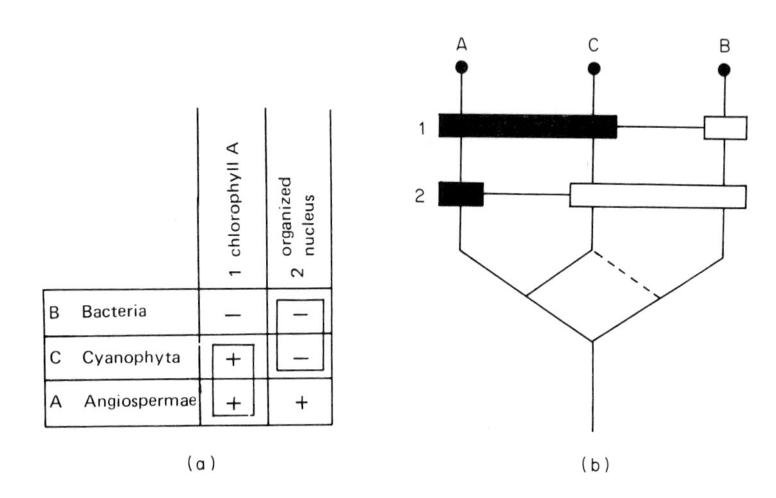

Fig. 6. Data matrix (a) and cladogram (b) showing the probably ancestry of the 3-taxon group Bacteria, Cyanophyta and Angiospermae.

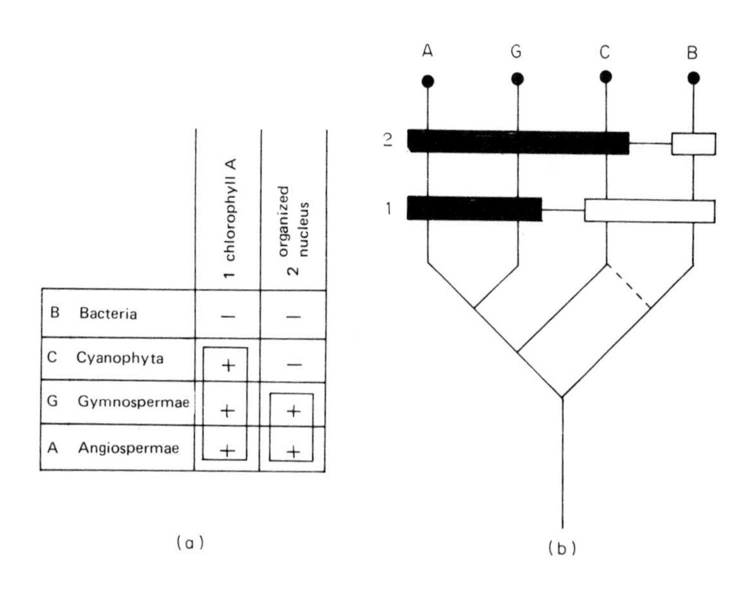

Fig. 7. Data matrix (a) and Cladogram (b) showing the probable ancestry of the 4-taxon group Bacteria, Cyanophyta, Gymnospermae and Angiospermae.

have had a relatively ancient origin, a view inconsistent with the Darwinian notion of descent by specialization. Furthermore, we must consider that chlorophyll A either evolved before the organized nucleus and it is a loss feature in the Bacteria, or it must have evolved twice to account for its parallel distribution in the Angiospermae and the Cyanophyta. It is certainly true that both the Bacteria and the Cyanophyta are ancient groups but the most parsimonious explanation for the distribution of the four character states would be to consider that the presence of chlorophyll A, is a shared derived feature for the Cyanophyta and the Angiospermae suggesting more recent common ancestry in terms of the sister Bacteria group. The organized nucleus in Fig. 6 is unique (autapomorhous) to the Angiospermae. The information content in this type of classification is very high and a number of different predictions can be developed from the model. For example, any organism with an organized nucleus is likely to possess chlorophyll A. Figure 7 shows the situation for the Gymnospermae which in this four-taxon group falls exactly into the predicted position.

2. The Position of Cicer

To consider how small molecule chemical characters might be used in angiosperm systematics we have chosen an example from the Leguminosae. In her paper entitled "The delimitation of the tribe Vicieae (Leguminosae) and the relationships of *Cicer* L.", Francis Kupicha (1977) gives a simple cladistic model for the relationships of the four tribes the Viceae, the Cicereae and the Trifolieae-Ononideae. It is a useful example to show how ancestor-derived polarities are established in various character states and demonstrates the value of chemical characters in testing existing phylogenetic hypotheses.

From a total of thirty-one characters, five two-state characters were found to be useful for evaluating the relative position of the Cicereae tribe with the Vicieae and the Trifoleae-Ononideae. Figure 8 gives the distribution of the ten character states in the four groups. Traditionally, *Cicer* has usually been related to the Vicieae because its stipules are not adnate to the petiole, it has hypogeal germination and also possesses tendrils (Fig. 8. A, D, E). However, it also shares two character states with the Trifolieae-Ononideae by the serrate leaves with craspedodromous venation and long-stalked glands

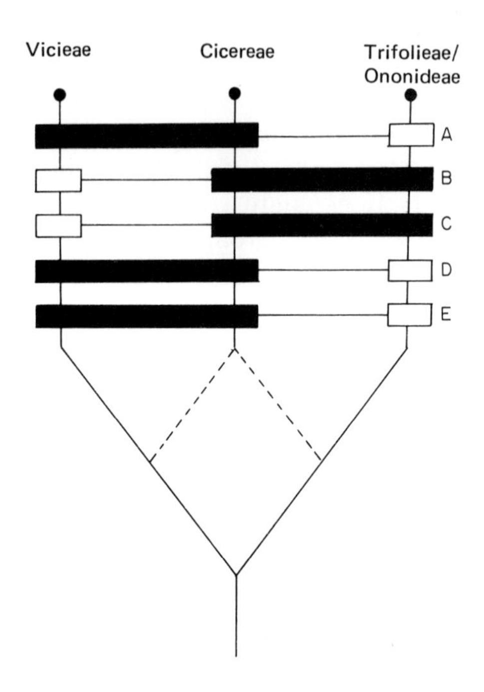

Fig. 8. The distribution of five 2-state characters in the 3-taxon group; Vicieae, Cicereae and Trifolieae-Ononideae. A, stipules adnate to petiole versus stipules not adnate to petiole. B, Leaves serrate with craspedodromous venation versus leaves not serrate without craspedodromous venation. C, Long-stalked glands present versus long-stalked glands absent. D, Hypogeal germination versus epigeal germination. E, Tendrils present versus tendrils absent.

(Fig. 8. B, C). In a strictly phenetic classification three character states versus two would be sufficient ground for grouping *Cicer* with the Vicieae. However, because a phylogenetic hypothesis cannot be resolved from this particular set of information, the same characters can either be checked out further in related groups or new sets of information can be gathered and analysed for the same group. Ideally, the sister group of this particular assemblage would be the obvious place to look first but as this is difficult to define, we have considered the character distributions in the remainder of the subfamily Papilionoideae in which the undefined sister group must reside (Fig. 9). Taking each character in turn: character state D, hypogeal

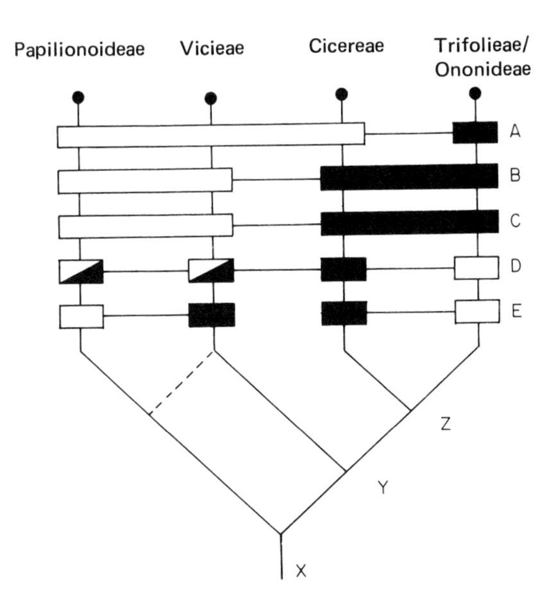

Fig. 9. Parsimony diagram for the possible phylogenetic relationship of the 3-taxon group; Viceae, Cicereae and Trifolieae-Ononideae after reinterpretating the ancestor–descendant polarities of the five 2-state characters by comparison to the Papilionoideae. See Fig. 8 for key to characters, and the text for explanation.

germination occurs frequently in all of the tribes, except the Trifolieae/Ononideae and is readily reversible with epigeal germination in many species of the Papilionoideae and the Vicieae. Its polarity cannot, therefore, be decided and it has no value as a taxonomic character at this level of universality. If we consider the two states of character A, stipules not adnate to the petiole versus stipules adnate to the petiole, the former condition is universal throughout the remainder of the Papilionoideae, the Vicieae and the Cicereae which means that stipules adnate to the petiole is best considered as derived, an autapomorphy, i.e. unique state for the Trifolieae-Ononideae. Long-stalked glands and serrate leaves with craspedodromous venation (characters B and C) are unique conditions for the Cicereae, Trifolieae and Ononideae. If we conclude that these are derived conditions and that the Cicereae are sister to the other two tribes, then only one difficulty is left – the interpretation of character E. Either tendrils have evolved twice, independantly for the Vicieae and the Cicereae or evolved once before branching point Y to become lost again after branching point Z. For this data there is no way of distinguishing between the two hypotheses.

From this diagram we are left with the hypothesis that the Cicereae share most recent common ancestry with the Trifolieae-Ononideae than with the Vicieae, i.e. whether the Vicieae are the sister groups to the three tribes or that its sister group resides in the Papilionoideae. In recent years there have been a number of papers devoted to the study of isoflavonoids and acetylenes, particularly those occurring as phytoalexins in the Leguminosae (Ingham, 1972; Ingham and Harborne, 1976). As an independent examination of the morphological hypothesis we have looked at the distribution of eight chemical characters (Table I). A serious refutation of the cladistic model in Fig. 9 could only come about if the Vicieae, or a group within the remainder of the Papilionoideae showed more recent common ancestry with either the Cicereae or the Trifolieae-Ononideae. Working on the basis that the presence of the eight chemicals is derived, their distribution is given in Fig. 10. The compounds Medicarpin and Maackian (Fig. 10, E, R) are common to all the groups. At the other extreme, the two compounds Pisatin and Wyerone (G, H) are unique to the Vicieae and have no systematic value except to emphasize the distinctiveness of that group. Pratensein (C) is clearly synapomorphic for the Cicereae and the Trifolieae-Ononideae, corroborating but not refuting the original common ancestry hypothesis based on morphological attributes. From the distribution of the three compounds Biochanin A, Formononetin and Diadzein (Fig. 10, A, B, D) it would appear that the Cicereae/Trifolieae/Ononideae form a monophyletic group, but having a more recent common ancestor with an, as yet, undefined group within the paraphyletic Papilionoideae rather than with the Vicieae.

3. The Caryophyllidae – a Suitable Case for Phylogenetic Systematics

Because we believe that the methods of phylogenetic systematics can provide better genealogical concepts than narrative methods, there

Table I. The Distribution of Eight Chemicals in Five Tribes of the Leguminosae

	A Bio-chanin A	B Formono-netin	C Praten-sein	D Daidzein	E Medi-carpin	F Maackiain	G Pisatin-	H Wyerone
Vicieae	—	—	—	—	+	+	+	+
Cicereae	+	+	+	+	+	+	—	—
Trifolieae-Ononideae	+	+	+	+	+	+	—	—
Remainder of Papilionoideae	+	+	—	+	+	+	—	—

+ Present
— Absent

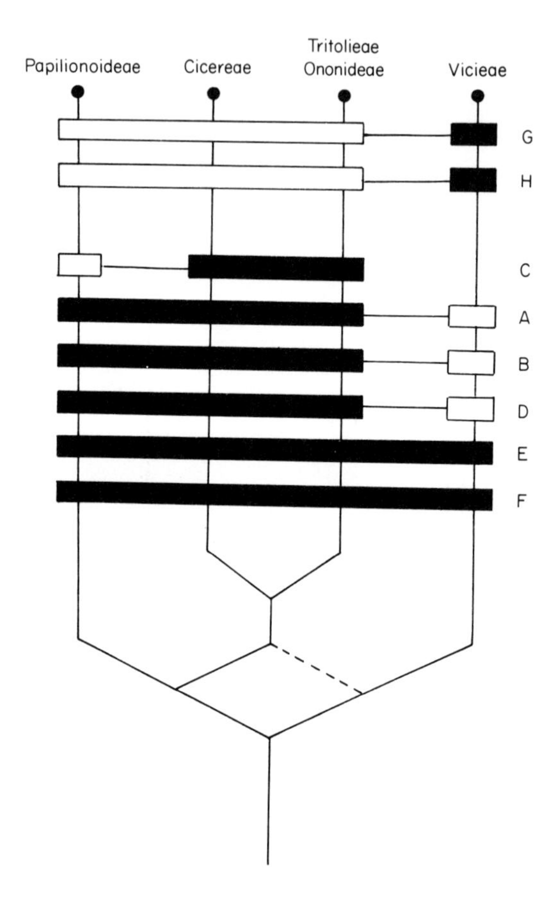

Fig. 10. Cladogram showing the distribution of eight chemicals of the Leguminosae in the 4-taxon group; Viceae, Cicereae, Trifolieae-Ononideae. Letters refer to the chemicals in Table I. Presence is considered derived (black squares and rectangles).

can be no better example than the relationships of the families of the order Caryophillidae to illustrate the difference between a traditional and a cladistic reconstruction. It is remarkable that, despite impressive amounts of comparative information produced over the last few years (see Mabry and Behnke, 1976, for review), the evolutionary relationships of the Caryophyllidae are still discussed entirely within a narrative framework.

As a starting point for discussion we have chosen Cronquist's classification (1968) to illustrate the traditional interpretation. We have no reason to criticize his scheme more than anybody else's, it is simply the one which figures most frequently in systematic discussion. In his diagram (Fig. 11, Cronquist, 1968: Fig. 4.5), the

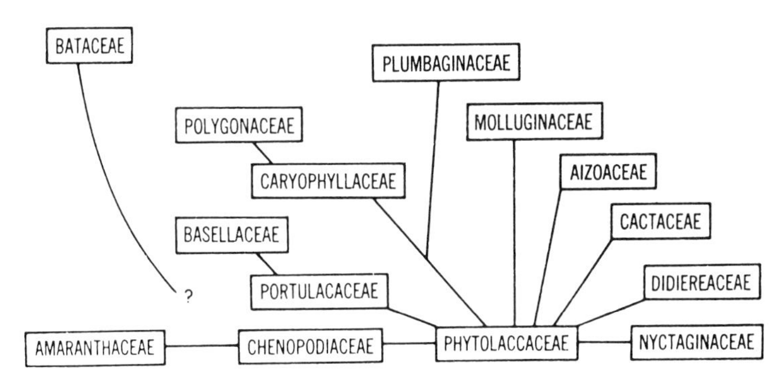

Fig. 11. Probable relationships of the families of the Caryophyllidae (reprinted from Cronquist, 1968).

probable relationships of the families of the Caryophyllidae are given as an "ideal" tree. This means that various concepts of resemblance are used to define "relationship", invoking "real" ancestors from the groups available for study. The arrangement is essentially a phenetic one because the concept of resemblance is not resolved and a mixture of relatively advanced and relatively primitive characters are used to express similarity. Ancestor-descendant polarities in homologous character state distributions are decided by considering that the Caryophyllidae take their origin in the Magnoliales subclass Magnoliidae probably in the Illiciaceae.

In phylogenetic systematics, all groups must be strictly monophyletic, a situation which can only be convincingly demonstrated by the identification of sister groups through the shared possession of derived features. We have extracted the available cladistic information and presented it in the cladogram of Fig. 12. For simplicity we had to assume that the families themselves are all monophyletic groups. In some cases this is reasonable (e.g. the Caryophyllaceae)

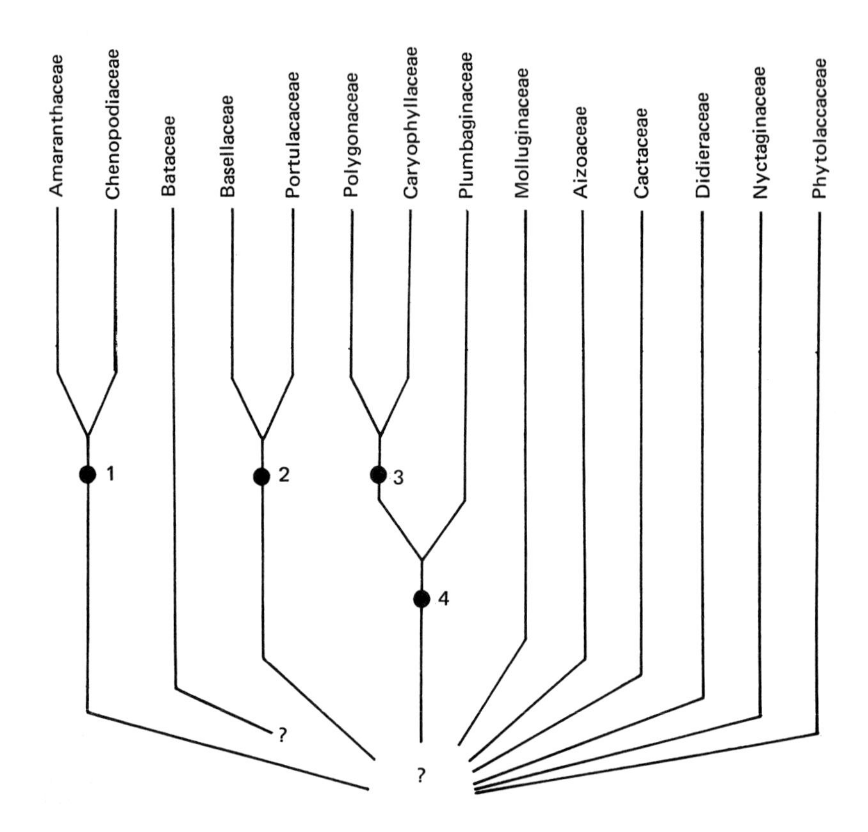

Fig. 12. Evolutionary tree showing the probable relationships of the families of the Caryophyllidae. Redrawn as a cladogram from Fig. 11. Unresolved branching points are given as question marks.

but in several cases this is not (e.g. Aizoaceae) since they are either paraphyletic excluding sister taxa placed within other groups, or they are polyphyletic, including taxa which are sister to groups within other families (e.g. the Phytolaccaceae). It is illogical to consider that modern families with species showing more recent common ancestry with each other to be ancestors of other modern groups. Thus, the Phytolaccaceae are placed at the top of the diagram (Fig. 12) rather than ancestral to all the other families. Should it be demonstrated that various taxa of the Phytolaccaceae have a number of different sister group relationships with other families in the order, then the family is not a natural group and should be disbanded. This point reflects a fundamental difference between narrative and analytical methods. An axiom common to all analytical methods is that the empirical data (whether they be OTUs, individuals, species or nucleotide sequences) are always terminal twigs in a reconstruction with hypothetical rather than real ancestors.

For the most part the relationships of the 14 families are unresolved. If n is taken to represent the number of taxa there should be $n-1$ branching points in a fully resolved cladogram, i.e. 13 branching points for 14 families. In Fig. 12, there are only four branching points. Obviously derived groups are easy to recognize as indicated by branches 1 and 2 for the four families Amaranthaceae, Chenopodiaceae, the Portulaccaceae and the Basellaceae. Branches 3 and 4 indicate sister group relationships for the Polygonaceae with the Caryophyllaceae and these two families with the Plumbaginaceae.

Since this group, as a whole, includes all of the Betalain-producing families we have to assume that the, as yet undefined, sister group of this assemblage is an anthocyanin-producing group. Thus, if all of the Betalain-producing families are a monophyletic group, either anthocyanin production has evolved twice from a Betalain sister group for the Polygonaceae/Plumbaginaceae/Caryophyllaceae branch and the Molluginaceae respectively, or that anthocyanins have secondarily evolved once and the Molluginaceae belong with the three other anthocyanin producing members. There is a third possibility suggested by Mabry (1973) that Betalains are a derived feature replacing Anthocyanins which due to its commonness in the Angiospermae must be considered relatively primitive.

In an attempt to resolve some of the branching points we have surveyed all of the available literature (see review in Mabry and Behnke, 1976) to find synapomorphic character states. These together with the Betalain character are summarized in Table II and included in the cladogram on Fig. 13. Character A, the presence versus absence of Betalains, whose presence is considered derived, means that the Caryophyllaceae and the Molluginaceae are sister groups to the Betalain-bearing families. There are three synapomorphies postulated here (E, F and H) for linking them, by considering that P-type plastids rather than S-type are derived (Behnke, 1976), the presence of perisperm versus absence is derived, and

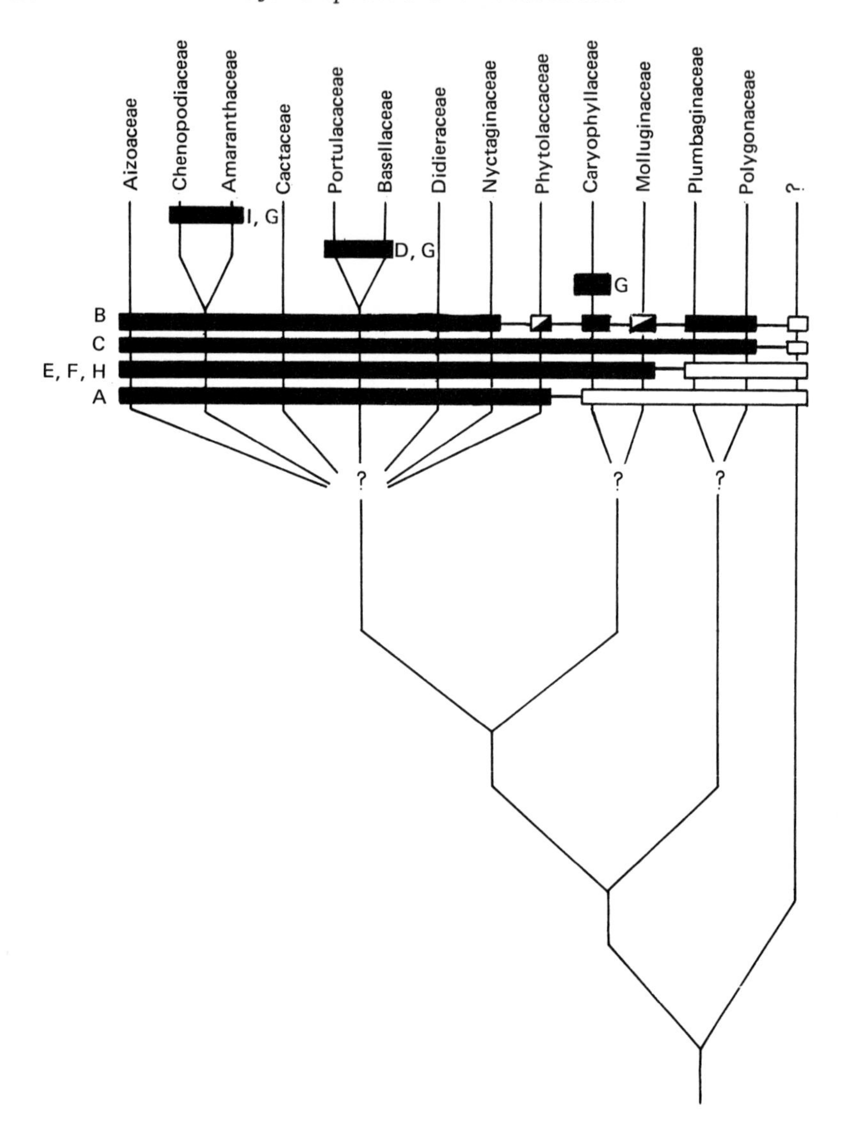

Fig. 13. Probable relationships of the Caryophyllidae as recognized on the basis of contemporary evidence. The letters refer to the character states of Table II.

that absence of endosperm versus the presence of endosperm is derived. The Plumbaginaceae and Polygonaceae can only be considered as the sister group by postulating that trinucleate versus binucleate pollen is derived (C). If any of these interpretations are

Table II. The Distribution of Nine Morphological Characters within the Caryophyllidae.

	Apotypic	Plesiotypic
A Betalain	present	absent
B Carpels	syncarpous	apocarpous
C Pollen	trinucleate	binucleate
D Calyx	2 sepals	3 or more
E Endosperm	absent	present
F Perisperm	present	absent
G Placentation	Free central or basal	basal and uniovulate, axile or parietal
H Plastids	P-type	S-type
I Crystalloids	absent	present

correct, then there are several awkward interpretations on other characters. If in Character B (apocarpy versus syncarpy), apocarpy is considered primitive its presence in the Molluginaceae, and the Phytolaccaceae is difficult to explain. If syncarpy is a relatively derived condition linking these 12 families together the idea that secondary apocarpy occurs, as originally suggested by Rohweder (1965a, b) for *Phytolacca* would hardly be a plausible parsimonious interpretation. The unusual distribution of the free-central or basal placentation means that either this character is trivial evolving on three separate occasions, the general phylogeny is wrong, or that it is primitive. This diagram summarizes what can be done on present evidence, the contemporary view of the relationships of the main families in the Caryophyllidae. A comparison with Cronquist's diagram (Figs. 11 and 12) shows, in fact, that despite an increase in data and as a result of using narrative interpretations the relationships of the Caryophyllidae are still unresolved, in all but four possible branching points.

It was our original intention to review the phytochemical work in the Centrospermeae, since one of us (PMR) has been studying the flavonoid biochemistry of the order for three years. During that time a considerable amount of phytochemical data on Alkaloids, Glycoflavones, Isoflavonols, 6-7 methylene dioxy flavonoids, steroids, saponins and Betalains of potential systematic value has accumulated. In examining the data we came to the conclusion that the only way that chemical characters would have any value at all would be if they had been searched for in every taxon, that they occurred in two or more taxa, and that by postulating presence versus absence we were dealing with homologous conditions. Significant results are given in Table III. Possible synapotypies (i.e. shared derived presence) were found in only six chemicals; Amaranthin, Phytollactin, Betanidin, Celosianin, Isoflavonoids, and 6-7 methylene dioxy flavanols. Their distributions in the Betalain families are given (presence meaning derived) in Fig. 14. Character A shows the presence of Betalains in the nine families. The presence of Celosianin (E) and Isoflavonoids (I) have so far been found only in the Chenopodiaceae and the Amaranthaceae. The next most closely related family appears to be the Cactaceae as indicated by the distribution of Amaranthin (B). Other chemicals tend to support the linking of the Amaranthaceae with the Chenopodiaceae a sister group with the Cactaceae (e.g. Phytolacctin (C)), although the presence of 6-7 methylene dioxy compounds links them with the Didieraceae, and Betanidin is found only in the Aizoaceae and the Cactaceae.

CONCLUSIONS

As already pointed out by Bremer and Wanntorp (1978) the debate about analytical methods in systematics has been going on for a long time although, on the whole, botanists have chosen to bypass it. Although it is possible to find a few botanical analytical enthusiasts, most work has in fact been carried out within a narrative framework. Since phytochemistry has largely played a supportive role in systematics the relevance of chemical data ia rarely assessed in a consistent way. As a result, chemical corroboration of existing classifications based on non-testable eclectic interpretations of anagenetic theories is the main justification for what has largely become a vast data-gathering enterprise. Haphazard data gathering might be a reasonable pursuit from a purely chemical point of view, but is extremely wasteful in systematics.

Hennig's phylogenetic systematics was one of the first formalizations of an analytical method which brought a deductive, scientific approach to comparative biology. By using explicit ground rules for

Table III. The Distribution of Betalains and other Chemicals in the Families of the Caryophyllidae.

Family	Betalains A	Amaranthin B	Phyto-llactin C	Betanidin D	Celosianin E	Indi-caxanthin F	Betanin G	Vulgaxanthin H	Iso-flavonoids I	6,7-methylene dioxy flavonoids J
Aizoaceae	+	−	+	+	−	−	+	?	−	−
Amaranthaceae	+	+	−	−	[+]	−	+	?	+	+
Basellaceae	+	?	?	?	?	?	?	?	−	−
Cactaceae	+	[+]	+	[+]	−	+	+	+	−	−
Caryophyllaceae	−	NA	NA	NA	NA	NA	NA	NA	−	−
Chenopodiaceae	+	+	[+]	−	+	?	+	+	+	+
Didieraceae	+	?	?	?	?	?	?	?	−	+
Molluginaceae	−	NA	NA	NA	NA	NA	NA	NA	−	−
Nyctaginaceae	+	−	−	−	−	+	+	+	−	−
Phytolaccaceae	+	−	−	−	−	−	+	?	−	−
Portulacaceae	+	−	−	−	−	+	+	+	−	−
Plumbaginaceae	−	NA	NA	NA	NA	NA	NA	NA	−	−
Polygonaceae	−	NA	NA	NA	NA	NA	NA	NA	−	−

+ present, [+] occasionally present, − not detected, NA not applicable, ? not known.

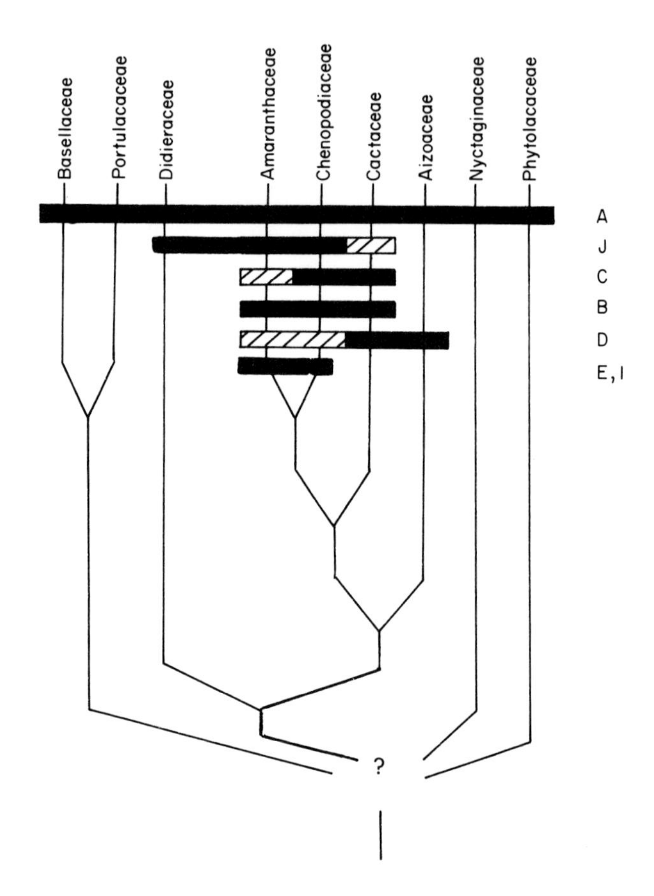

Fig. 14. The distribution of six chemical characters within the Betalain families. The letters refer to the character states of Table III.

the compilation of cladistic theories it removes subjectivity and provides falsifiable hypotheses of genealogical relationship. Chemical characters have to be treated in the same way as any other source of characters and although they are usually more difficult to assess, as compared to morphological features, they have the same potential as primary raw data or independent falsifiers of existing hypotheses. However, until such time when some agreement is reached as to the scientific validity and purpose of comparative biology, chemosystematics will remain a narrative rather than an analytical undertaking.

ACKNOWLEDGEMENTS

We would like to thank Dr. J. Ingham for advice, Bob Press for drawing the figures, the photographic studio at the British Museum (Natural History) for making the slides and Mrs M. Humphries for typing the manuscript.

REFERENCES

Arber, E. A. N. and Parkin, J. (1907). On the origin of Angiosperms *J. Linn. Soc. Bot.* **38**, 29–80.

Behnke, H.-D. (1976). Ultrastructure of sieve-element plastids in *Caryophyllales* (Centrospermae). Evidence for the delimitation and classification of the order. *Pl. Syst. Evol.* **126**, 31–54.

Bremer, K. and Wanntorp, H. E. (1978). Phylogenetic systematics in botany. *Taxon* **27**, 317–329.

Camin, J. H. and Sokal, R. R. (1965). A method for deducing branching sequencies in phylogeny. *Evolution* **19**, 311–326.

Cohn, A. (1853). Untersuchungen über die Entwicklungsgeschicte Mikroskopischer Algen und Pilze. *Nova Acta Acad. Caesar. Leop. Carol.* **24**, 103–256.

Cronquist, A. (1968). "The Evolution and Classification of Flowering Plants". Nelson, London.

Darwin, C. (1859). "On the Origin of Species". Murray, London.

Farris, J. S. (1970). Methods for computing Wagner trees. *Syst. Zool.* **19**, 83–92.

Farris, J. S., Kluge, A. G. and Eckardt, M. J. (1970). A numerical approach to phylogenetic systematics. *Syst. Zool.* **19**, 172–189.

Felsenstein, J. (1978). Cases in which parsimony or compatibility methods will be positively misleading. *Syst. Zool.* **27**, 401–410.

Gardiner, B. G., Janvier, P., Patterson, C., Forey, P. L., Greenwood, P. H., Miles, R. S. and Jefferies, R. P. S. (1979). The salmon, the lungfish and the cow: a reply. *Nature* **277**, 175–176.

Hennig, W. (1966). "Phylogenetic Systematics". University of Illinois Press, Urbana.

Hill, C. and Crane, P. (in preparation). "The Origin of the Angiosperms".

Ingham, J. L. (1972). Phytoalexins and other natural products as factors in plant disease resistance. *Bot. Rev.* **38**, 343–424.

Ingham, J. L. and Harborne, J. B. (1976). Phytoalexin induction as a new dynamic approach to the study of systematic relationships among higher plants. *Nature* **260**, 241–243.

Kluge, A. G. and Farris, J. S. (1969). Quantitative phyletics and the evolution of anurans. *Syst. Zool.* **18**, 1–32.

Kupicha, F. K. (1977). The delimitation of the tribe Vicieae (Leguminosae) and the relationships of *Cicer* L. *J. Linn. Soc. Bot.* **74**, 131–162.

Mabry, T. J. (1973). Is the order Centrospermae monophyletic? *In* "Chemistry in Botanical Classification. Nobel Symposium 24" (G. Bendz and J. Santesson, eds), pp. 275–285. Academic Press, New York and London.

Mabry, T. J. and Behnke, H.-D. (eds) (1976). Evolution of Centrospermous Families. A Symposium held on July 8, 1975, during the XIIth International Botanical Congree Leningrad. *Pl. Syst. Evol.* **126** (1), 1–105.

Moore, G. W., Barnabas, J. and Goodman, M. (1973). A method for constructing maximum parsimony ancestral amino acid sequences on a given network. *J. Theor. Biol.* **38**, 459–485.

Patterson, C. (1980). Methods of Paleobiogeography. *In* "Vicariance Biogeography; a Critique" (G. Nelson and D. E. Rosen, eds). Columbia University Press, New York. In press.

Rohweder, O. (1965a). Centrospermen-Studien. 1. Der Blutenbau bei Uebelinia Kiwuensis T.C.E. Fries (Caryophyllaceae). *Bot. Jahrb. Syst.* **83**, 406–418.

Rohweder, O. (1965b). Centrospermen-Studien. 2. Entwicklung und morphologische deutung des Gynociums bei *Phytolacca. Bot. Jahrb. Syst.* **84**, 509–526.

Rosen, D. E. and Shuh, R. T. (1976). (Book review) "Flowering Plants: Evolution above the Species Level". *Syst. Zool.* **24**, 504–506.

Round, F. E. (1965). "The Biology of the Algae". Arnold, London.

Sneath, P. H. A. and Sokal, R. R. (1973). "Numerical Taxonomy". Freeman, San Francisco.

Stanier, R. Y. and Van Niel, C. B. (1941). The main outlines of bacterial classification. *J. Bact.* **42**, 437–466.

Szalay, F. S. (1977). Ancestors, descendents, sister groups and testing of phylogenetic hypotheses. *Syst. Zool.* **26**, 12–18.

Taktajhan, A. L. (1959). "Die Evolution der Angiospermen". Gustav Fischer, Jena.

18 | The Compatibility of Occurrence Patterns of Chemicals in Plants

G. F. ESTABROOK

Department of Botany and University Herbarium, Michigan University, Ann Arbor, MI 48109, USA

INTRODUCTION

Ornduff *et al.* (1974) state that the value of biochemical characters for assessing the systematic relationships of plants has been demonstrated clearly. Among the objectives of some plant systematists is to estimate evolutionary relationships of plants, and then to structure a classification that reflects this estimate. Here I will discuss some principles that can be used to reason with the occurrence patterns of chemicals in plants so that it becomes clear how they are used as evidence to support or refute estimates of evolutionary relationships or classifications based on them. To illustrate my points, I will draw examples from data generously supplied by O. R. Gottlieb of the Instituto de Quimica, USP, Brazil. When referring to his data, names of taxa and of chemicals are deliberately omitted and the estimates of relationships presented in the examples below do not necessarily reflect his professional opinion. I will also discuss unpublished data supplied me by T. O. Duncan, Department of Botany, UC Berkeley, CA, for which taxonomic or chemical names are not given here. The excellent flavonoid study of Ornduff *et al.* (1974) provides a third source of examples for this discussion.

Systematics Association Special Volume No. 16, "Chemosystematics: Principles and Practice", edited by F. A. Bisby, J. G. Vaughan and C. A. Wright, 1980, pp. 379–397, Academic Press, London and New York.

Please refer to their work for details of taxonomic and chemical identity.

BASIC ASSUMPTIONS

There are many reasonable ways to approach the analysis of chemical data for systematic purposes, but for any approach to contribute most effectively to scientific debate and understanding, the situations to which it is intended to apply should be clearly circumscribed. The compatibility approach to the analysis of chemical data is based on the principle of the logical consistency of hypotheses. To apply the ideas of compatibility analysis the following conditions should be met. A collection of populations, subspecies or species etc. is under study. This collection will be called the study group. The entities in the study group will be called evolutionary units, EUs. Compatibility analysis is done to try to achieve a better understanding of the evolutionary relationships among the EUs in the study group. We must assume that each EU evolved from a single ancestral EU (not necessarily available for study) so that the evolutionary relationships among the EUs in the study group can be represented with a tree diagram. For example, this condition rules out EUs that are hybrids. This can be potentially troublesome as many plant taxa are known to be of hybrid origin, and it is possible that unbeknownst to the worker, one or more of his EUs arose in this way.

Each EU has been surveyed for the presence or absence of chemicals of interest to the worker, and there is no "missing data". Furthermore, the presence or absence of a particular chemical in an EU is typical of virtually all its individuals, so that each EU clearly does or does not have each chemical. Within-EU variability of a chemical makes it difficult to "trust" that chemical as a basis for between-EU comparisons.

A third basic assumption requires the data gathering technology to reveal the presence of a chemical in an EU if and only if it is there. Differential presence of chemicals related to ontogenetic stage, environmental causes, deterioration or loss during preservation, time of day the collection was made and many other possible factors unrelated to the EU's capacity to make them, undermine confidence in the data for comparative purposes.

To summarize the assumptions preliminary to a compatibility analysis of occurrence patterns of secondary chemicals in plants:

(1) each EU evolved from a single ancestral EU so that the evolutionary history of the study group has a tree diagram;
(2) the presence of a chemical in an EU is typical of (virtually) all its individuals;
(3) the data gathering techniques do reveal the presence of a chemical in an EU if it is there.

In some studies, breaches of these assumptions would be so great that a compatibility analysis would be invalid. In such cases, other data analysis techniques are likely to be invalid as well.

AN IDEAL OCCURRENCE PATTERN

What should be the properties of the pattern of presence and absence of a chemical among EUs for it to be ideally related to the evolutionary history of the study group? There are many reasonable concepts in answer to this question, some more restrictive than others. The concepts of compatibility analysis state that a chemical's pattern of presence and absence during the evolution of a study group of EUs is ideally related to their evolutionary relationships if either one or the other of the following two conditions is met.

(1) The ancestor of the study group lacked the chemical and it arose during the evolution of only one EU (possibly not available for study), and all the descendants of this EU inherited the presence of this chemical. This is illustrated in Fig. 1a.
(2) The ancestor of the study group possessed the chemical and it was lost during the evolution of only one EU (possibly not available for study), and all the descendants of this EU inherited the absence of this chemical. This is illustrated in Fig. 1b.

Fig. 1c, d, e, f illustrate various violations of this ideal relationship between the presence or absence of a chemical and the evolution of the study group.

A portion of an evolutionary tree (such as the portion consisting of plants capable of producing a particular chemical) is said to be *convex* if it is possible to get from any place in it to any other place in it along line segments that all lie in that portion. Notice that when the occurrence pattern of a chemical is ideally related to the evolutionary tree, then the portion of the tree that can make the

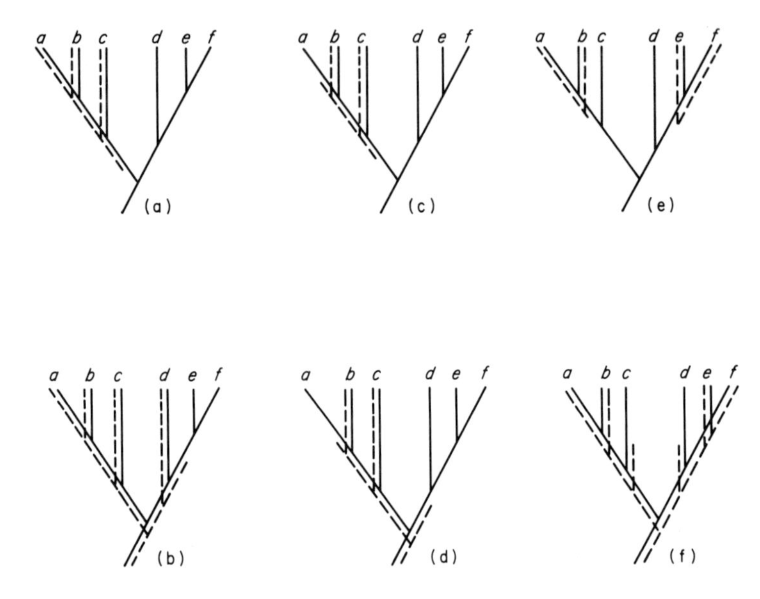

Fig. 1. Possible relationships between the presence/absence patterns of a chemical and the evolutionary history of a group of EUs. Dashed line indicates presence.

chemical and the portion of the tree that cannot make the chemical are both convex.

COMPATIBILITY OF HYPOTHESES

Each chemical that has been examined for the presence or absence in the EUs corresponds to the hypothesis that its occurrence pattern is ideally related to the evolutionary history of the study group. If such a hypothesis were true, then the chemical would be gained or lost in exactly one place in the tree diagram representing the evolution of the study group. If the tree were cut at that place, then it would fall into two convex pieces: all the EUs that possessed the chemical would be in one piece; all the EUs that lacked the chemical would be in the other piece. Such a hypothesis can be stated diagrammatically, as shown in Fig. 2 by encircling the names of EUs that possess the chemical in one circle and the names of EUs that lack the chemical in another. These two circles are then joined with a line bearing the name of the chemical. If the hypothesis is true, then this line can represent the place in the evolutionary tree diagram where the chemical arose or was lost.

If two such chemical hypotheses are considered together, there may be EUs, call them *a, b, c* and *d*, with one chemical present only in *a* and *b* and the other chemical present only in *a* and *c*. In such a case the two hypotheses cannot both be true; they logically contradict each other so that at least one must be false. The situation is illustrated in Fig. 2 where the hypothesis that the occurrence pattern of chemical 23 is ideally related to the evolution of the study group *is incompatible with* the hypothesis that the occurrence pattern of chemical 17 is ideally related to the evolution of the study group.

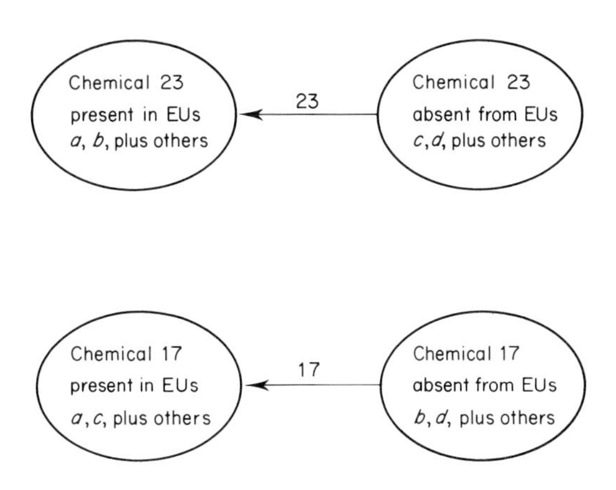

Fig. 2. The hypothesis that the occurrence pattern of chemical 23 is ideally related to the evolution of the study group *is incompatable with* the hypothesis that the occurrence pattern of chemical 17 is ideally related to the evolution of the study group.

If for two chemicals four EUs cannot be found to play the roles of *a, b, c* and *d*, then the hypotheses that their presence/absence patterns are each ideally related to the evolutionary history of the study group do not contradict each other (on the basis of the available chemical data). In this case the occurrence patterns of the two chemicals are said to be *compatible*. Notice that a chemical that occurs in just one EU, or in all but one EU, has an occurrence pattern that is compatible with that of any other chemical.

Ideas relating to the compatibility of evolutionary hypotheses have been in the literature for many years. Wilson (1965), Camin and Sokal (1965) and LeQuesne (1969) were among the first to state clearly related concepts. Estabrook (1972, 1978) discusses them in review. Mathematical properties that enable consideration of three or more evolutionary hypotheses simultaneously, and that provide the basis for algorithms for the computer analysis of data were established by Estabrook *et al.* (1975, 1976a, b), McMorris (1975, 1977), and Estabrook and McMorris (1977). Estabrook and Meacham (1979) have shown how to determine the compatibility of characters for which no *a priori* assumptions concerning the direction of evolution have been made. This method is used to produce the undirected estimates of evolutionary relationships presented below.

CLINCH

Cladistic Inference by Compatibility of Hypotheses, CLINCH, is a computer program written by Kent Fiala of the Zoology Museum of the University of Michigan, to carry out compatibility analysis. The examples to follow were constructed using this program under the MTS operating system on the six megabyte Amdahl computer at this University of Michigan in Ann Arbor, MI. I shall use the data of O.R. Gottlieb to illustrate the use of this program.

The presence or absence of each of 64 chemical skeleta in each of 15 EUs (genera in a subfamily) were determined in the laboratory of O.R. Gottlieb. These data were arranged in a matrix in which each row corresponds to an EU and each column to a chemical skeleton. Where each row and column intersect *A* indicates presence and *B* indicates absence. Of the 64 chemical skeleta 24 were present in more than one but less than 14 EUs. These 24 were analysed by CLINCH to determine the compatibility of their occurrence patterns.

Among the various information printed by CLINCH is a matrix summarizing the compatibility, 1, or incompatibility, 0, for the hypotheses associated with every pair of chemicals. Table I illustrates a portion of this compatibility matrix. For example, the hypothesis that the pattern of chemical 9 is ideally related to the evolutionary

Table I. Portion of a Compatibility Matrix

2	1							
3	1	1						
4	1	1	1					
5	1	1	1	1				
6	1	1	1	1	1			
9	1	1	0	0	0	0		
11	1	1	0	0	0	0	1	
13	1	1	0	0	0	0	1	0
	1	2	3	4	5	6	9	11

history of these genera is compatible with that for chemicals 1 and 2, but incompatible with that for chemical 3.

To understand better the patterns of agreement and disagreement among hypotheses corresponding to chemicals, the various maximal groups of mutually compatible hypotheses are discovered and revealed. (Maximal means not contained in a strictly larger group of mutually compatible hypotheses.) Groups of mutually compatible hypotheses are called *cliques* (from the analogous graph-theory term). All the presence/absence patterns of the chemicals (whose hypotheses are) in a clique are simultaneously consistent with some estimates of evolutionary relationships for the EUs in the study group. All these estimates can be summarized in a diagram. When the members of a clique are specified to CLINCH, it prints out instructions for drawing the diagram that summarizes the estimates of relationship with which all presence/absence patterns of chemicals in the clique are simultaneously consistent. The largest clique for Gottlieb's data has in it 14 of the 24 chemicals. (This is an unusually large fraction of the total number of different chemicals to be in a single clique.) EUs equivalent for these 14 chemicals are placed in the same circle, and ancestor–descendant circles are jointed with a line bearing the names of the chemicals that are gained or lost there. Figure 3 shows the diagram that summarizes the estimates of relationship for these 14 chemicals. Such diagrams are called *partial estimates* because they leave unresolved many details of relationship among the EUs.

CHOICE AND RESOLUTION OF ESTIMATES

The previous example is unusual in that an exceptionally large number of chemicals showed mutually compatible presence/absence

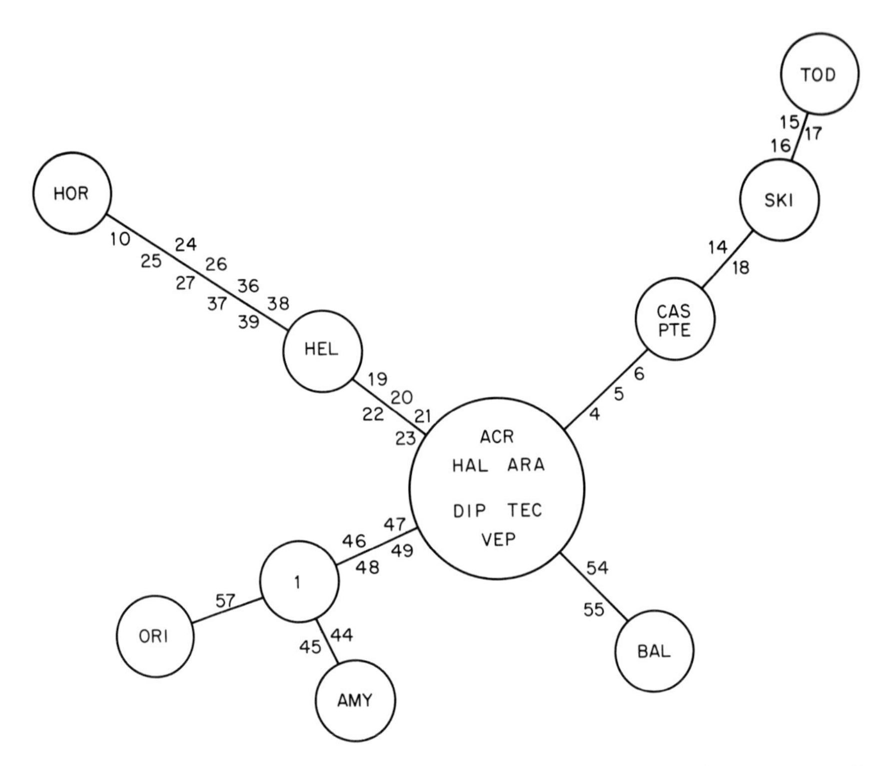

Fig. 3. First example data from O.R. Gottlieb showing undirected partial estimate of evolutionary history, based on largest collection of compatible chemical characters.

patterns. Often many fewer than half the chemicals participate in the largest cliques, and there may be several large cliques differing in size by only one or two chemicals. In cases such as this, it may be appropriate only to discuss the patterns of consistency such as they are, or to compare them with other data. If it is desirable to try to choose a single (or only a few) best partial estimate of evolutionary relationships, one approach is to choose as a basis for a partial estimate those chemicals that are in *all* the largest cliques; this is the approach of Estabrook *et al.* (1978). Another is to choose a single chemical whose presence/absence pattern distinguishes two large groups of EUs and is compatible with that of many other chemicals; this is illustrated by Estabrook and Anderson (1978).

A third approach is made possible by the work of C.A. Meacham (unpublished) who has determined a means of calculating a probability that a group of chemicals has compatible presence/absence patterns at random. Meacham's model of randomness is that proposed by Sneath *et al.* (1975) in which the frequencies of EUs that possess a chemical remain the same.

Data by O.R. Gottlieb for another study of 16 genera analysed for the presence or absence of 64 chemicals provides a good example of this technique. Figure 4 shows the partial estimate based on the

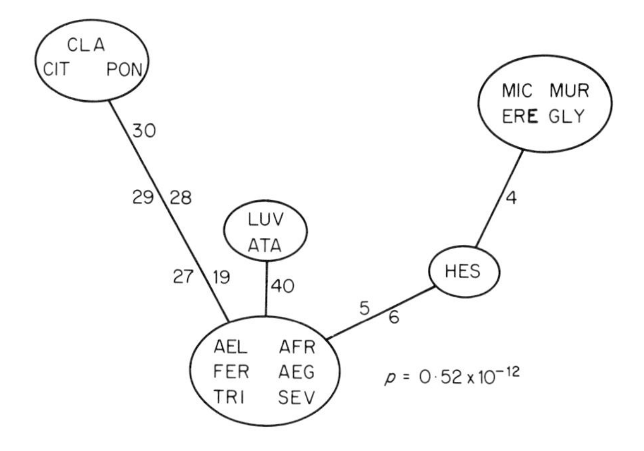

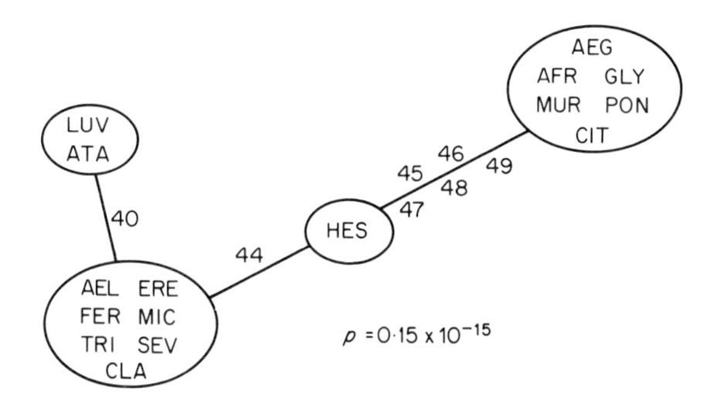

Fig. 4. Second example data from O.R. Gottlieb showing undirected partial estimates of evolutionary history based on largest (above) and least probable (below) collections of compatible chemical characters.

largest clique (nine chemicals) and that based on the clique least likely to have occurred at random (seven chemicals). In this example, I have chosen the least likely clique as a basis for the primary partial estimate of evolutionary relationships. Based on this partial estimate, the genus can be broken into those EUs that lack chemicals 45–49, and those EUs that have chemicals 44–49. EU HES is common to both pieces and provides a basis for linking them back together. Figure 5 shows (lighter lines) the results of compatibility analysis within each piece where estimates were based on the largest, which were also the least likely, cliques. These secondary estimates have

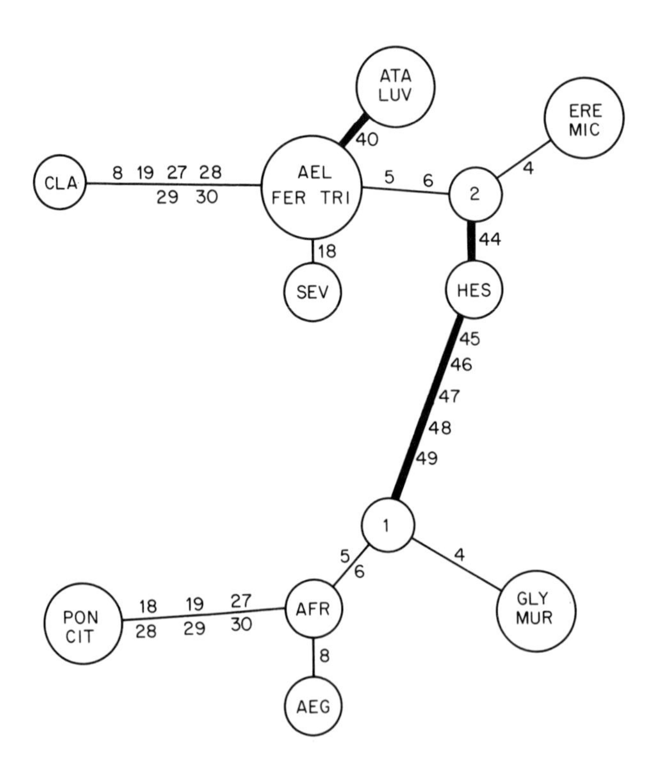

Fig. 5. Second example data from O.R. Gottlieb showing a more refined undirected partial estimate of evolutionary history based on combined secondary analysis of the two major phyletic lines revealed in the primary estimate based on the least probable collection of compatible chemical characters, shown here with broad lines.

been reunited to provide a more resolved partial estimate of evolutionary relationships. As might be expected in a case with two large and very different cliques, a great deal of parallelism is required to explain the data.

WHEN TO ABANDON THE ANALYSIS

Firm guidelines to decide when a clique constitutes a credible basis for making a partial estimate of evolutionary relationships do not exist at present. Those with statistical experience realize that the probability that a group of chemicals form a clique at random is not the most desirable number on which to base a decision. Better would be the probability that a clique less likely than the least likely actually observed, would occur at random among all the chemicals assayed for the study collection. Unfortunately, I do not yet know how to calculate this number. But less rigorous considerations can be made.

T. O. Duncan has kindly provided me with data on the presence and absence of 14 two-way paper chromatography spots in 87 populations representing 20 taxa in the *Ranunculus hispidus* species complex. These data were analysed for compatibility and all cliques were printed by CLINCH. Table II gives some results of Meacham's probability program. Spot 5 has somewhat more and spot 11 has considerably fewer compatibilities than expected at random, but otherwise the numbers expected are close to those observed. The least likely clique (5, 8, 10) $p = 0.85 \times 10^{-1}$ would occur at random about once in 12 times. With 14 spots there are 364 distinct triples that could have been cliques. It seems relatively clear that the presence/absence of these spots is unlikely to give an improved picture of the evolutionary relationships among these populations of the *Ranunculus hispidus* species complex. When these chemicals are identified, they could fall into classes whose occurrence patterns were quite compatible and consistently supported an estimate of evolutionary relationships, in spite of the negative result with preliminary spot data. This further supports the importance of making accurate chemical determinations.

EVALUATING A CLASSIFICATION

Frequently, surveys for secondary plant compounds are made

Table II. Probability Analysis of 14 TBA × HOAC Spots found in 87 Populations representing 20 Taxa in the *Ranuculus hispidus* Species Complex

Spot	Number of Compatibilities	
	Expected	Observed
1	1.59	1
2	0.85	1
3	0.67	1
4	1.23	2
5	6.35	9
6	0.45	1
7	1.02	2
8	1.16	3
9	5.04	4
10	4.17	4
11	3.13	0
12	1.02	0
13	1.09	1
14	0.45	1

for populations representing the taxa in a group of plants that has recently been revised taxonomically. In such instances it is reasonable to ask how, and to what extent, the occurrence patterns of the chemicals surveyed are consistent with the classification. To determine this we need to make precise the concept "the occurrence pattern of a chemical is consistent with a classification". Of the many possible concepts that could be defined, I shall present one related to compatibility. For this purpose the flavonoid data for the genus *Lasthenia* (Compositae) of Ornduff *et al.* (1974) make good examples for several reasons. Their work is well known by plant chemosystematists; the chemistry has been well studied by Bohm *et al.* (1974), and in some cases there is some idea of the biosynthetic pathways involved; a biosystematic study revealing chromosome numbers and intertaxa fertility has been done by Ornduff (1966).

A compatibility analysis of the data given by Ornduff *et al.* (1974) was made in which I included, together with the occurrence patterns of the flavanoids, the section memberships of the EUs. In this study there were some problems with the basic assumptions. Ornduff (1966) presents good evidence to suggest that EUs *k, o*

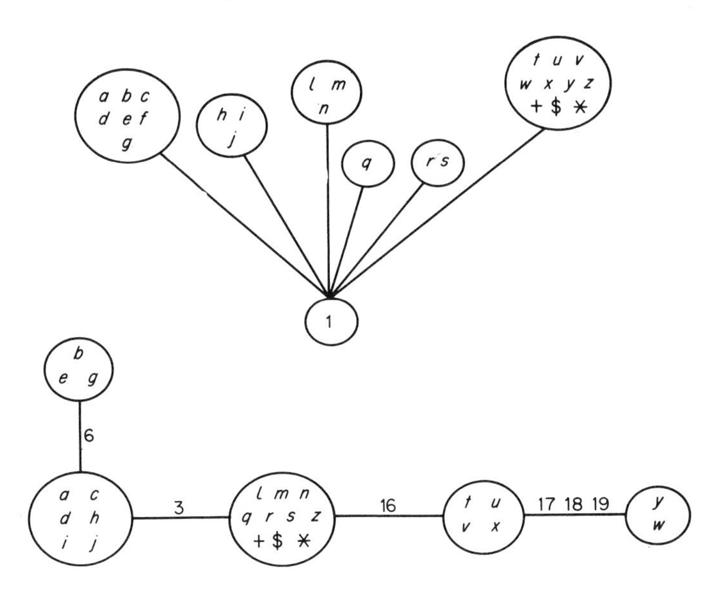

Fig. 6. Flavonoid presence/absence data for populations representing the taxa in *Lasthenia* (Asteraceae) from Ornduff *et al.*, 1974. Cladistic character above represents the restrictive hypothesis that every section is monophyletic. Below is the partial estimate of evolutionary history based on flavonoids whose presence/absence pattern is consistent with this hypothesis.

and *p*, representing *L. microglossa* and *L. ferrisiae* are of hybrid origin, so these were omitted from the compatibility analysis. Further, EU *f*, representing *L. macrantha* ssp. *bakeri* showed within EU variability for several flavanoids. It might have been better to omit *f* from the study group as well, but I kept it and scored it present for the only partly present flavonoids. These problems should be borne in mind when evaluating the conclusions suggested by the examples to follow.

The data, coded as above, were subjected to compatibility analysis. Character 24 represents the hypothesis that each section of the genus is monophyletic, i.e. contains all the descendants of its most recent common ancestor (Fig. 6, top). Compatibility analysis reveals that flavonoids 3, 6, 16, 17, 18 and 19 are consistent with this hypothesis and mutually compatible as well. Fig. 6 (bottom) shows the partial estimate based on the occurrence pattern of these flavonoids.

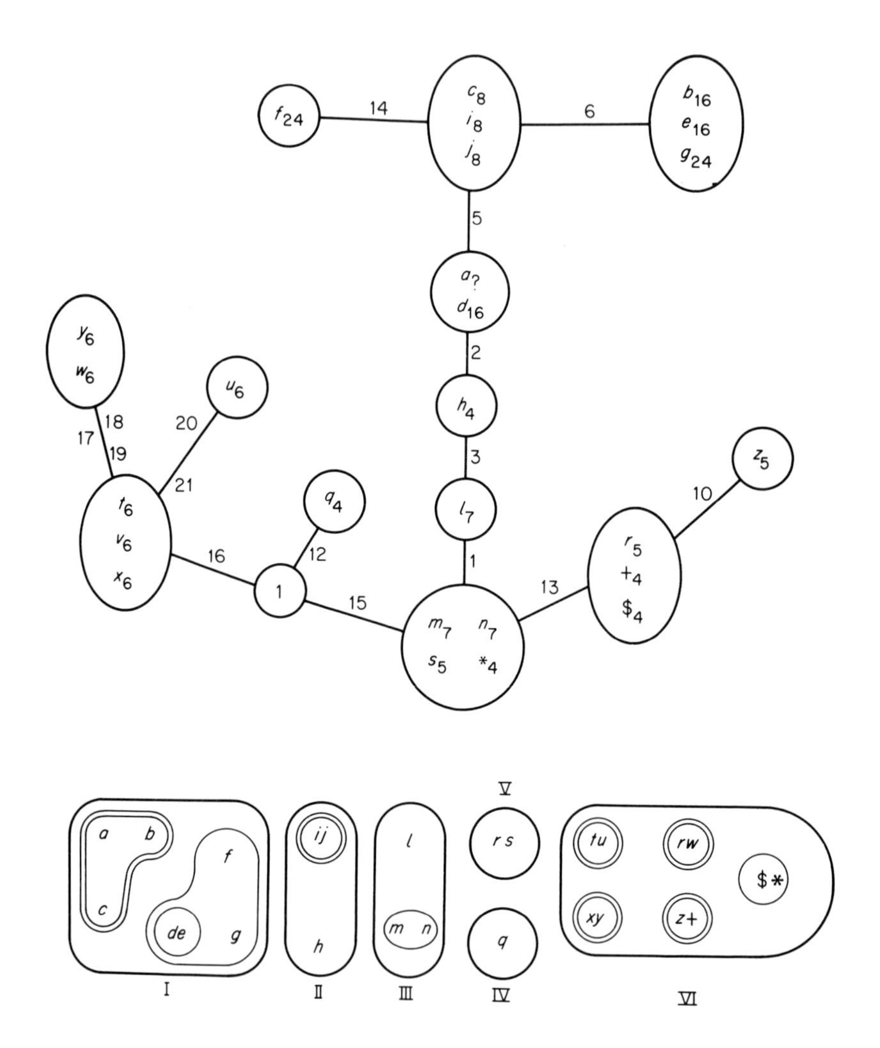

Fig. 7. Flavonoid presence/absence data for *Lasthenia* (Asteraceae) from Ornduff *et al.*, 1974. Twenty-one flavonoids in 26 populations represents 6 sections and 14 species. Undirected estimate of evolutionary history based on the largest collection of compatible chemical characters (above); the taxonomy of Ornduff (1966) (below).

There was a single largest clique with 11 out of the 16 flavonoids whose occurrence patterns corresponded to contradictable hypotheses. The estimate of evolutionary relationships based on these 11 flavonoids is shown in Fig. 7, where for interest's sake, chromosome numbers are included. There is clearly a high level of compatibility among the occurrence patterns of these flavonoids. The taxonomy of Ornduff (1966) is shown diagrammatically in the lower part of Fig. 7. All the patuletin containing EUs are to the left of flavonoid 15; all the patuletins (15–21) are compatible with each other, and they are compatible with most of the other flavonoids as well. Thus the partial estimate of relationships among EUs to the left of 15 (Fig. 7) is the same for all large cliques. EUs corresponding to sections I and II are those that lie above flavonoid 3. To attempt a resolution of relationships in sections III–VI, EUs to the right of 16 and below 3 were subject to secondary analysis, as shown in Fig. 8 (top). In the context of this smaller group of EUs, flavonoid 13 can be consistently replaced by 7, 9 and 11.

Requiring taxa to be monophyletic is undesirably restrictive and can result in an awkward and impractical taxonomy. A good taxomony should reflect in some way the best available estimate of evolutionary relationships, but it should also be free to reflect estimates of amount of evolutionary difference among EUs as well. Thus the relation between the evolutionary tree and the taxonomy for a study group should be, simply, that all taxa are convex on the tree. Recall that a taxon is convex on a tree means that it is always possible to pass along the unique path of phyletic line segments from one member to another without ever leaving the taxon.

Possible more complete resolutions of the primary estimate (left), and of the secondary estimate (right), are shown in Fig. 8 below each respectively. These more completely resolved estimates were structured to preserve as much as possible of the convexity of the taxonomy, yet still to be consistent with the partial estimates from which they were derived. In neither resolution can both section V and section VI be simultaneously convex. In the resolution of the secondary estimate, the "slight" modification of sliding *r* and *s* together to be a monophyletic group would permit both sections V and VI to be convex. Such "sliding" would be inconsistent with the occurrence pattern of flavanoid 11. However, the closeness of *r* and *s* in this resolution suggests that sections V and VI are "almost" convex on it.

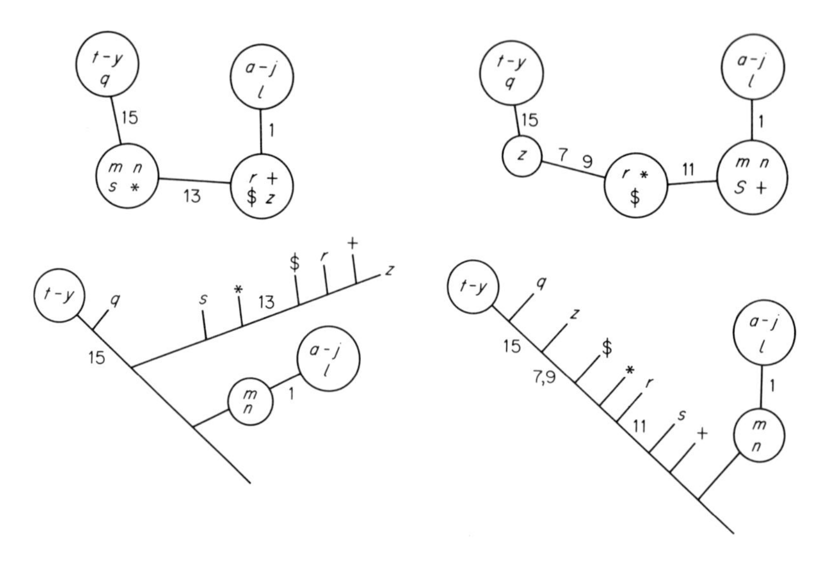

Fig. 8. Flavonoid presence/absence data for Lasthenia (Asteraceae) Ornduff *et al.,* 1974. Flavonoid 13 is replaced by Flavonoids 7, 9 and 11 in the more refined estimate to the right. Possible further refined estimates are respectively shown below each.

BIOSYNTHETIC PATHWAYS

More closely related to the genetics of the organisms whose evolutionary relationships we would estimate are the enzymes that make possible the chemical reactions that produce pools of chemical compounds that can be detected in plants. Occurrence patterns of chemicals in plants could be explained in a variety of ways when considerations of biosynthetic pathways and the kinetics of the enzymes on them are made. Bohm *et al.* (1974) present an estimate of the biosynthetic pathways for the major classes of the flavonoids that occur in *Lasthenia.* The presence and absence patterns of the four anthochlors found in *Lasthenia* are given in Table III. The occurrence patterns of one of the four anthochlors is incompatible with that of the other three. Here I would like to share a fantasy to illustrate how (1) the occurrence patterns of *enzymes* on the biosynthetic pathways that produce these anthochlors could all be compatible; (2) these compatible occurrence

Table III. Presence and Absence of Anthochlors in *Lasthenia*

	EU	Name	B	O	M	S
	abc	*L. chrysostoma*	X	X	X	
	d	*L. macrantha* (typical) A	X	X	X	
I	e	*L. macrantha* (typical) B	X	X	X	X
	f	*L. macrantha ssp. bakeri*	X	X	X	X
	g	*L. macrantha ssp. prisca*	X	X	X	
	h	*L. debils*	X		X	X
II	i	*L. leptalia* A	X	X	X	
	j	*L. leptalia* B	X	X	X	X
III–VI	l	*L. chrysantha*	X			
	mn+	Rest of genus (15 EUs)				

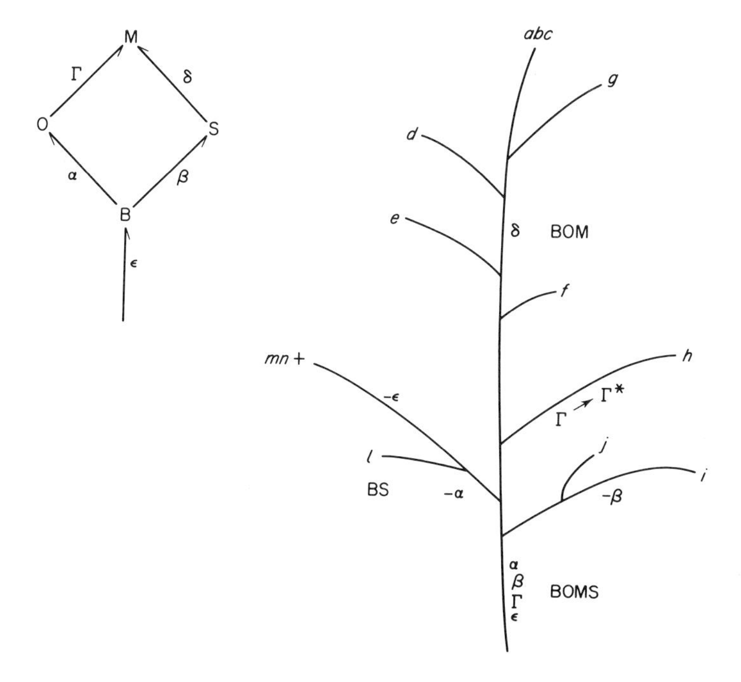

Fig. 9. (Left) Biosynthetic pathways for the anthochlors in *Lasthenia* (Asteraceae) from Bohm *et al.* 1974, with entirely fanciful kinetics (α, fast; Γ, slow; Γ*, fast; β, slow; δ, fast). (Right) Entirely fanciful estimate of evolutionary history for the taxa of Table III.

patterns could give a partial estimate of evolutionary relationships; (3) this partial estimate could have a consistent refinement on which all of Ornduff's taxa (subspecies–section) would be convex. This fantasy is illustrated in Fig. 9 with a hypothetical evolutionary tree on the right and Bohm's pathways on the left. Here Greek letters refer to enzymes. When epsilon is present, there is a pool of anthochlor *B* from which the rest of the anthochlors are made. The kinetics of alpha are rapid so that *O* is made faster than gamma can convert it to *M*. Beta is slow so that when delta is absent *S* can be detected, but when delta is present *S* is consumed and becomes too scarce to be detected. Gamma star is a fast version of gamma with the same effect on *O*. The presence of alpha, beta, gamma and epsilon is considered primitive in the evolutionary tree shown to the right of Fig. 9. We need to account for the loss of *S* by two different mechanisms in two different places. Beta is lost in the evolution of *i* and delta is gained in the evolution of the ancestor of *abc, g* and *d*. Likewise *O* must be lost twice, once when gamma star replaces gamma in the evolution of *h*, and again when alpha is lost during the evolution of the ancestor of *mn*+ and *l*.

Although this fantasy is purely hypothetical and almost certain to be incorrect, I present it to illustrate the kinds of explanations that we might be able to look forward to from chemosystematics as measuring technology improves, if we continue our debate with well defined, clearly exemplified concepts and methods.

REFERENCES

Bohm, B. A., Saleh, N. A. M. and Ornduff, R. (1974). The flavonoids of *Lasthenia* (Compositae). *Am. J. Bot.* **61**, 551–561.

Camin, J. H. and Sokal, R. R. (1965). A method for deducing branching sequences in phylogeny. *Evolution* **19**, 311–326.

Estabrook, G. F. (1972). Cladistic methodology: a discussion of the theoretical basis for the induction of evolutionary history. *Ann. Rev. Ecol. Syst.* **3**, 427–456.

Estabrook, G. F. (1978). Some concepts for the estimation of evolutionary relationships in systematic botany. *Syst. Bot.* **3**, 146–158.

Estabrook, G. F. and Anderson, W. R. (1978). An estimate of phylogenetic relationships within the genus, *Crusea* (Rubiaceae) using character compatibility analysis. *Syst. Bot.* **3**, 179–196.

Estabrook, G. F. and Meacham, C. A. (1979). How to determine the compatibility of undirected character state trees. *Math. Biosci.* **46**, 251–256.

Estabrook, G. F. and McMorris, F. R. (1977). When are two qualitative taxonomic characters compatible? *J. Math. Biol.* **4**, 195–200.
Estabrook, G. F., Johnson, C. S., Jr and McMorris, F. R. (1975). The idealized concept of the true cladistic character. *Math. Biosci.* **23**, 263–272.
Estabrook, G. F., Johnson, C. S., Jr and McMorris, F. R. (1976a). An algebraic analysis of cladistic characters. *Discrete Math.* **16**, 141–147.
Estabrook, G. F., Johnson, C. S., Jr and McMorris, F. R. (1976b). A mathematical foundation for the analysis of cladistic character compatibility. *Math. Biosci.* **29**, 181–187.
Estabrook, G. F., Strauch, J. B., Jr and Fiala, K. L. (1977). An application of compatibility analysis to the Blackith's data on orthopteroid insects. *Syst. Zool.* **26**, 269–276.
LeQuesne, W. J. (1969). A method of selection of characters in numerical taxonomy. *Syst. Zool.* **18**, 201–205.
McMorris, F. R. (1977). On the compatibility of binary qualitative taxonomic characters. *Bull. Math. Biol.* **39**, 133–137.
McMorris, F. R. (1975). Algebraic analysis of uniquely derived characters. *In* "Proceedings of the Eighth International Conference on Numerical Taxonomy" (G. F., Estabrook, ed), pp. 399–415. Freeman, San Francisco.
Ornduff, R. (1966). A biosystematic survey of the goldfield genus *Lasthenia* (Compositae: Helenieae). *Univ. Calif. Publ. Bot.* **40**, 1–91.
Ornduff, R., Bohm, B. A. and Saleh, A. M. (1974). Flavonoid races in *Lasthenia* (Compositae). *Brittonia* **26**, 211–220.
Sneath, P. H. A., Sackin, M. J. and Ambler, R. P. (1975). Detecting evolutionary incompatibilities from protein sequences. *Syst. Zool.* **24**, 311–332.
Wilson, E. O. (1965). A consistency test for phylogenies based on contemporaneous species. *Syst. Zool.* **14**, 214–220.

19 | Problems in Interpreting Asymmetrical Chemical Relationships

J. C. GOWER

Rothamsted Experimental Station, Harpenden, Herts AL5 2JQ, England

Abstract: Systematists are familiar with methods for analysing systematic relationships expressed as similarities or taxonomic distances but may be less familiar with methods for analysing asymmetry. Two examples are discussed of the statistical analysis of asymmetric chemical relationships:

(1) First passage times that are a measure of mutation rates between pairs of amino acids.

(2) An analysis of the ability of yeasts to utilize 23 compounds (mainly sugars) with the objective of trying to determine metabolic pathways.

Pictorial (graphical) representations of the data are obtained that fit well but considerable chemical knowledge is needed for proper interpretation.

INTRODUCTION

Taxonomists are familiar with a variety of methods for analysing square tables, or matrices, whose values are symmetrically disposed ($d_{ij} = d_{ji}$). Similarity, dissimilarity and taxonomic distance are everyday examples. Yet not all comparisons are of this kind; for example the number of characters possessed by one species that are not possessed by another is an asymmetric relation and immunological examples have been mentioned in this book. Sometimes a specific parametric model that incorporates asymmetry can be

Systematics Association Special Volume No. 16, "Chemosystematics: Principles and Practice", edited by F. A. Bisby, J. G. Vaughan and C. A. Wright, 1980, pp. 399–409, Academic Press, London and New York.

specified and its terms estimated by standard statistical methodology such as maximum likelihood, but here I shall be concerned only with a method of a more general nature that makes no specific model assumptions, but allows the data to be exhibited in an informative way. Such methods have been found useful in many fields of application and the examples I give here suggest that the method will be of interest and use to systematists using chemical information. The theory may be found in Gower (1977), see also Constantine and Gower (1978), and I shall not report it in detail here. But some account of the method must be given before the examples become intelligible.

Suppose D is a square matrix whose values compare, in some way, the row and column items and whose rows and columns are classified by the same things. The relationship is not necessarily symmetric ($d_{ij} \neq d_{ji}$). Any square matrix may be expressed, uniquely and trivially, as the sum of a symmetric matrix M and a skew-symmetric matrix N, thus

$$D = M + N \tag{1}$$

where $m_{ij} = \frac{1}{2}(d_{ij} + d_{ji})$ and $n_{ij} = \frac{1}{2}(d_{ij} - d_{ji})$.
The symmetric matrix M may be analysed by some appropriate standard method but if the analysis is taken no further, as is often the case, there is an implicit assumption that the departures from symmetry are uninteresting, representing experimental error or noise. Yet the skew-symmetric part may have structure of real interest and some method is needed to examine this possibility.

Gower (1977) proposed an analysis of the skew-symmetric part N, that is somewhat related to a principal components analysis, but differs from it in many respects. As in components analysis better fits are obtained by using increasing numbers of dimensions but, unlike components analysis, all solutions must be in an even number of dimensions. Thus the smallest number of dimensions that can be used is two, the next smallest is four etc. We shall concentrate here on two-dimensional fits with the form

$$n_{ij} = \lambda(u_i v_j - v_i u_j). \tag{2}$$

The vectors $(u_1, u_2, \ldots, u_n)$ and $(v_1, v_2, \ldots, v_n)$ may be plotted, as in components analysis, to give n points P_i with coordinates (u_i, v_i) $i = 1, 2, \ldots, n$. Such diagrams have to be interpreted

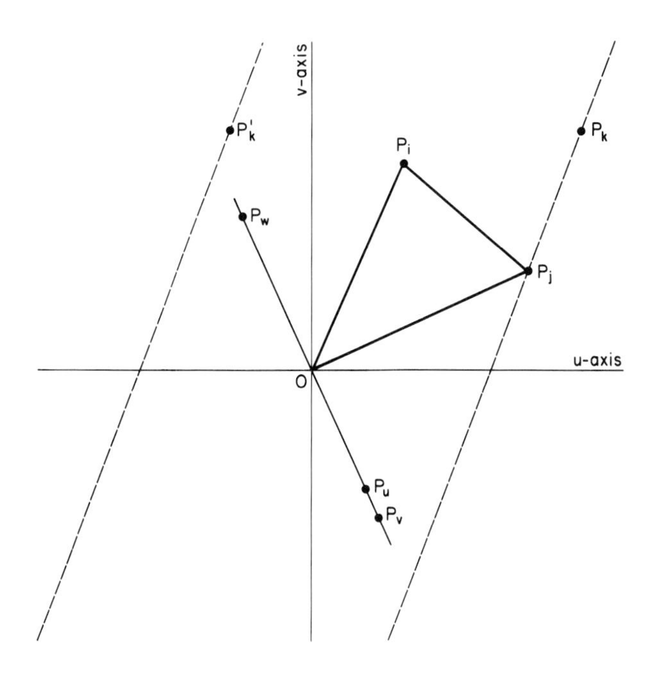

Fig. 1. Geometry of asymmetry space. The area of triangle OP_iP_j approximates n_{ij} and the area OP_jP_i approximates $-n_{ij}$. The dotted line containing P_k is the locus of all points such that $n_{ik} = n_{ij}$ and the dotted line containing P'_k is the locus of all points such that $n_{ik} = -n_{ij}$. All pairs of points on any line through the origin subtend zero area triangles at the origin; thus the estimated values of n_{uv}, n_{uw} and n_{vw} are all zero.

carefully for (2) does not represent a distance, but is proportional to the area of the triangle with vertices P_i, P_j and the origin O. Areas, not distances, are the main interpretive tool and skew-symmetry is exhibited in this representation because the signs associated with areas of triangles depend on whether their vertices are traversed in a clockwise or an anti-clockwise sense. Points which are close together will generate small areas so should approximate small values in the skew matrix, but so will pairs of points that are distant but colinear with the origin. From Fig. 1 we see that all points P_j that have the same skew-symmetry with P_i lie on a straight line parallel to OP_i. The locus of points with skew-symmetry of opposite sign is another straight line equi-distant from OP_i and parallel to the first line but on the opposite side to OP_i. Thus in this space colinearities correspond to the property that all points equi-distant from P_i in a distance-space lie on a circle. The area of any triangle $P_iP_jP_k$ approximates $n_{ij} + n_{jk} + n_{ki}$. If this is small it could be because all these values are small or because large positive and negative values cancel out and this depends on whether the triangle is located near the origin or not.

There is one case of special interest, that occurs in both examples below. When v_j $(j = 1, 2, \ldots, n)$ is constant h (say), (2) becomes

$$n_{ij} = \lambda^*(u_i - u_j) \tag{3}$$

where the constant h has been absorbed with λ to give λ^*. This is still a two-dimensional form, but because the coordinates in one dimension are constant, it is essentially a one-dimensional solution and will be referred to as such. When this occurs there is a great simplification, as all the points $P_i (i = 1, 2, \ldots, n)$ are colinear on the line $v_i = h$. All triangles have this same constant altitude and their areas are therefore proportional to their bases of lengths $(u_i - u_j)$ or $(x_i - x_j)$, as is shown in the notation of Fig. 2.

An ideal analysis should combine the separate analyses of the skew-symmetric and symmetric parts of D and Gower (1977) examines a few specific models to show how this can be done.

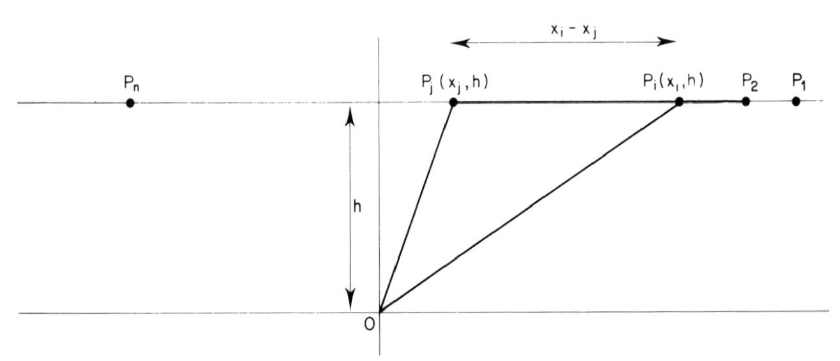

Fig. 2. One-dimensional skew-symmetry. The points $P_i (i = 1, 2, \ldots, n)$ with coordinates (x_i, h) are colinear on a line distance h from the origin. The area of triangle OP_iP_j is $\frac{1}{2}h(x_i - x_j)$ so that n_{ij} is proportional to $x_i - x_j$.

With a one-dimensional skew matrix this combination becomes especially simple as only the quantities u_i ($i = 1, 2, \ldots, n$) have to be incorporated and the examples we are now ready to examine show how this can be done.

EXAMPLES

1. An Example of a Model for the Evolution of the Genetic Code

The first example does not concern observational data, but only quantities calculated from simple assumptions intended to model the possible evolution of the genetic code. Jorrê and Curnow (1975a, b) give the details so I shall provide only a brief sketch here. The model was of Markoff chain form with the 64 codons taken as the basic states of the chain. One and only one codon-base might mutate into another base and all such mutations are equally likely and hence are equal to 1/9 (three possible mutations at each of the three codon bases). Mutations to a stop code are regarded as lethal. From such a matrix it is easy to compute the probability p_{AB} of a chain of mutations causing the replacement of amino acid B by amino acid A before a stop code is met. Also the average number of steps m_{AB} taken to pass from A to B can be calculated (technically, the mean first passage time). This is a measure of the ease of mutation from amino acid A to B. Clearly neither p_{AB} nor m_{AB} are likely to be symmetric relationships. Jorré and Curnow gave a table of the values $p_{AB} \times m_{AB}$ and Curnow and Jorré (1978) of p_{AB} and m_{AB} separately.

The skew-symmetric part of the table with values $p_{AB} \times m_{AB}$ turns out to have a strong linear form (Fig. 3) which can be combined with an ordination of the symmetric part (Fig. 4) to produce well defined contours representing the superimposed asymmetry.

The first thing to say about Fig. 4 is that it gives a very good approximation to the 400 numerical values of Jorré and Curnow's table. Thus it describes the data well and also in an easily assimilable form. MET is at the bottom of a trough with the other amino acids arranged around its slopes at varying levels. Mutations uphill take more steps than those downhill. This characterization shows that the diagram is a good representation of the data, but can it help with interpretation?

The model is concerned with codon mutations so it seems sensible

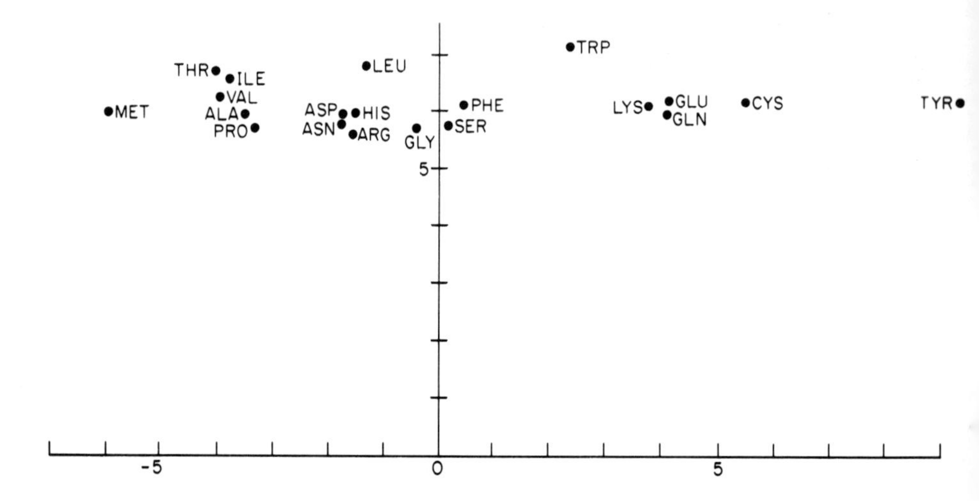

Fig. 3. One-dimensional skew-symmetry exhibited in the analysis of the quantities $p_{ij} \times m_{ij}$ of Curnow and Jorré (1975a).

to examine the diagram in conjunction with the codons coding for the amino acids (see Fig. 4). It is now clear that those amino acids that can replace each other with a single mutation lie close together, as they should, but it is not clear why some simple mutations occur "on the level" and others "on the slope". The probabilities p_{AB} are likely to be confusing things so I have analysed the skew part of the actual first passage times given by Curnow and Jorré (1978). Again an excellent one-dimensional fit was obtained which is shown in Fig. 5 with the codons. There is now a clear trend from the amino acids with six codons down to those with a single codon. Further those amino acids coded only be codons that can easily mutate to "stop" occur late in the list. Mutations of an amino acid with many codons to one with few can happen in many more ways than the reverse. The "stop" code cuts off certain routes and puts those amino acids at the bottom of the trough at a disadvantage.

This example is based on artificially generated data and merely illustrates properties known to be built into the model. Nevertheless it has been shown that a complicated process can generate data from which a simple interpretable structure can be found. It would be

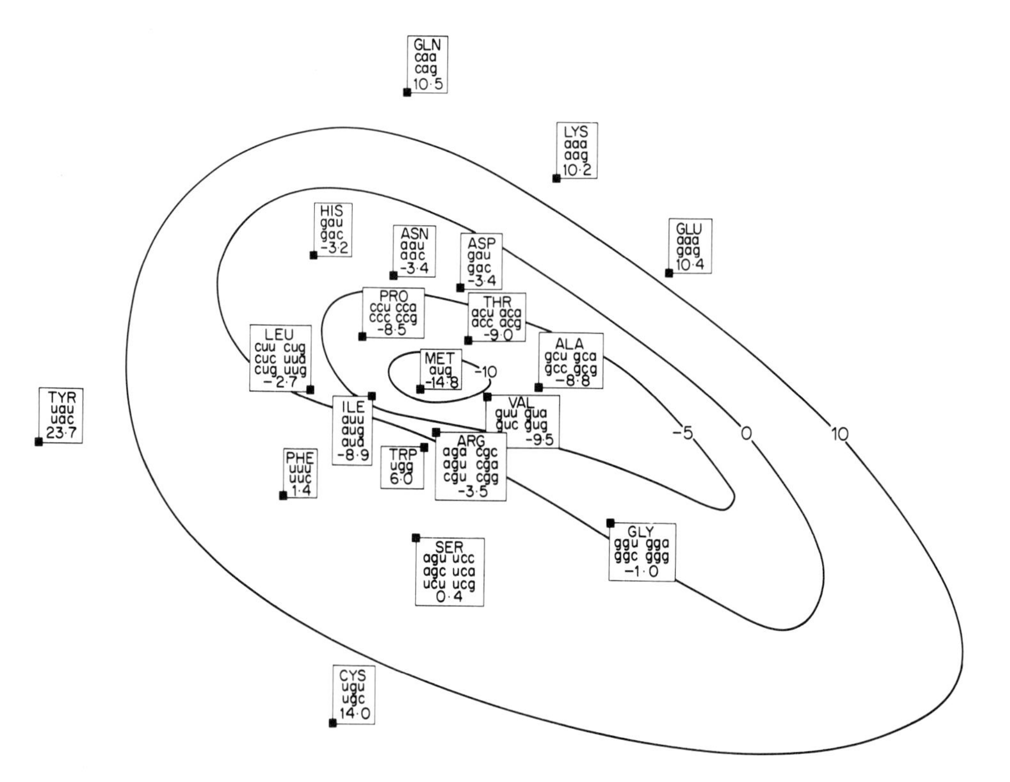

Fig. 4. Combination of skew-symmetry and symmetry in the analysis of the quantities $p_{ij} \times m_{ij}$ of Curnow and Jorré (1975a). The distances between pairs of amino acids approximates the values $p_{ij} \times m_{ij}$, using a non-metric multidimensional scaling analysis. Each box contains the abbreviated amino acid name, followed by the codons that code for that amino acid and finally the value of the one-dimensional skew-symmetry given by Fig. 3. Contours of the skew-symmetry values give the bowl-shaped trough that is shown.

worth trying the method on experimental data of a similar kind to see if any patterns occur, and if they do whether they differ from those given by the model.

2. An Example of the Utilization of Sugars and other Substances by Yeasts

In this example 29 compounds, mainly sugars, are contrasted in

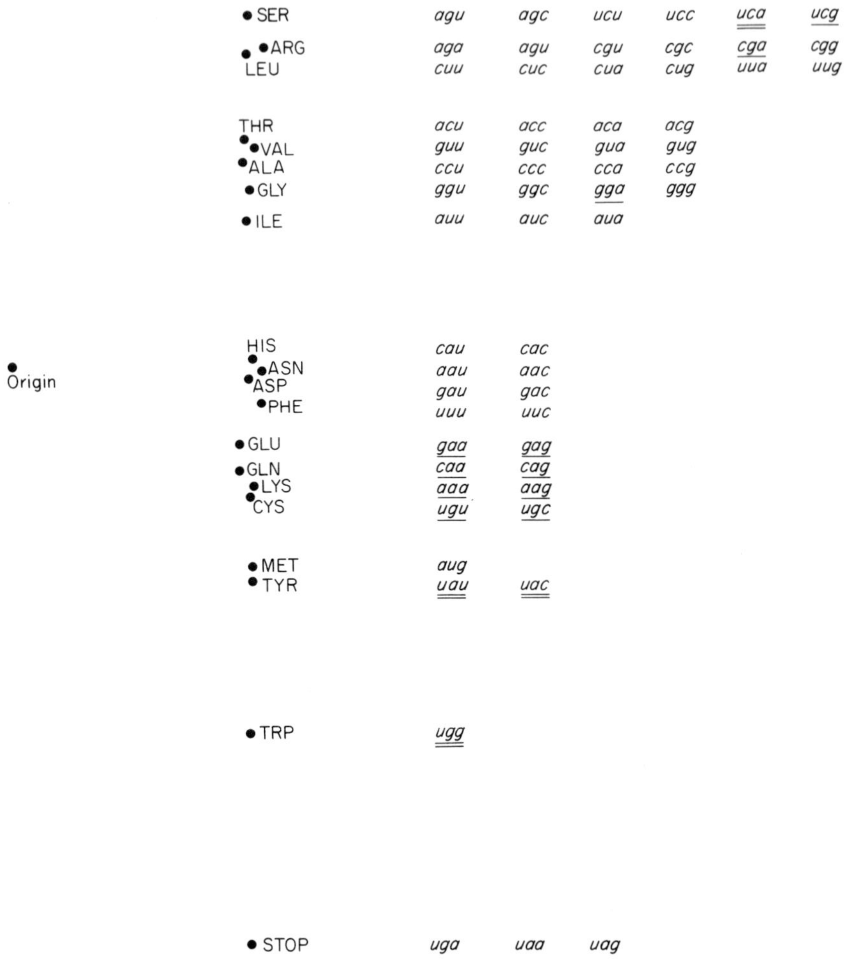

SER agu agc ucu ucc uca ucg
ARG aga agu cgu cgc cga cgg
LEU cuu cuc cua cug uua uug

THR acu acc aca acg
VAL guu guc gua gug
ALA ccu ccc cca ccg
GLY ggu ggc gga ggg
ILE auu auc aua

HIS cau cac
Origin
ASN aau aac
ASP gau gac
PHE uuu uuc

GLU gaa gag
GLN caa cag
LYS aaa aag
CYS ugu ugc

MET aug
TYR uau uac

TRP ugg

STOP uga uaa uag

Fig. 5. One-dimensional skew-symmetry exhibited in the analysis of the quantities m_{ij} of Curnow and Jorré (1978). A single underline means that there is one way in which the set of codons can mutate to STOP with one mutation, a double underline indicates mutation to STOP with one mutation in two different ways.

terms of the ability of 497 species of yeasts to use them for growth. Barnett (1976) gives 2×2 tables for many of the different pairs of compounds showing the number of yeasts that use both (a_{ij}) one (b_{ij} or c_{ij}) or neither (d_{ij}) compounds and discusses how tables with low values of either or both of b_{ij} and c_{ij} suggest possible metabolic pathways. His arguments are reminiscent of those used by LeQuesne (1969) to order characters according to their evolutionary development and hence construct cladistic trees. Here we are mainly concerned with representing the *a, b, c, d* values diagramatically in such a way that interesting 2×2 tables can be detected by eye. To do this a simple result given by Banfield and Gower (1980) is used. They point out that the quantity $x_i = a_{ij} + c_{ij}$ is invariant to the choice of j and $x_j = a_{ij} + b_{ij}$ is invariant to the choice of i. In fact x_i is merely the number of positive characters possessed by the ith unit. It follows that the quantities $b_{ij} - c_{ij}$ may be calculated from $x_j - x_i$ and so necessarily have the simple one-dimensional skew symmetric form referred to above in equation (3). This is a mathematical consequence of the definition of the terms of a 2×2 table and is not a property of the data. It is interesting that while the simple matching dissimilarity coefficient $b_{ij} + c_{ij}$ requires $n - 1$ dimensions for exact representation $b_{ij} - c_{ij}$ never requires more than 1 dimension. This one dimension can usually be superimposed on an ordination of some symmetric similarity coefficient of the (a, b, c, d)-values.

Figure 6 shows an ordination using a distance $(b_{ij} + c_{ij})^{1/2}$ which is equivalent to a principle coordinates analysis based on a simple matching coefficient, or equivalently a principal components analysis of the original binary data. The skew symmetric x_i-values have been superimposed and contoured by the parallel lines shown. The salient features of Fig. 6 is how the number of yeasts using the different substances increases from right to left. A subsidiary feature is the separation by the wavy line, between the upper and lower parts of the diagram. This separation is in terms of Table XVI (glucosides, D-galactose and L-rhamnose) and Table XVII (pentoses, alditols, D-galactose and L-*xylo*-hexulose) given by Barnett (1976) and so presumably has some biochemical significance. D-galactose is a substance common to both tables and appears satisfactorily close to the boundary between the two regions in Fig. 6.

Because *a, b, c, d* cannot be negative this kind of diagram has some interesting properties. First, close pairs of points must have b and c both near zero and this property is true of almost any sensible choice of taxonomic distance δ_{ij}, e.g. $\delta_{ij} = (b_{ij} + c_{ij})$, $\delta_{ij} = (b_{ij} + c_{ij})^{1/2}$, $\delta_{ij} = b_{ij} + c_{ij} + d_{ij}$ etc. Because b and c are near zero, the associated x-values, $x_i = a_{ij} + c_{ij}$ and $x_j = a_{ij} + b_{ij}$

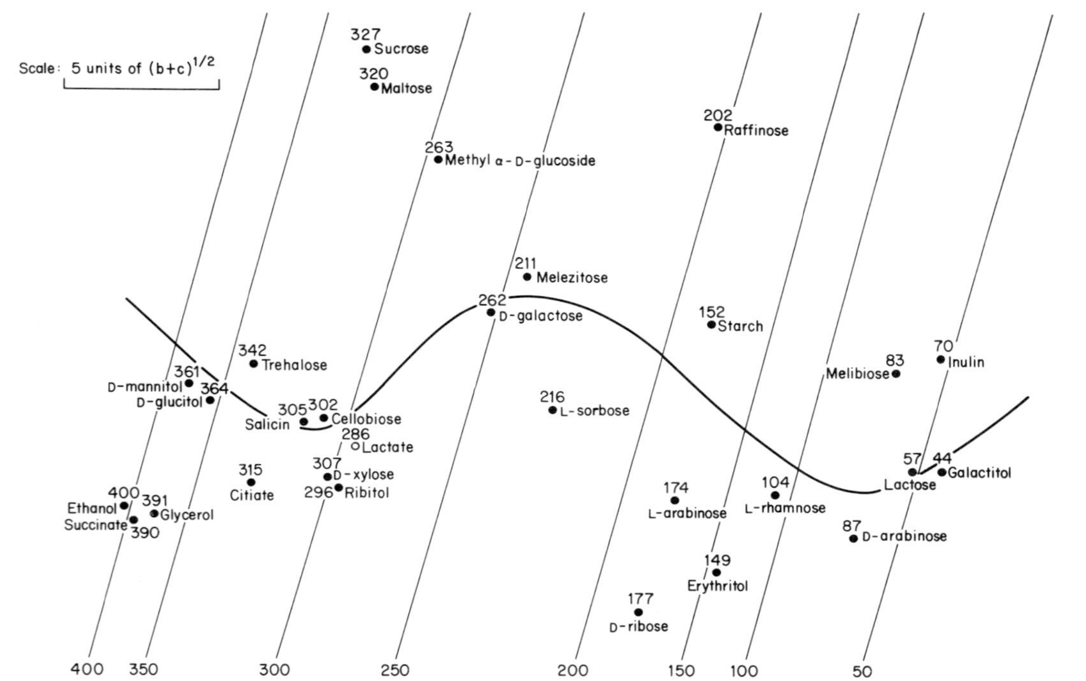

Fig. 6. Principal coordinates analysis of relationships between pairs of sugars and substance are estimated by the quantities $b_{ij} + c_{ij}$ giving the number of yeasts (out of 484) that utilize either substance i or j or both. One-dimensional skew-symmetry is given by the number of yeasts that use each substance, and these numbers have been plotted on the figure and give rise to the parallel contour lines. The wavy line separates substances given in Tables XVI and XVII of Barnett (1976). This figure is slightly modified from Banfield and Gower (1980).

should both be nearly equal to a_{ij}, otherwise there must be distortion at this point of the ordination. It follows that if a set of points cluster, all pairs of the cluster will tend to have small values of b and c so all members of the group should have x-values that tend to cluster around the same x-values of a. The converse is not true; items with the same x-values need not be close together.

If the ordination is accurate, the knowledge of δ_{ij}, x_i and x_j allows all (a, b, c, d)-values for all 2×2 tables to be reconstructed. Thus the simple addition of n numbers permits the recovery of much information normally lost in an ordination. A table for which $b_{ij} = 0$ will have $\delta_{ij} = c_{ij}$ and $x_i - x_j = c_{ij}$, two equal quantities that can be detected on the diagram and used in an attempted construction of metabolic pathways. Banfield and Gower (1980) discuss in more detail the occurrence of such tables with this data.

CONCLUSION

The above are far from being exhaustive analyses. They are intended only to indicate the possibilities of handling this kind of chemical data. A proper analysis requires the collaboration of a chemist or biochemist.

REFERENCES

Banfield, C. F. and Gower, J. C. (1980). A note on the graphical representation of multivariate binary data. *Appl. Statist.* In press.

Barnett, J. A. (1976). The utilisation of sugars by yeasts. *Adv. Carbohydrate Chem. Biochem.* 32, 125–234.

Constantine, A. G. and Gower, J. C. (1978). Graphical representation of asymmetry. *Appl. Statist.* **27**, 297–304.

Curnow, R. N. and Jorré, R. P. (1978). Mean first passage times between the amino acids and the evolution of the genetic code: corrections to an earlier paper. *Biochimie* **60**, 197–202.

Gower, J. C. (1977). The analysis of asymmetry and orthogonality. *In* "Recent Developments in Statistics" (J. Barra *et al.* eds), pp. 109–123. Amsterdam.

Jorré, R. P. and Curnow, R. N. (1975a). A model for the evolution of the proteins, cytochrome *c*: mammals, reptiles, insects. *Biochimie* **57**, 1141–1146.

Jorré, R. P. and Curnow, R. N. (1975b). The evolution of the genetic code. *Biochimie* **57**, 1147–1154.

LeQuesne, W. J. (1969). A method of selection of characters in numerical taxonomy. *Syst. Zool.* **18**; 201–205.

20 | Analysis of Chemical Data: a Summing Up

P. H. A. SNEATH

Department of Microbiology, Leicester University, Leicester LE1 7RH, England

In summing up the contributions on the analysis of chemical data, the most important point is that taxonomists now recognize that chemical information does deserve good data analysis. This was not always so. In the early days of chemotaxonomy such data were often suspect simply because they were chemical and not morphological. Despite a more enlightened attitude to chemical information, time has also brought a second realization, which is that the interpretation of chemical data is no more simple than that of morphology. The contributions in this volume abundantly illustrate this point.

But before considering these complexities, we may ask why we should have expected chemical data to be different from any other kind of information. Various answers to this have been given in the past. First, it has been said that chemical data are in some sense more fundamental, or more directly an expression of the genome, whereas morphological characters subsume the effects of many genes. Secondly, it has been felt that by having a closer relationship to the genome, chemical characters would in some way afford sharper taxonomic information. Again, some have hoped that chemotaxonomy would avoid many of the problems of homology, or of the plasticity of the phenotype, or of adaptive significance of genes,

Systematics Association Special Volume No. 16, "Chemosystematics: Principles and Practice", edited by F. A. Bisby, J. G. Vaughan and C. A. Wright, 1980, pp. 411–417, Academic Press, London and New York.

which had posed difficulties with more traditional taxonomic characters. Many of these statements were never very convincing, and it would seem unlikely that they can assist very much in the interpretation of chemistry of living creatures. The complexities of DNA now being uncovered – satellites, amplified sequences and so on – show that even when we study the physical form of the genome itself we are still involved in difficulties, as acute as those found in morphology.

There are, however, a few special aspects of chemical data that deserve our attention. First, it is important to bear in mind the different implications of data on small molecules and on the large informational macromolecules like DNA and proteins. The occurrence of a small molecule, such as a flower pigment, represents a relatively small quantity of information, reflecting perhaps the presence of a few enzymes, or differences in a few nucleotides in a gene. The same is true of the simple presence or absence of large molecules like isoenzymes. The amino acid sequences of proteins, however, and the nucleotide sequences of nucleic acids, represent relatively large quantities of information. In this case the information may be the fine detail of a single gene (as with a protein sequence) or it may reflect the detail of numerous genes, as in DNA studies (although often in variable or unknown amounts). Serological cross-reactions are perhaps best viewed as reflections of fine gene structure, although it is now clear that the correspondence may be poor, particularly if one uses mixtures of antigens.

A second important property of chemical data is their very wide range of specificities. One can have, on the one hand, serological systems that distinguish between the smallest taxonomic entities that can be readily studied and, on the other hand, one can use protein sequences to compare organisms from the most diverse taxa. Although it may not always be possible to discover (or interpret) chemical systems of desired specificity, this wide scope of chemotaxonomy is notable. Possibly the most impressive aspect is the power of protein sequences to see, as it were, deep into the past and across the largest gulfs of taxonomy. Thus, there can be few systems that can be used to assess relationships between organisms as diverse as wheat and silkworm, or man and yeast, in the way that the amino acid sequence of cytochrome *c* can do. Another remarkable aspect is the fidelity with which the fine structure of the single gene that codes for cytochrome *c* reflects the main relationships of the entire genome: phylogenetic trees based on this one protein unambigously represent the main evolutionary groupings that have been deduced from other biological and geological evidence.

Thirdly, although chemical characters offer many new approaches to systematics, they are relatively opaque to intuitive analysis. Tables of the occurrence of chemical substances, or of serological cross-reactions, or lists of protein sequences, are difficult to make sense of by inspection. We have not yet gained the wide experience of the comparative morphologist, and I believe that we must in any event develop better numerical methods to analyse chemical data. Such methods as we have are often inadequate, as several contributors have discussed at length.

The determination of homology, in any of its usual meanings, presents peculiar difficulty in chemotaxonomy – though it must be remembered that there are numerous problems also with morphology. Is the presence of a particular chemical compound in two genera to be regarded as homology? We seldom know it is produced by the same biochemical pathway in the two genera: more often we simply assume this is so. When a compound is absent, it may be due to defects in quite different parts of a pathway, but without considerable effort we cannot be sure of this. Unlike a morphological structure, it is not easy to see whether a biochemical pathway is missing, and draw the appropriate conclusions. Even with protein sequences, the investigation of disputed homology is difficult. Nevertheless it is to the credit of biochemists that they have started to distinguish different aspects of homology which before were confused, so that terms like isology and paralogy will in due course help to clarify the field and also, it is to be hoped, disciplines outside chemotaxonomy.

Uncertainty about whether joint absences of a chemical should count toward similarity lead the chemotaxonomist into problems of how to handle negative matches. Part of this may be due to the tendency to divide quantitative measurements into "negative" and "positive" regions. Allied to this is the problem of the best scales to employ for such measurements. It is, of course, fundamental to understanding chemotaxonomy to distinguish scales that are analogues of character states of one organism (e.g. intensity of flower pigment in a species) from those that are analogues of a resemblance of relationship between two organisms, and which are equivalent to similarity coefficients (e.g. extent of serological cross-reaction between two species). But the success of chemotaxonomy is now leading to an interest in many new forms of scaling of both kinds, and the contribution by Gower, on some mathematical consequences of asymmetric relations is an example of these new developments.

One persistent problem, to which no comprehensive solution seems to be in sight, is what weight to give to chemotaxonomy in comparison to other evidence. It is, of course, possible to combine chemical characters with other characters and give each character equal weight whatever its origin, as is the usual practice in numerical taxonomy. But the main difficulty is to combine separate classifications when it is not possible to say how much taxonomic information is contributed by one of them, for example a serological classification. In such situations we must regard the classes of data as separate lines of evidence; concordance reinforces our trust in the reasonableness of the proposed taxonomic conclusions (albeit in only a rough-and-ready way), and discordance reduces our confidence in them.

It is not part of my task to enter into the merits of different aims of systematic analysis. The type of method should be appropriate to the aims of one's study, and we are entitled to choose the aims we wish, whether phenetic, cladistic, or diagnostic. The question to which we should address ourselves in chemotaxonomy is whether chemical data presents any particular problems for one or other of these aims. It does not seem to me that there is any strong evidence here: any method can use chemical data. Nevertheless, two of the contributions on analysis of chemical data relate directly to cladistics, so brief comment is appropriate.

Estabrook shows elegantly the limitations and prior assumptions for cladistic analysis; whatever strategy is adopted these are critical to the reconstruction of phylogeny. It is particularly important to realize that no cladistic method can avoid making assumptions about the way in which evolution generally proceeds, and that the use of chemical data in no way obviates the need to make these assumptions. He has also made a useful start on how one estimates the probability that different cliques of characters would be cladistically compatible, and this leads to consideration of the reliability of the cladogram under the assumptions of that particular cladistic method.

The contribution of Humphries and Richardson investigates the application of the cladistic methods of Hennig. This chapter is particularly revealing I feel, because unlike the usual morphological Hennigian studies, it is not at all clear at first sight which chemical character states can be plausibly regarded as ancestral or derived. It is evident that such decisions are not easy, yet if they are largely incorrect, the cladogenies will be incorrect. It is also evident that one can choose chemical characters to support almost any hypothesis about the phylogenetic relationships of major groups – bacteria, blue-green and other algae, fungi, higher plants – if one picks only a few characters. The question raised by this chapter, therefore, is why certain chemical characters should be considered more reliable or important than others for reconstructing the phylogeny. More broadly, the extent to which Hennigian studies are based (albeit unconsciously) on prior phenetic analyses (albeit intuitive ones) is here called in question: we must ask ourselves to what extent phenetic conclusions are a prerequisite to deciding if character states are ancestral.

The present and welcome trend toward investigating the reliability of systematic conclusions is seen in several other contributions to analysis of chemical data. Before discussing these it is important to realize that since all numerical analyses are based on finite numbers of characters, they are subject to statistical sampling error. This is true whether or not we believe we may reach stability in numerical relationships if we use enough characters, or whether we believe observable characters can be treated as a quasi-random set of variables. This does not imply that increasing the number of characters without limit or proper choices will lead to perfect estimates of taxonomic relationships – any more than increasing without thought the number of meteorological stations will lead to perfect weather prediction. But to controvert the statement that sampling error is present would be to assert that *estimates* are independent of the number of observations. If so, at the extreme, one or two characters, or one or two meteorological stations, would be as good as twenty or a hundred. The use here of the word "estimate" correctly expresses the situation with regard to systematic relationships. Furthermore, sampling error applies equally to cladistic, phenetic and diagnostic methods; it applies to any method of selecting variable observations that does not presuppose one knows the correct answer in advance, as is known to every statistician.

This principle is relevant to the two papers on cladistics that have just been mentioned. It implies that whatever the cladistic method, the resulting cladogram cannot be called true or false. Instead it has some probability of being true, or of lying within certain limits about the true one. To controvert the argument it would be necessary to show that random reduction in the number of characters would always maintain the topology of the cladograms (excluding indeterminate cases). This cannot happen, because reduction to one character will lead to different minimal cladograms unless all characters are completely correlated; the proof is thus complete.

Problems of this kind have led to renewed interest in studies of congruence, on which there are important contributions by Harris and Bisby, Mickevich, and Gottlieb. It is evident we need a great deal more work on congruence in all its aspects. As a general principle, congruence has the same function as the principle of maximum correlation of evidence in the philosophy of science. The more things agree together, the more confidence one has that they represent a true or useful scientific hypothesis. But the principle of congruence is not necessarily straightforward in its detailed application to systematics. It is therefore good to see new work in this area.

There are, however, several difficulties in application of the concept of congruence. First, relatively few numerical congruence comparisons between classifications have been made in which sampling error has been taken into account. One cannot expect better congruence than one would obtain from random subsets. The scanty evidence available suggests that incongruence between classifications from character subsets of biological significance (e.g. adults *versus* larvae) is only a little larger than that expected from randomized subsets. Secondly, it is essential to use split-data designs familiar in statistics with randomized subsets if progress is to be made. More methodological studies like those of Mickevich are needed to compare methods on standard data sets, where one is testing congruence between analytic methods. Thirdly, the magnitude of congruence may be greatly influenced by replication of the organisms considered. For example, if three species are almost equidistant in phenetic space, one can readily appreciate that if one chooses a single specimen from each species the relations obtained from two subsets of characters may differ appreciably. The correlation between the phenetic distances might be poor and one might conclude that congruence between the classifications was unsatisfactory. Yet if one takes two specimens from each species, the intraspecies distances are likely to be very small in both subsets, and this could lead to a dramatic increase in the correlation between the two phenetic classifications. There will evidently be a complex interplay between sampling, taxonomic structure and congruence.

These effects of sampling and structure of organisms may explain some of the conflicting impressions from the contributions of Gottlieb and of Harris and Bisby. Certain levels of taxonomic rank in angiosperms, for example, may have relatively little structure in the sense of tight clusters of taxa, and there may instead be a haphazard but fairly even scattering of taxa in phenetic space. This is the impression given by some of the angiosperm orders, whose composition and distinction from nearby orders often seems problematical. In contrast, many families seem quite distinct from nearby families and contain separate genera each composed of a tight cluster of species. The amount of apparent congruence could here depend greatly on the level of ranks that were studied.

In conclusion, we should not be discouraged by inconsistencies in chemotaxonomic data. Unexplained findings can be the seeds of new discoveries. New modes of evolution, new patterns of taxonomic structure, new insights into ecology, may be suggested by findings that are at first sight merely tiresome aberrancies. To make new discoveries we will need new analytic methods, both chemical and mathematical. The contributions in this volume illustrate how these can be developed and we can look forward to a bright and challenging prospect.

21 | Principles and Practice in Chemosystematics: a Summing Up

K. A. JOYSEY

University Museum of Zoology, Downing Street, Cambridge CB2 3EJ, England

When I first planned to attend the Southampton Symposium on Chemosystematics it was my intention to listen and to learn, rather than to hold forth as a speaker. Subsequently, I was invited to provide a summing up at the end of the meeting. My verbal presentation at the appointed time was, of course, somewhat superficial. When I tried to write down a more substantial version for publication it soon became evident that, unless I could refer to the full text of the papers (which was not possible), my comments would appear to be even more superficial. Accordingly, the paragraphs which follow have been tidied up from a tape recording of what I actually said. I hope that they will be read in the spirit in which they were offered, that is, first impressions of verbal deliveries without access to the full texts published in this volume. If any of the other contributors should feel that they have been misrepresented then I wish to apologize for any misunderstanding on my part.

I believe that I was chosen as the victim to provide this summing up because it was thought that I had a general interest in the subject matter but no particular axe to grind. Although I have tried hard to be neutral it seems proper that I should declare my interest; I am interested in any new information which may contribute to our knowledge and understanding of evolution. Hence, my motives in

Systematics Association Special Volume No. 16, "Chemosystematics: Principles and Practice", edited by F. A. Bisby, J. G. Vaughan and C. A. Wright, 1980, pp. 419–425, Academic Press, London and New York.

attending the Symposium were not strictly systematic, but I hope to indicate what the field of Chemosystematics looks like when viewed through the eyes of a broad-minded palaeozoologist who is currently involved with the evolutionary interpretation of amino acid sequence data. I have tried to pick out some key points and have done so largely in a chronological sequence following the excellent structure of the Symposium.

At the start, Cronquist assured us that he wished to incorporate chemical evidence so long as it did not disturb conventional concepts too much. He is, of course, correct in stating that parallel evolution is rampant at both the morphological and the chemical level but he gave the impression that the newer chemical evidence must be concordant with traditional lines of evidence in order to gain acceptance; he is not the only one who has expressed that view. I could not help wondering why he was less inclined to discount morphological evidence when it failed to be concordant with chemical evidence but I have to admit that, to some extent, I am tarred with the same brush. I tend to feel the same way when our own amino acid sequence data for myoglobin appears to suggest that a mammalian order should be split; one feels protective towards the conventional order. Of course, it all boils down to the weight of evidence for each hypothesis; I hope (but I am not sure) that I am here using the term weight in its colloquial sense, rather than in the sense of character weighting.

Cronquist also stated that the validity of methods used in analysing amino acid sequence data is not immediately obvious; I would agree with that, even though I have been involved in such analyses. He claimed that such studies can be tested only by their results, and that the different methods have been judged largely by their success in producing plausible phylogenetic schemes. This also has the ring of truth, but such claims should not be carried too far. If I can tell a parable, on the way to lunch on the first day we all went across the lawn to a door which was locked and then, like a flock of sheep, we all went to another door which again turned out to be the wrong one. Cronquist happened to be leading the flock and as we arrived at the second door he appositely commented that this was a case of the blind leading the blind.

Wright, using a quotation from Simpson, drew special attention to the importance of drawing on all available lines of evidence but he did not give us explicit guidance on how they should be integrated. He reminded us of the hazards of jargon (e.g. maximum parsimony equals greatest economy) and he advocated that care must be taken to obtain an adequate sample of the species under study. This is important whether we are dealing with chemical or morphological criteria. I am very conscious that some of the myoglobin mammalian sequences with which I have been involved were based on material from a single individual. We have taken shelter in the fact that "abnormal" myoglobins (using the term by analogy with "abnormal" haemoglobins to indicate variant forms within the human species) are very rare in man, only four examples being known among thousands studied. It was, therefore, rather unsettling to go along to the poster session on the first evening and see an exhibit by Ferguson, Taggart and Mason demonstrating, by immunological methods, an example of myoglobin polymorphism in a salmonid fish species.

Turning now to the session on secondary substances, Harborne introduced the theme that it was essential to know something of function before the systematic usefulness of characters could be evaluated. He cited examples from the phytoalexins which inhibit the development of fungal spores, and he found congruence with systematic results based on anthocyanins which function as flower pigments, and also with the ability of cells to cope with nicotinic acid when in tissue culture. Indeed, he provided a wealth of information and then he shook some of us with his comment that such characters were of best taxonomic value when not under what he called natural selection, which I interpret to mean not of adaptive significance. Harborne illustrated this point by reference to the fact that quite closely related plants may have blue flowers designed for bee pollination or red and yellow ones intended for the humming birds. Estabrook tried to take up this important point in discussion but there was insufficient time and the point has not been developed since; I wish it had been.

Mabry stressed the importance of alternative patterns of biosynthetic pathways and their use in judging relationships. He claimed that homogeneity of chemical race suggested a more recent origin for a group than heterogeneity of chemical structure, on the basis that time was required for heterogeneity to develop. But in Bosbach's contribution to the poster session, and subsequently in Hurka's lecture, we learned that heterogeneity could be brought to the surface in *Capsella* just by digging the ground over.

Howse and Bradshaw told us that termites and ants communicate by chemical messages and on the scale of activity of these little insects it is perhaps not surprising that "language" differences occur over distances of a few miles. Can you imagine trying to sort out human relationships on this basis? The problem is equivalent to that of trying to get agreement between physical anthropologists and social anthropologists and, to make matters worse, it emerged that the capability of the insects to produce a particular chemical language is partly dependent upon their diet. It was all very salutory.

Turning to Charlwood and Bell, the technical expertise and the data handling equipment now available at Kings College, London, is clearly fantastic and we eagerly await masses of results when they have got the last bits installed. I could not help feeling that if the same outstanding talent for data handling was applied to morphological characters they would also make great advances. In the discussion Mickevich asked the key question, namely, what is the basis for decisions on homology? I think that this was the most important point of the day but it was promptly brushed under the carpet, having been sidetracked by a brief discussion on transduction. The problem of homology was not picked up again until Sneath mentioned it in his review of the analysis of chemical data.

Dando and Southward's exhibit in the poster session on the barnacle *Chthamalus* drew attention to another important point; enzymatic discrimination led to the recognition of morphological characters which had not previously been noticed and these new morphological criteria were then found to hold good for species discrimination. This integration of different criteria, such that each has predictive value in relation to the other, stands in contrast to the incongruence of different lines of chemical evidence which were stressed by Harris and Bisby at the opening of the session on the analysis of chemical data.

Turning now to the session on proteins and other large molecules, I hesitate to review the reviews given by Hurka on botanical enzymes and Rollinson on zoological enzymes. They both stressed the practical and technical pitfalls, and we heard about variation due to age, sex, physiological state of the material, period of storage, and even about new polymorphisms which can be discovered merely by changing the gel. I took special note of Hurka's comment that, for phylogenetic studies, different enzymes were found to be informative for different groups and that different lines of evidence gave crossing classification, which is in agreement with the findings of Harris and Bisby. I also took special note of Rollinson's belief that there is no correlation between rates of evolution as measured by morphological and chemical criteria. I recall that this subject was discussed at our Birmingham meeting on Chemotaxonomy and Serotaxonomy in 1967 and, although it has not been discussed in this volume, I believe that there is still considerable diversity of opinion on this matter.

Berry demonstrated natural selection in action on mouse haemoglobins, citing differences in the relative frequency of alleles between younger and older mice living under harsh conditions. For one of their enzymes systems, he has shown that selection acts in the opposite direction in males and females. He pointed out that when we detect a variant we do not know whether we are looking at something that is, of itself, important or looking at a section of chromsome containing something else which is important. Reflect on that statement in relation to those who stress the importance of understanding function, and those who purport to avoid using characters of adaptive significance. Any character may be linked with other characters of adaptive significance and yet we may not have been able to crack the correlation. Berry also mentioned that a divergence pattern for mouse populations based on chemical data did not correlate with one based on morphological data.

Gray told us some fascinating stories of species phylogeny in plants based on amphiploidy, which made the zoologists thoroughly jealous particularly because he was followed by Miller, Webb and Thorpe's account of their studies on gobioid fish. They found that a classification based on haemoglobin seemed to bear absolutely no relationship to one based on osteology, and neither could be matched with one based on the patterns of sensory papillae; once again this rings a bell in relation to the paper by Harris and Bisby.

Boulter told us that it was quite fortuitous that the amino acid sequences of cytochrome *c* did fit the generally accepted pattern of vertebrate evolution. This was particularly fortunate because some molecules which have been studied subsequently did not fit so well. He advocated that when parallelism is common, as it is in myoglobin, then it is better to choose another protein; I found this rather dispiriting. I think that we would agree that reliable phylogenies cannot yet be deduced from this sort of data, partly because of the shortcomings of the data itself and partly because of the unacceptable assumptions which are involved in the data handling. Nevertheless, the resulting cladograms form a useful basis for discussion and provide clues as to where further investigation might be fruitful. In particular, discrepancies draw our attention to points of conflict between the evidence from different sources and may sometimes lead us towards a change of opinion.

Dover did not have time to make us familiar with the necessary jargon of the relatively new field of DNA comparisons; we need to see it in print in order to have time to digest his message but the subject clearly deserves a whole symposium. Such work probably reflects our closest approach to the comparison of actual genomes and it raised the theoretical point as to whether we would wish to regard a gene as "present" if it were indeed detectable but switched off. At the character level, if a previously suppressed gene were to be switched on again after 200 million years of silence, should we regard the character as being homologous or should we dismiss it as parallel? Such thoughts are relevant to Estabrook's contribution which was included in the session on the analysis of chemical data.

Returning to the second day, Cristofolini and Friday presented divergent views. Cristofolini showed what can be achieved if you have faith in what you are doing, for example, he accepted that immunological distance broadly correlates with amino acid distance. In contrast, my colleague Friday has become somewhat disillusioned and lost faith; I think that this is no bad thing so long as it leads to a constructive reappraisal. He admitted that there is a broad correlation between immunological and sequence distance but emphasized that there are very considerable anomalies and, in a positive way, he has set out to measure those anomalies. Friday emphasized the difficulties inherent in using immunological evidence as a basis for systematic conclusions but I am sure that he is not going to throw away the baby with the bath water.

I understand that it was Sneath's responsibility to summarize the third day session on the analysis of chemical data and this he has done already. It remains for me to ask where do we stand in relation to the Birmingham meeting of 1967, and what progress have we made? First, regarding our principles, we no longer have biochemists claiming that they possess the ultimate answer to all our problems, largely because we no longer think in terms of a battle between morphological and chemical criteria. The advocates of each new method now recognize that all lines of evidence have to be integrated. Secondly, our practices have been developed enormously. We have got an enormous amount of new data, we have greatly improved techniques for getting even more data, and we have greatly improved facilities for handling that data, but we still cannot agree on what to do with it. This observation is not special to chemosystematics, for it applies just as much to morphological characteristics. We still have need to develop our principles, and we are actively engaged in doing so. For example, if I can trespass again on the subject matter of the last session, Mickevich told us that homoplasy does not affect parsimony based analysis. As I see it, the adoption of the parsimony procedure itself is likely to reduce homoplasy artificially. Perhaps we are going in circles, and what we really need is a model for evolution.

In this respect, the Systematics Association is planning a Symposium in Cambridge (1980) on the problems of phylogenetic reconstruction. In the present Symposium there has been an emphasis towards the phenetic approach whereas in Cambridge the emphasis will be phylogenetic. I hope that we will be able to move towards an understanding of that elusive model for evolution.

Author Index

Subject Index